Learning the Nuclear: Educational Tourism in (Post)Industrial Sites

BALTISCHE STUDIEN ZUR ERZIEHUNGS-
UND SOZIALWISSENSCHAFT

BALTIC STUDIES IN EDUCATIONAL
AND SOCIAL SCIENCES

Herausgegeben von Gerd-Bodo von Carlsburg,
Natalija Mažeikienė und Airi Liimets

Edited by Gerd-Bodo von Carlsburg,
Natalija Mažeikienė and Airi Liimets

BAND 36

Zu Qualitätssicherung und Peer Review der vorliegenden Publikation

Die Qualität der in dieser Reihe erscheinenden Arbeiten wird vor der Publikation durch die Herausgeber der Reihe geprüft.

Notes on the quality assurance and peer review of this publication

Prior to publication, the quality of the work published in this series is reviewed by editors of the series.

Natalija Mažeikienė (ed.)

Learning the Nuclear: Educational Tourism in (Post)Industrial Sites

Bibliographic Information published by the Deutsche Nationalbibliothek
The Deutsche Nationalbibliothek lists this publication in the Deutsche
Nationalbibliografie; detailed bibliographic data is available online at
http://dnb.d-nb.de.

Library of Congress Cataloging-in-Publication Data
A CIP catalog record for this book has been applied for at the
Library of Congress.

This monograph is written in the framework of the research project 'The
Didactical Technology for the Development of Nuclear Educational Tourism in
the Ignalina Nuclear Power Plant (INPP) Region (EDUATOM)'. The project has
received funding from European Regional Development Fund
(project No 01.2.2-LMT-K-718-01-0084) under grant agreement with the
Research Council of Lithuania (LMTLT).

The monograph is approved for publishing by the Council of the Faculty of
Social Sciences, Vytautas Magnus University, Lithuania (Ref. No. 54-1).

Reviewers : Prof. Dr. Velta Lubkina, Rezekne Academy of Technologies, Latvia
Assoc. Prof. Dr. Maja Ćukušić, University of Split, Croatia.

ISSN 1434-8748
ISBN 978-3-631-84163-1 (Print)
E-ISBN 978-3-631-84734-3 (E-PDF)
E-ISBN 978-3-631-84735-0 (EPUB)
E-ISBN 978-3-631-84736-7 (MOBI)
DOI 10.3726/b18090

Open Access: This work is licensed under a Creative Commons CC-BY 4.0
license. To view a copy of this license, visit
https://creativecommons.org/licenses/by/4.0/
© Natalija Mažeikienė, 2021.

Peter Lang – Berlin · Bern · Bruxelles · New York · Oxford · Warszawa · Wien

This publication has been peer reviewed.

www.peterlang.com

Cover image: Lisi Raskin, Ignalina Heights Promotional Material, 2020. A3, poster.

'Ignalina Heights' by artist Lisi Raskin is a fictional real estate development that imagines a future residential neighborhood on the grounds of the decommissioned Ignalina Nuclear Power Plant in Lithuania. The promotional poster consists of a photograph that captures an ominous swell in the lake whose waters were once used to cool the reactor core. The slogan refers to a half-life or a period of decay of radionuclides. In the case of the waste from Ignalina, the half-lives of various isotopes range from hundreds to millions of years, eclipsing the history of human civilization.

Table of Contents

List of Abbreviations 9

Natalija Mažeikienė
Introduction. Nuclear Tourism as an Emerging Area of Learning about Nuclear Energy 11

Ilona Tandzegolskienė
Revisiting Educational Potential of the Industrial Heritage Tourism: Ruhr Area in Germany and Ignalina Power Plant Region in Lithuania 19

Linara Dovydaitytė
The Pedagogy of Dissonant Heritage: Soviet Industry in Museums and Textbooks 65

Ineta Dabašinskienė
Place and Language Transformations in a Post-Soviet Landscape: A Case Study of the Atomic City Visaginas 111

Eglė Gerulaitienė and Natalija Mažeikienė
Energy Tourism at Nuclear Power Plants: Between Educational Mission and Retention of "Safety Myth" 135

Natalija Mažeikienė and Eglė Gerulaitienė
Chernobyl Museum as an Educational Site: Transforming "Dark Tourists" Into Responsible Citizens and Knowledgeable Learners 175

Magdalena Banaszkiewicz
Fun in the Power Plant. Edutainment in the Chernobyl Exclusion Zone Tourism 225

Lina Kaminskienė
What We Find Outdoors: Discovering Nuclear Tourism Through Educational Pathways 239

Judita Kasperiūnienė
Innovative Technological Solutions in Virtual Nuclear Education 273

Odeta Norkutė and Natalija Mažeikienė
Energy Literacy in Geography Curriculum: Redefining the Role of
Nuclear Power in Changing Energy Landscapes 313

List of Figures 375

List of Tables 379

About the Authors 381

List of Abbreviations

ACM	Association for Computing Machinery
AGR	Advanced Gas-Cooled Reactors
ANSTO	Australian Nuclear Science and Technology Organisation
AR	Augmented reality
BBC	British Broadcasting Corporation
CEE	Central and Eastern Europe
CEZ	Chernobyl Exclusion Zone
CO2	Carbon Dioxside
EDF	Électricité de France S.A
ELTA	Lithuanian News Agency
EU	European Union
GI	Graphical interface
GIS	Geographic Information System
GT	Geospatial Technologies
HBO	Home Box Office company
IBL	Inquiry-based Learning
ICOMOS	International Council on Monuments and Sites
INPP	Ignalina Nuclear Power Plant
IT	Information Technologies
LL	Linguistic Landscape
Lithuanian SSR	Lithuanian Soviet Socialist Republic
MW	Megawatt
NASA	National Aeronautics and Space Administration
NPP	Nuclear Power Plant
OECD	Organisation for Economic Co-operation and Development
OPAL	Online performance and learning
PBL	Problem-based Learning
PPE	Personal Protective Equipment
RBMK	Reaktor bolshoy moshchnosty kanalny Реактор Большой Мощности Канальный (in Russian)
RES	Renewable Energy Sources
SL	Service-learning

STEAM	Science, Technology, Engineering, Arts and Mathematics
STEM	Science, Technology, Engineering and Mathematics
TASS	Telegraph Agency of the Soviet Union
TICCIH	The International Committee for the Conservation of the Industrial Heritage
UNESCO	United Nations Educational, Scientific and Cultural Organization
UNSCEAR	United Nations Scientific Committee on the Effects of Atomic Radiation
USA	United States of America
USSR	Union of Soviet Socialist Republics
VLE	Virtual Learning Environment
VR	Virtual reality
WHO	World Health Organisation
WNA	World Nuclear Association

Natalija Mažeikienė

Introduction. Nuclear Tourism as an Emerging Area of Learning about Nuclear Energy

Sites of nuclear energy research, development and testing of nuclear weapons, atomic energy reactors or places of nuclear disasters are becoming attractive tourist destinations. The authors of this book discuss the educational potential of nuclear tourism and learning about nuclear power in informal and non-formal learning settings. This monograph is an outcome of a research project EDUATOM devoted to the elaboration of the virtual nuclear tourism route in Ignalina Power Plant region in Lithuania[1].

On the one hand, the popularity of nuclear tourism is related to the development of energy tourism, which has long been a niche and expert-based tourism, attracting engineers, scientists, high school students studying nuclear physics, chemistry, engineering, and recently has been undergoing a transformation by opening up to new groups of tourists – students of all ages and citizens of different groups. On the other hand, the rise of nuclear tourism has been related to the cultural processes of heritagization when the history of nuclear energy research and the atomic energy industry have become an issue of atomic heritage that serves for the identity building of nations and communities. Such heritagization of the atomic past in the U.S. is exemplified by exhibitions and museums established in atomic cities, at nuclear weapons complexes, and other venues to construct the memory on the era of the atomic bomb development under the Manhattan Project during the Cold War period (Mollela, 2003).

A positive approach towards atomic heritage referring to atomic energy industry represents an idea of scientific achievements and promises of the atomic era. On the contrary, atomic heritage related to nuclear disasters (The Chernobyl Museum, Chernobyl Exclusion Zone, etc.) and disastrous use of nuclear weapons (Hiroshima Peace Memoria in Japan, the Bikini Atoll Nuclear Test Site in the Marshall Islands, Kakadu National Park in Australia) represents

---

1   The research project 'The Didactical Technology for the Development of Nuclear Educational Tourism in the Ignalina Nuclear Power Plant (INPP) Region (EDUATOM)' was funded by the European Regional Development Fund according to the supported activity 'Research Projects Implemented by World-class Researcher Groups' under Measure No. 01.2.2-LMT-K-718 grant (No. 01.2.2-LMT-K-718-01-0084/232).

a dystopian account by representing the dark and difficult pasts (Storm et al., 2019). Both utopian and dystopian visions of the use of the atom created at these heritage sites appeal to a broader public imaginary of nuclear and radioactive dangers and disasters (Ibid.).

Heritagization of the atomic past and development of tourism at nuclear power plants is incorporated into the broader process of creating value of industrial heritage, when the memory of the industrial past in post-industrial society begins to be nurtured and industrial landscapes, buildings, and artifacts start to be treated as valuable cultural objects that must be preserved. In this process, nuclear power plants with closed reactors and damaged landscapes become objects of industrial heritage (Storm, 2014).

In the chapter *Revisiting Educational Potential of the Industrial Heritage Tourism: Ruhr Area in Germany and Ignalina Power Plant Region in Lithuania,* the co-author of this book, **Ilona Tandzegolskienė** discusses how educational industrial tourism in the (post)-industrial landscapes becomes a transformative experience, preserves memory, promotes urban development, and identity building. Narrative practices in educational tourism, art projects, and entertainment activities connect landscape and industrial facilities with memory and human experience of local community and tourists. These practices of interpretation of the past and present become a means of constructing a new post-industrial identity of the community. However, this process of remembrance and creating a heritage has a contradictory nature since the industrial past and industrialization are associated in many cases with negative painful processes of the obstructive and devastating impact on the landscape, natural environment, social development, and local identity. Scholars use the metaphor of wound and scar to express this negative element in the nature of industrial heritage (Storm, 2014). Authors analysing the industrial past in Lithuania (Drėmaitė, 2002, 2012) reveal negative meanings ascribed to industrialization which is associated with Soviet legacy. Industrialization falls into the category of dissonant heritage, inconvenient, and unwanted past.

The co-author of this book, **Linara Dovydaitytė,** dedicates the chapter *The Pedagogy of Dissonant Heritage: Soviet Industry in Museums and Textbooks* to heritagization of the nuclear industry in Lithuania and reveals features of the memory work on the nuclear past. In the post-Soviet politics of memory (including public pedagogy and educational discourse), the Soviet industry is treated as a difficult legacy since it is associated not only with modernization but also related with Soviet occupation, environmental issues, and negative impact on the social and cultural identity of citizens. INPP and nuclear industry in Lithuania are considered in the public political discourse as a Soviet nuclear project, and in this sense, the past of the nuclear industry does not fall into the category of valuable heritage. At the same time, conflictual interpretation of the Soviet industrial past leads to problematic development of the identity of post-industrial society since the work and life of former industrial communities are not being interpreted as valuable and memorable. That is why the contradictory nature of heritagization of the industrial past poses challenges for tourism and public pedagogy (museums and other educational sites). The authors of this book emphasize the importance of combining the critical thinking approach with empathy to local communities and formal workers of the industry.

In this regard, **Ineta Dabašinskienė,** the author of the chapter *Place and Language Transformations in a Post-Soviet Landscape: A Case study of the Atomic City Visaginas,* poses a question how after the closure of the INPP the unique multilingual and multicultural profile of the atomic town Visaginas can become a valuable resource for the education and tourism which would contribute to producing an economic value and building a new positively affirmed post-nuclear identity.

Nuclear tourism is analysed in the book as a specific case of energy tourism. On the one hand, excursions and activities of the Visitor Centres are aimed at developing STEM, energy literacy and environmental skills; on the other hand, loyalty of energy companies' consumers has been formed. Furthermore, energy companies conduct corporate branding and public relations through tourism, seek to shape positive attitudes of energy consumers and citizens towards energy sources and energy companies. The co-authors of this book, **Eglė Gerulaitienė and Natalija Mažeikienė,** present a critical assessment of nuclear tourism in the chapter *Energy Tourism at Nuclear Power Plants: Between Educational Mission and Retention of the "Safety Myth",* discussing the features of nuclear tourism at atomic reactors. Visitors to nuclear power plants participate in the educational process by gaining knowledge in various fields about the operation of nuclear power plants, participating in STEM education, and improving energy literacy. Alongside all this education, pronuclear indoctrination takes place, when nuclear tourism becomes a means of persuasion and purposeful communication of the nuclear industry, to form pro-nuclear attitudes and positive opinion about the nuclear energy industry and specific companies. Nuclear power plants use tourism to demonstrate security practices and procedures, strengthening the image of a reliable and safe industry. Another important development in nuclear tourism is the transition from expert-based to experience-based tourism, whereas nuclear reactors, like other industrial and energy tourism objects, attract tourists due to their specific physical qualities – exceptional grandeur, unusual appearance, and shape. When visiting large-scale industrial facilities, tourists experience special strong feelings – admiration for the majesty of industrial 'cathedrals'. The experience of tourists in nuclear reactors is twofold – on the one hand, visitors are aware of the dangers posed by radiation, and this causes a special thrill. On the other hand, the safety procedures organized at nuclear power plants involve tourists in 'security theatre' performances, which also create special feelings and experiences for visitors and explains the attractiveness of this tourist destination.

In addition to excursions organized by energy companies and nuclear power plants, other nuclear tourism destinations also attract tourists' attention. **Natalija Mažeikienė and Eglė Gerulaitienė,** co-authors of the chapter ***Chernobyl Museum as an Educational Site: Transforming 'Dark Tourists' into Responsible Citizens and Knowledgeable Learners,*** analyse the educational potential of the Chernobyl Museum as a cultural interpretation of nuclear disaster. Whereas expositions of nuclear power plants reflect an optimistic narrative presenting the nuclear energy as a future technology, antinuclear critical discourse on the unsafety of atomic industry and nuclear accidents (50 Miles, Chernobyl, Fukushima) is represented by museums, art projects, and tourist facilities which are not connected to the atomic industry. These sites raise questions about the real costs of nuclear energy – how much it 'costs' in terms of human health and life, evaluating its impact on the environment and future generations.

These 'dark' sites of nuclear energy disasters are memorials of the unsafety of nuclear energy, the danger to humanity and nature. Nuclear disaster-related museums and tourist destinations are a unique way to culturally construct a nuclear disaster, and it becomes a result of collective imagination and memory work. Nuclear disaster sites, museums, and memorials as a variety of dark tourism present people's suffering and victimhood.

In this sense, a narrative on the mentioned objects differs from the optimistic story told by power plants, which radiates safety and reliability. Such places as the Chernobyl Museum has become a valuable spot to learn – it introduces the visitor to a structural approach to disaster, tells a story of how organizations and communities mobilized response efforts to disasters. That is why it has turned into a precious source for civic education and studying history. The Museum constructs and depicts the nuclear community, covers themes of nuclear geography, represents a critical historical approach to Soviet-regulated nuclear science and nuclear energy, which led to the disaster. In addition to the educational potential, the nuclear disaster expositions are designed to awaken the visitors' existential experiences – to feel sublime – the ungraspable existential feeling of horror. The artistic installations at the Chernobyl Museum evoke deep philosophical and religious thoughts, contemplations, and feelings.

Nuclear disaster tourism combines an educational impact, which embraces rich knowledge from history, geography, sociology, nuclear energy, biology, and the environmental sciences. At the same time, the cultural construction of the Chernobyl catastrophe appeals to a broader area of nuclear imaginary dealing with dystopian post-apocalyptic images of nuclear disasters that were created in cinematography, literature, artworks (i.e. Andrei Tarkovsky's movie 'Stalker', Svetlana Alexievich's famous book on Chernobyl). It constructs prerequisites to establish stronger ties between non-formal learning in museums and tourist destinations and formal learning by using the intertextuality approach – by combining resources in outdoor education with school curriculum texts and using fiction, documentary and feature films.

Magdalena Banaszkiewicz, while discussing intensive touristification of the Chernobyl Exclusion Zone in the chapter *Fun in the Power Plant. Edutainment in the Chernobyl Exclusion Zone Tourism* reveals how entertainment is created by appealing to the nuclear tourism imaginary stimulated by the global popular culture (i.e. video game 'S.T.A.L.K.E.R.', HBO series *Chernobyl*). Nuclear tourism in the Chernobyl Zone seeks to create visitors' specific experience, which becomes a mixture of thrill, sense of risk, and excitement. Magdalena Banaszkiewicz describes new approaches in tourism when tourism based on principles of pleasure and relaxation 3S (sun, sea, sand) gives up position to 3E (entertainment, education, excitement) and 3F (fun, friends, feedback). Combining entertainment with education, tours to The Chernobyl Exclusion Zone, can be turned from the ethically controversial endeavour of dark and toxic tourism to activities which 'can provide a strong educational experience, raising awareness about the current environmental issues and the polluted environmental conditions around us' (Di Chiro (2000), cit. by Yankowska and Hannam, 2014, p. 937).

The authors of this book discuss how cooperation between educators in the tourism sector and those in formal education can take place. In the chapter *What We Find Outdoors: Discovering Nuclear Tourism Through Educational Pathways,* **Lina Kaminskienė** analyses the concept of outdoor education, deliberating the possibilities of using educational resources outside the school. Educational nuclear tourism includes a specific form of non-formal education which creates an educational potential for visitors when links with formal education are strengthened through the implementation of contextual learning, place-based education, and region-focused curriculum. The concept of outdoor education is applied in the sites of nuclear tourism through school journeys, field trips, and other events and educational activities. According to Kaminskienė, outdoor education in nuclear tourism sites could be organized through place-based education which incorporates concepts of experiential education, community-based education, and education for sustainability.

A variety of educational strategies, such as problem-based learning, action research, landscape analysis, cultural journalism, and many others, contributes to learning on specific topics and issues related to the nuclear power (the role of nuclear energy in the broader energy landscape, energy literacy and current challenges in nuclear energy use, problems of nuclear waste disposal, nuclear disasters as a sign of unsafety and insecurity of nuclear energy, impact on communities living near nuclear sites). The place is conceptualized in place-based education not only as a real physical and social environment, it can be conceived as imaginative or virtual space. Therefore, the nuclear topic can be included in the school curriculum not only as a real place of physical visits to nuclear power plants and places of nuclear disasters.The role of the virtual environment in learning about nuclear is revealed by **Judita Kasperiūnienė** in her chapter *Innovative Technological Solutions in Virtual Nuclear Education.* The author analyses how learning about the atom is integrated into formal and informal education. The knowledge and abilities to understand and make decisions about the use of nuclear energy in various spheres of human life became especially relevant after the Chernobyl catastrophe and other nuclear accidents (i.e. Fukushima). Kasperiūnienė discusses nuclear/atomic literacy, learning about nuclear in non-formal and informal education, which takes place in education laboratories, museums, exhibitions, and virtual spaces by using new emerging technologies, such as virtual and augmented reality, computational dynamics, and virtual world. Development of nuclear literacy in contemporary educational environments of virtual reality and mixed reality tours is based on visualization, immersion, and interactivity. In nuclear literacy development, growing attention is given to the use of game-based learning, where experiential learning takes place through playing and using elements of challenge, fantasy, and curiosity. According to Kasperiūnienė, virtual, mixed, and augmented reality tours create a simulation of an existing location with the help of audio and video technologies that are very valuable in nuclear education, STEM, and nuclear tourism when recreating a realistic representation of reality and presenting views to inaccessible and restricted areas in nuclear power plants or sites of nuclear disasters. In the digital guides and tours, technology (hardware and software) is combined with narration and storytelling. Geolocation technologies and geolocation storytelling create an emotional experience, arouse curiosity and empathy, and foster critical thinking. In nuclear formal and informal education, additional educational opportunities are created by Serious Games, which combine learning with play and entertainment, help to gain knowledge, and develop skills while solving educational tasks and overcoming challenges.

In the chapter *Energy Literacy in Geography Curriculum: Redefining the Role of Nuclear Power in Changing Energy Landscapes,* the co-authors **Odeta Norkutė and Natalija Mažeikienė** reveal the significance of connecting informal and non-formal learning activities in educational tourism with the formal school curriculum to make educational tourism destinations attractive and useful to students and teachers. In the process of developing educational nuclear tourism routes and educational platforms on nuclear topics, geography curriculum could become an important school subject that covers nuclear energy–related topics. Geography is called a school 'subject of survival' and subject 'for the future'. An aim to promote environmental and energy literacy, provide social competences, turns school geography into a central school subject which could create an educational response to energy issues in relation to climate change and pollution issues. While learning on the economic, environmental, and social aspects of the use of nuclear energy in the formal curriculum, pupils will find relevant and very helpful additional and complementary knowledge and experience on the issue in non-formal and informal settings of nuclear tourism.

It is obvious that attempts of educational specialists and tourism providers to promote educational nuclear tourism will be fruitless without strengthening junctions with formal education, without connecting settings of informal learning with other physical, virtual and imagined places of learning. At the same time, learning about nuclear energy emerges as a transformative experience when it is derived from and nourished by a wider cultural nuclear discourse and nuclear imaginary in cinematography, literature, media, entertainment and popular culture, virtual reality. In that regard, the role of educators and educational institutions still needs to be defined and envisaged.

References

Di Chiro, G. (2000). Bearing witness or taking action? Toxic tourism and environmental justice. In R. Hofrichter (Ed.), *Reclaiming the environmental debate: The politics of health in a toxic culture* (pp. 275–300). Cambridge, MA: MIT Press.

Drėmaitė, M. (2002). Pramonė kaip paveldo objektas. *Kultūros paminklai,* 9, 110–118.

Drėmaitė, M. (2012). Industrial Heritage in a Rural Country. Interpreting the Industrial Past in Lithuania. In M. Nisser, M. Isacson, A. Lundgren, & A. Cinis (Eds.), *Industrial Heritage around the Baltic Sea* (pp. 65–78). Uppsala: Uppsala Universitet.

Molella, A. (2003) Exhibiting atomic culture: the view from Oak Ridge. *History and Technology*, 19:3, 211–226, DOI: 10.1080/0734151032000123954

Storm, A. (2014). *Post-Industrial Landscape Scars*. New York: Palgrave Macmillan.

Storm, A., Krohn Andersson, F., and E. Rindzevičiūtė (2019). Urban nuclear reactors and the security theatre. The making of atomic heritage in Chicago, Moscow, and Stockholm In: Heike Oevermann, and Eszter Gantner (Eds.). *Securing Urban Heritage: Agents, Access, and Securitization* (pp. 111–129). London: Routledge.

Yankovska, G. and Hannam, K. (2014). Dark and toxic tourism in the Chernobyl exclusion zone. *Current Issues in Tourism*, 17:10, 929–939.

Ilona Tandzegolskienė

Revisiting Educational Potential of the Industrial Heritage Tourism: Ruhr Area in Germany and Ignalina Power Plant Region in Lithuania

Abstract: The main topic of the chapter is related to the changes that took place during the historical stage of the industrial period. The changes are analysed through the preservation and revitalisation of objects, the perspective of landscape reconstruction, and the renewal of the urban city. The main keywords in the chapter are related to the preservation of industrial heritage, the meaning of historical events, the determination of the value of remembrance based on the theory of the scar, the importance of identity, and the principles of educational tourism. In this work, the case study method is applied. The underlying principles of this method allow to devise certain characteristic strategies in constructing the new face of an industrial heritage object, post-industrial landscape, or urban city. The case study discusses the example of the Ruhr area in Germany through which certain trends on how problematic and specific objects are transformed into part of the cultural heritage and places of interest are shown. The obtained results are presented in connection with the challenges posed by the closure of the Ignalina Nuclear Power Plant (INPP) and the search for the new identity in the city of Visaginas. The study report also presents elements of a sustainable and long-term industrial heritage analysis scheme including the uniqueness of the site, its significance to society, value potential, approaches to demonstrating objects based on constructivist philosophy, educational tourism, and new learning experiences.

Keywords: industrial heritage, Post-industrial landscape, Educational tourism, Learning experience, Ruhr area Germany, Ignalina power plant region, atomic town Visaginas.

Introduction

The classical conception of cultural heritage, which dominated approximately until the mid-20th century, focused on tangible and intangible objects that were maintained and developed by inheritance from older generations of the society. Capelo et al. (2011) note that such an interpretation of this conception is no longer appropriate for the current reality and invite to perceive heritage as something continuously created and recreated which comes from the past, yet has a strong interaction with the present. Therefore, the context of the concept of heritage has expanded from architectural structures to the terrain, from nature reservations to the landscape, from ethnographic rural images to images of industrial cities, and from the demonstration of "primitive equipment" to powerful industrial giants.

The objects of industrial heritage are becoming a part of the cultural heritage today, and the revival or giving a "second life" to these objects is in many respects and interesting phenomenon. This change is associated with environmental perspectives, economic development, social and cultural revival, and empowerment of intellectual potential. Along with the main functions such as enhancing the attractiveness of a public space, assuring the structural sustainability of a city or a region, "awakening" of the creative potential of the population, or reducing the unemployment rate in the region, revival of industrial heritage sites also focuses on cherishing the intangible values such as involving museums into the preservation of the memory about the Industrial Revolution epoch. According to Alanen and Melnick (2000), museums can preserve the interaction between cultural development and industrial heritage which leads to the creation of new artistic-cognitive spaces; furthermore, the transformation elements demonstrated in the factories and plants are linked to gaining historical knowledge through education. Storm (2014 analyses the objects of industrial heritage and post-industrial landscape using the scar metaphor that encourages us not to forget this object by reliving the past in the present moment, as well as to construct the future by exposing us to the values through the process of revitalization. Hence, the revitalisation of industrial objects encourages us to investigate and document the history of the industrial revolution, to record the growing balance between the vegetation and anthropological elements in the places of landscape reconstruction, and to create a new conception of the urban image focusing on its development as a cultural, cognitive, and leisure centre. It also encaptures social change, the scale of development, the need for the transformation, the rethinking of functionality, the details of the structure, and the stages of crisis resolution. When constructing remembrance of the industrial heritage or creating post-industrial landscape spaces, the following benefits of educational tourism are notable: motivation to travel, perceive, and learn through realising urban cultural development and industrialisation through various objects and landscapes.

The main theme of this chapter is related to the changes that occurred in the period of the Industrial Revolution which lead to widespread discussions about the reconstruction of industrial heritage, post-industrial landscape, and urban revitalisation as an opportunity to raise awareness of the value of remembrance and the significance of historical events through educational tourism. The aim of this chapter is to overview the primary changes in the industrial heritage and post-industrial landscape in relation to the need of post-industrial society to preserve and maintain this specific heritage, as well as to highlight the elements of educational tourism that underlie the principles of sustainable and long-lasting cognitive and learning process.

Consequently, the case study method, which is successfully used in social sciences, architecture, and landscape architecture planning, will be applied in the present work. In the current research, the case of Ruhr area in Germany is presented as a case where problem-causing specific objects can be identified for reconstruction to make them a part of cultural heritage. The example of the Ruhr area in Germany is expected to help highlight certain strategies relevant to the INPP region and the city of Visaginas.

Theory Part 1: Spatial and Social Changes of Industrial Heritage and (Post)-Industrial Landscape

Changes in the Conception of Heritage in the Industrialisation Period

The conception of heritage is relevant in this chapter since the object under analysis represents the challenges of the Industrial Revolution and captures the moment of preservation and remembrance. In order to understand the relevance of today's industrial heritage and post-industrial landscape to culture, education, and history, it is not enough to refer to the classic conception of this notion. Nowadays, this notion also encompasses monuments, legacy of agrarian culture, natural, cultural, and post-industrial landscape that are significant for a group or society in general, i.e. heritage becomes factories, factory sites, mines, gigantic industrial machines, and even dwellings or public spaces meant for workers (Jensen, 2000; Čepaitienė & Mikailienė, 2017). De la Torre and Mason (2002) explain heritage value as a set of positive characteristics or qualities perceived by certain individuals or groups about cultural objects or sites where these aspects are not immutable in a changing context. They highlight the right of every person to participate in history writing, emphasising heritage as having social and political implications. It is noteworthy that the conception of heritage includes both material aspects associated with monuments, artefacts, natural structures and buildings; and intangible, non-material aspects, associated with authentic lifestyle in a particular area such as customs, special skills, or abilities; national cuisine, songs, and dances (Storm, 2014; Copic & Tumaric, 2015; Čepaitienė & Mikailienė,

2017). Capelo et al. (2011) reinforce this idea by stating that heritage exists in the material physical sense as an object – a building or a landscape while at the same time these material objects are given meaning through memories, dispositions, or imagination. This conception has been considerably developed in recent decades and has expanded from individual to urban architectural structures, from industrial contexts to the natural ecosystem and landscape; consequently, the significance of heritage and cultural value has increased.

With such a broad understanding of heritage, difficulties arise in the definition of the object and the separation and evaluation of the aspects of the classification which are related to geographical, chronological, and typological elements of value. Alexander (2010) and Storm (2014) note that the initial perception of heritage is equated with the conscious desire of our predecessors to tell us something and to leave us something as an important phenomenon or an object. However, this definition is increasingly used when seeking to focus on a particular gift or even oppositely – a burden (we proudly carry it in the case of a gift or quietly bear it in the case of a burden), so that we can "leave" it for the future generations. Storm (2014) also maintains that it is becoming increasingly hard to clearly define a heritage object due to the difficulty to define and capture logical links between human experience and physical artefacts. The author (ibid.) emphasises that heritage is ever-changing and seen as an activity or a process which is more focused on intangible things, such as authenticity, identity, and artistic or technological uniqueness. Bearing this in mind, heritage is constantly being redefined in relation to the context of the area, the history being told, the value system, and current changes. Thus, it is important to talk about heritage in the context of the values and attitudes of the current population while not forgetting unrelenting aspect of the value of identity. Interestingly, the concept of heritage, which is recorded in historical documents of the epoch, preserved in drawings, presented through narratives, or illustrated in photographs and immortalised in artistic projects, is also identified with something positive; namely, it is enriching people's lives in that to some extent it is partially offering some sense of immortality, Similarly to Storm (2014) Neittleingham (2018) refers to the moment of restoration and revitalisation of the heritage object as a time "scar". While these objects become immortalised and associated with longer publicity, the primary purpose of the objects is destroyed and the identity value is altered. It is noteworthy that certain "unwanted heritage" or even "dark" heritage exists which narrates conflicting situations and memories related to the "victim-offender" or "powerful-powerless" relationships.

Therefore, when presenting the philosophical approach to the conception of heritage, several conceptions are distinguished. One of them is the essentialist conception of heritage which underlines the natural and innate value of an object. Another is the "humanist view" which emphasises the universal, transcendental, objective, and unconditional characteristics of heritage. Finally, the constructivist approach is based on the premise that the identity lies not in things but rather in relations, and it is social relationships and modern values that determine the identity of cultural and social objects while clarifying the significance of the past in the present time (Alexander, 2010).

Active analysis of industrial heritage objects started around 1960–1970, when the focus was on the achievements of technological processes, accumulated experience of the local community, and stories told by workers. Built in the 19th century, industrial heritage buildings with large windows mostly built close to the sea or lake or in urban areas are now reconstructed into restaurants, schools, exhibition halls, museums, and places for cultural events. However, there is also talk about the scars of the ruined post-industrial landscape of the 20th century such as abandoned factory complexes, shutdown power plants, monolithic buildings, and workers' canteens with broken windows.

Furthermore, the new conception of economic utility and sustainability has led governments to reconsider the cultural and historical value of these forgotten objects that have lost their functionality and devastated landscapes. Thus, the re-adaptation of industrial heritage buildings and the post-industrial landscape to modern needs has recently become of interest. The redesigned living environment and the uniqueness of the objects are apparent in such design solutions as preserved authentic and visible steel girders, plastered walls, original and wide windowsills, and open spaces. Many famous projects such as Bankside Power Station in London, opened in 2000 as an art museum (Art Museum – Tate Modern), or lofts in the former mill in Gothenburg in Sweden are perfect examples and design solutions of the industrial past. Industrial heritage objects due to their contrasting appearance with creatively managed landscape attract new residents or new visitors, and in turn not only start serving modern life but are also associated with financial prosperity.

Nowadays, industrial heritage raises a great deal of interest due to its exotic otherness and creates an aura of mysticism and romance. The number of abandoned, obsolete, and unused post-industrial infrastructure sites and industrial heritage objects testifies how the industrial revolution has encouraged mankind to move towards new technologies and technological advances neglecting the agricultural economy. It means that the true purpose of the land is changing towards the idea that land is a place where one can absorb "resources" and "take" what humanity needs. These objects, structures, and landscapes that bear witness to the industrial and technological past have acquired a certain cultural, social, and historical value both at the regional and the urban scale. In this case, industrial heritage includes waterpower, river navigation canals, locks, mills, gasworks, post-industrial sites, warehouses, railways and stations, harbours, or mining sites (Krinke, 2001; Alexander, 2010; Sutestad & Mosler, 2016).

It is noteworthy that "industrial ruins" were not immediately perceived as beautiful or highly valued heritage objects since they were associated with excessive consumption of resources and urbanisation in the industrialisation period and reminded of devastating or even dangerous places. In terms of heritage value, such places or objects are often too stigmatising to be included as relevant and valuable points of interest (Loures, 2008; Alexander, 2010; Storm, 2014). Loures (2008) and Loures, et al. (2017) maintain that a negative public perception of objects created during the Industrial Revolution still exists even if they are no longer functional and their origins are no longer in line with the contemporary aesthetic, ecological, and value-based conceptions.

According to Storm (2014), nuclear power plants with closed reactors and damaged landscapes can also be added to the industrial heritage list, such as the nuclear power plants at Ågesta in Sweden, Calder Hall nuclear power plant in England, INPP in Lithuania and which are currently closed and being dismantled. However, these types of objects attract visitors because of the "nuclear fear" and sense of danger or because of the utopian signs of the past.

Some authors (Copic & Tumaric, 2015) claim that the analysis of industrial heritage and identification of (post)-industrial sites was initially only of interest to "amateurs" and "enthusiasts," yet today it has become an interesting topic widely investigated by various researchers in sciences and arts. Not only is the historical, technological, social and architectural value, but also the scientific, creative, and educational value of these objects is actively discussed (Loures, 2008; Alexander, 2010; Storm, 2014). Mason and Avrami (2000) and Loures (2016) also invite to take a broader look at this phenomenon either from the ecological or educational point of view referring to the relevance of its social, cultural, and economic value. Meanwhile, Copic and Tumaric (2015) recommend analysing the historical and architectural uniqueness of these objects and landscapes as the interest in these areas is continuously growing.

The Use of Scar Metaphor in Defining the Relevance of Heritage

When analysing the sites of (post)-industrial landscape, Storm (2014) employs the leitmotif of "scars" and distinguishes three categories of industrial heritage: the first one is related to the canonised understanding of heritage scars where the place used for industrial purposes is redefined and reused in setting its new goal; the second one is referred to the ruined post-industrial landscape scars which are associated with abandoned, collapsing, or unused places that are simultaneously considered romantic and disgraceful; and the third category underlines undefined post-industrial landscape scars which are not considered significant to be remembered or preserved. Storm (2014, 8) notes: "<...the idea of the scar challenges understandings of heritage as regards the relationship between the mental and the physical. A scar is something you live with. It is a bodily experience, a physical memory. If shown, it is also evocative and can trigger narration: Where did you get that one? <...> A scar, on the other hand, is an organic metaphor. It is not a tool for human beings, but an integrated part of human experience."

There is more to be gained from this description of the conception than from predefined categories. Remembrance, in this case, requires consolidation, whereas the scar metaphor becomes a conceptual tool for holding and protecting memories. Scar, wound, and other organic metaphors used to define heritage are used to show large-scale transformations that have been associated with gigantic constructions and painful changes in the landscape or a terrain (e.g. construction of hydroelectric or gigantic factories). From the point of view of industrial heritage, the use of the scar metaphor also implies the perception that well-being is inevitably costly. It requires a "sacrifice" as industrial growth and the desire to live better alter the image of a location, ruin and devastate nature. The scar does not disappear anywhere. It is present, yet it can be shaded, hidden, or consciously forgotten.

However, according to some authors (Loures et al., 2017), visual value has a stronger impact when assessing the (post)-industrial landscapes and industrial heritage objects, whereas historical remembrance is not always preserved, i.e., historical amnesia occurs. This is due to the fact that fit-for-new life and domesticated industrial objects and landscapes have often been restored and rearranged by those who do not have a personalised memory of the area or the object. Consequently, the historical remembrance and identifying memories of such areas have been lost. In fact, many (post)-industrial landscapes and industrial heritage objects simply disappear or are "destroyed", which can be referred to as a scar because the identity of those who lived and worked in that place is simultaneously destroyed. One could talk about the connectedness with the place and the community that lived there since the significance of identity depends on social interaction and constant dialogue with others (Bairašauskienė, 2018).

Meier and Aytekin (2019) have conducted 222 open-ended interviews in six regions of Europe noting that there is a sense of nostalgia for the past among the people who have lived and worked in an altered industrial landscape. This is manifested through the perception that their jobs have been destroyed, and their sacrifice will no longer be appreciated by anyone. Both the pointlessness of the former work process and the hope for the new investments, which are linked to the opportunities of revitalisation, are visible. Consequently, people living in such place begin to look at the process of change from a distance, with a certain detachment, which is called a panoramic view. In turn, panoramic view requires both physical and emotional distance which leads to the reflection on the object/location and the search for a new self. Historically, the transformations of industrial cities, industrial sites, and post-industrial landscapes in the US and the UK have taken place by altering their use and investing in new and spacious residential dwellings, replanted and scenic landscapes, or planning and installing museum spaces, art galleries, and leisure and entertainment areas.

Meanwhile, the demolition of industrial complexes and the devastation of landscapes in Eastern Europe have freed quite a number of cities and locations from the utopian rhetoric of the Soviet heritage, yet this has also posed a new question: "How could new non-utopian spaces be created from objects that represented the pride, industrial growth, and high self-esteem of people who worked and lived there?" According to Drėmaitė (2002), there is a lot of debate in Lithuania about the industrial legacy of the Soviet period not only in the psychological sense when talking about foreign workers who arrived, low quality production, poor organisation, non-ecological conditions, and hard work, but also because of the emergence of uncharacteristic to Lithuania unified architecture. This period is considered a negative experience and is referred to as the scar that is not consciously analysed or remembered. This is tantamount to an irreversible historical loss of remembrance where cameras and photographs are unable to capture and preserve the existing memory since the buildings and memories have been destroyed.

Another important aspect is that the focus is often given to an individual building rather than to the landscape and environment as potential heritage. In this case, the natural surface of the land and the buildings located there, as well as the particularity of the region, results in the proposition how to strategically plan the presentation of the object, covering both the uniqueness of the buildings and the landscape. Such thinking could enhance the creative potential, the design, and strategic planning capabilities of the region seeking to protect the historical remembrance of the Industrial Revolution. The genre of photography is introduced as an appropriate means of conveying information that helps to capture the decline of material assets and the restoration of nature in those places. The romanticisation of industrial heritage and the environment acquires a different outlook through photography, where the moments and changes of the past are captured by technology. Everyday routines and a gaze into a daily life of a working person dominate in the photographs, hence conveying the life and story of the working people. This genre suggests a new idea that the daily life and "injustices" of the past and present are hidden in the museums and restored industrial heritage objects, photo reports, or films in order to avoid the discussions about uncomfortable things.

However, some artists and photographers purposefully try to bring back the image of the labourer to the landscape or industrial heritage objects in order to create a visual illusion of a working person. To summarise, the discussion focuses more on the breakdown between the material things and location: the importance of the tools they hold, the machines they operate, or the living environment that represents the fragments of their home life is valuable only as long as the living place is used, worked on, and lived in (Meier & Aytekin, 2019). Dortmund Zollern Colliery can serve as an example because its remembrance is associated with natural resilience and hard-working but proud people. The scar here is perceived through the simplification and redefinition of the hard work of a labourer in the mines and the suppression of work problems by the architectural evaluation of the area. In this case, the identity in photographs and installations is strengthened by emphasising local authenticity, storytelling, and the preservation of the working culture.

Another reason for the existence of the scar could be named – the invisibility and uncertainty of industrial heritage objects, which may be due to their remote location or perception as unimportant. This can be exemplified by the city of Visaginas and the INPP since these objects have neither a clear definition nor cultural value and currently are more imaginable and implicit. The decommissioning of the first reactor at the INPP in 2004 and the second one in 2009 had a major impact on the fate of the city since the majority of the city's population worked in the plant, and the city itself was built to service the nuclear plant (Šliavaitė, 2010). Hence, it has become inconspicuous as to what the purpose of the city and the surrounding landscape around the closing INPP should be today, and what value should be attributed to it. At the moment, it is difficult to associate it with recreational zones and eco-tourism.

Therefore, the time perspective also determines the prospects of revitalisation and re-use since industrial heritage objects and the ruined landscape are identified as scars of uncertainty. Since they are left in an uncertain position, it is difficult to remember their past but also foresee the future. Hereby, both tactical and visual perceptions of the perspectives of the object and the location remain unclear and sink into "oblivion". This can be attributed to the Lithuanian industrial period which is perceived as obscure and devastating to the unique nature of the country and at the same time identified with a particular negative experience at a certain historical period, namely, the Soviet times. Hence, the heritage and landscape of the industrialisation period can hardly find their way to the construction of Lithuania's historical narrative and national identity since the heritage of mansions and industrial objects of interwar Lithuania are more precious to Lithuanians (Drėmaitė, 2002).

In addition to the records in the historical sources that may help understand the impact of industrial heritage at the regional and national level, it is also important to talk about the social context which includes the sources of the society's culture and way of life. This may explain the underlying causes of industrial growth and changes in people's lives that have led to this particular type of change, as well as structural, economic, and social changes. Due to the negative experiences, this historical stage in the country (Lithuania) is not actively researched and described. Nevertheless, material objects, such as archival data, buildings, locations or landscapes, and intangible objects, such as workers' stories, lifestyles, or folklore of that period, may answer the question asked a little earlier about creating new non-utopian spaces and identifying the value of industrial heritage. It is notable that the restoration and preservation of identity is possible through sharing the narrative, where nostalgic memories of the past and community involvement in preserving the historical value can help create a historical moment of this remembrance. Nostalgia is not the only element of the narrative that revives identity – sadness and indifference can also become the narrative storyline of identity restoration.

The Trends of Urban Change Dependent on Industrial Heritage Objects or Post-Industrial Landscape

In addition to gigantic buildings and landscapes, urban spaces are often considered as objects of industrial heritage, and tourism is a powerful tool to inspire their reconstruction and conservation. Murphy and Boyle (2005) agree that culture has "become" a part of the strategy of numerous cities where tourism is developed. Analysing relevant case studies, a lot of cities across the UK (Glasgow, Liverpool, Newcastle and Birmingham) and other countries (Antwerp, Bilbao, Genoa, Rotterdam) have been identified as cultural cities. This idea marks a radical development of the concept of "culture" and contributes to educational tourism, heritage, and entertainment (Murphy & Boyle, 2005; Ismailova et al., 2015; Copic & Tumaric, 2015). The urban tourism strategy and the related issues including the purpose, scale, pace, and appropriate impact of the change on the region/locality and policy encourage the creation of a new product that attracts domestic investment, and thereby, demonstrate the economic growth of the city/area. It is worth noting that tourism oriented towards the development of cultural values has become a major theme of international tourism and became an important focus in the reconstruction of urban areas. According to some authors (Murphy & Boyle, 2005), it is highly significant that the development of industrial heritage objects and urban tourism has been linked to the economic and educational goals. Consequently, it leads to raising investment and revitalisation of urban areas. As Meethan (1996) notes that cities "took advantage of cultural capital" title because it saved closed factory buildings and unused areas from demolition and turned them into tourist and educational attractions while developing services sector.

Murphy and Boyle (2005) analyse cities that symbolise and demonstrate how industrial cities can change, such as Glasgow, Bilbao, Newcastle, Rotterdam, Porto, and introduce a conceptual model of the post-industrial city. The authors (ibid.) further draw attention to the tendencies of the development of tourism culture in these cities emerging from the need for tourist accommodation services, development of leisure and retail services, creation of attractive heritage, revival of forgotten and neglected places, and the organisation of festivals and other cultural events. When the support was received from politicians and the new initiative was strongly supported by the local community, the cities of Liverpool and Bristol have become as an example to follow. The ethnographic study conducted by Meier and

Aytekin (2019) demonstrated that social identity and community building of the local people is just as important for the transformation of industrial objects. This reveals the importance of the political decisions that promote urban development and economic growth, attraction of private investors, collaboration of public and private sectors, local government involvement, and the community support for the conception. Moreover, great importance is attached to the creation of new attractions, to the enhancement of the attractiveness of information/tourism centres, to the search for marketing solutions, wellness and leisure centres, to the analysis of possibilities for more museums and gallery facilities, to the thematic prerequisites for organisation of festivals and other events, and to the promotion of retail sales and real-estate development. Special emphasis is placed on the need for individuality in urban change and development since a successful beach revitalisation, organisation of festivals, or construction of a conference centre may not always be justified. Therefore, it is suggested that project development should help preserve something that is unique and specific to the area while remembering cultural signification avoiding "uniformity and recurrence" in other cities (Alexander, 2010; Murhpy & Boyle, 2005; Rodríguez-Ferrándiz, 2014; Ismailova et al., 2015).

In summing up the first part, it may be concluded that industrialisation objects testify to the industrial development, urban, or even regional development, and therefore have an indispensable historical value in several respects, such as: (1) it is a typical and important proof of industrial development and industrialisation processes; (2) it tells about social identity and belonging to the community, and it shows the daily work of a working person signifying his/her social value; (3) it has technological and scientific value in terms of production, engineering, engineering, architecture, and planning; (4) it can have a high aesthetic value for the quality of architecture, design, and planning, which may become an inspiration in the future from the creative point of view; (5) it is a unique intangible heritage that can be recorded in archives, through experience, through elements of the industrial landscape, industrial machinery and even their layout in the room, and through urban spaces in which they exist; (6) authenticity, as the survival of specific object typologies or landscapes, adds special value and should be carefully assessed and preserved. The first three subthemes of the theoretical part can be summarised in one figure (see Fig. 1) which demonstrates the process of objects' analysis and the most important stages of transformation.

Educational Potential of Industrial Heritage Tourism

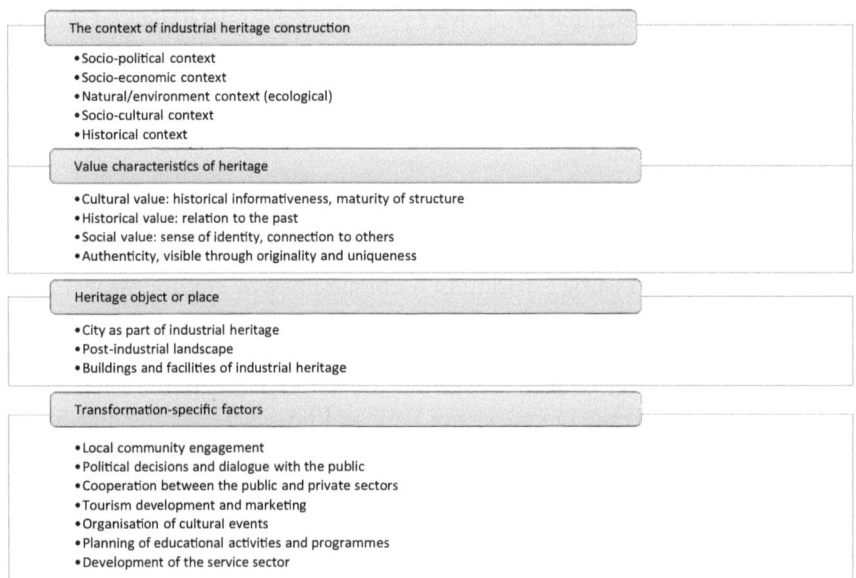

Fig. 1: Process and analysis of industrial heritage construction (compiled by I. Tandzegolskienė)

Theory Part 2: The Possibility of Transformative-Experiential Learning Developed within Educational Tourism When Interacting with the Objects of Industrialisation Process

A modern tourist chooses the goal and type of the travel not only seeking to flee or escape from something, but also in search of intrinsic motivation that incorporates self-awareness and the need to gain new experience. While discussing the objects of industrial heritage, there are three most frequently distinguished conceptions: industrial tourism, post-industrial tourism, and industrial objects of cultural tourism. For instance, Szromek and Herman (2019) distinguish between industrial and post-industrial tourism. The former is concerned with tourism related to the cognition of the operating industrial companies through educational programmes developed by the companies, while the latter is associated with travelling to places where production has been terminated. However, both terms are frequently used as synonyms simply meaning – trips to the objects and places of industrial heritage. These can refer to museums, parks, or other infrastructural objects which involve and motivate by their authenticity and give a possibility to understand and analyse the location by employing the accumulated historical legacy and heritage. They perform the function of exhibition rather than production. When defining educational tourism, not only industrial places but also cultural objects and nature objects are included.

Therefore, some authors (Rithie et al., 2003; Richards, 2011; Akabakar et al., 2014; McGladdery & Lubbe, 2017; Harazneh et al., 2018) propose expanding the conception of industrial tourism by referring to it as a transformative experience that is acquired while learning in a unique place or environment. In the field of educational tourism, Smith (2006) expresses an idea that heritage is more intangible. Even though tangible objects and places can be identified as heritage, it is more relevant to talk about the act of remembrance related to the present, the meaning of experience here and now, whereas the places and objects themselves are historical and cultural means that can help you understand and master it. It shows that trying to understand the value of an object or a locality, it is important to invite tourists/visitors to feel and experience what is significant. Storm (2014) notes that when an industrial heritage object or a post-industrial landscape is presented, people mainly interpret it through "seeing" and "feeling". In this way, a tourist/visitor is guided by the aesthetic criteria of cognising the history or the place, where the eyes and the "re-absorption of experience" become very important. It is also confirmed by the idea of sociologist Urry (1995) that heritage objects or places under analysis attract tourists and visitors because they can experience the place.

All of the provided examples show that scientific literature presents different definitions of educational tourism. Richards (2011) claims that educational tourism is a consequence of cultural tourism fragmentation, and today educational tourism is a separate niche having nothing to do with volunteering, language teaching, or creative tourism. Ankomah and Larson (2000) maintain that educational tourism consists of the following sub-types: cultural tourism, historical tourism, eco-tourism, rural tourism, heritage tourism, and student exchanges between educational programmes. Ritchie et al. (2003) provide a conception of educational tourism that covers an anthropological perspective based on motivation factors and dependent on the age group of tourists. According to the author (ibid.), educational tourism is best defined by a combination of these two words "willingness to learn". Following this conception, acquisition can become a primary or a secondary motivator to travel depending on how the learning process occurs – formally (using the services of an expert or a guide) or non-formally (autonomously or as a result of intrinsic motivation). The author proposes sub-segments of educational tourism, such as educational tourism of adults and seniors, school and university/college students; he also mentions motivation problems, the impact of educational tourism, and the necessity of regional development. Meanwhile, Sharma (2015) focuses on the topics that can be developed depending on the region and defined as subtypes of educational tourism: historical tourism, heritage tourism, archaeological tourism, wildlife tourism, sports tourism, farm/agro tourism, pilgrimage tourism, rural tourism, eco-tourism, cultural tourism, culinary tourism, film tourism. Yet another conception comes from Sie, Patterson and Pegg (2016:108) who distinguish three characteristic features in their definition of educational tourism: "It (1) is an organised trip that provides choices for self-directed learning; (2) leisure travel with an expert-led educational element; and (3) is a trip where intrinsically motivated and like-minded travellers contribute to authentic and personal experiences."

The provided main definitions of educational tourism presuppose that a unified definition of the concept itself does not exist, yet some tendencies can be identified: the need for thoughtful travel planning, learning, and reflection; a motivated choice of an object or place associated with formal or non-formal education and the age group; and authentic and individually created experiences.

Analysing the conception of educational tourism, considerable attention is given to the discussion of motivation. Abukabar et al. (2014) provide Cohen's conception formed in 1972, according to which tourists are divided into two subgroups by employing various expressions of motivation: the explorer and the drifter or the individual and mass tourist. The explorer and the drifter seek new experiences and novelty, while the individual or the mass tourist seek acquaintance or in other words familiarity. This is also associated with the possibility to travel to nearby regions, for instance, within 24 hours, or go on more distant trips in one's own country or abroad (up to one month) including international exchange travels to another country for several months or semesters (Maga & Nicolau, 2018). Based on the set goals, this chapter will focus on the aspiration to learn to acquire new experience while travelling and visiting heritage objects raised in the period of industrial revolution, whereas the themes of international exchange and studies will not be expanded. Rehberg (2005) expanded the division of groups of motivating factors by adding such ones as "achieving something positive", "quest for the new", and "quest for oneself". The first group puts more emphasis on ethical values; the second focuses on friends, culture, and new experience; and the third one is more concerned with personal interests and motives, mainly professional, career, or intellectual purposes.

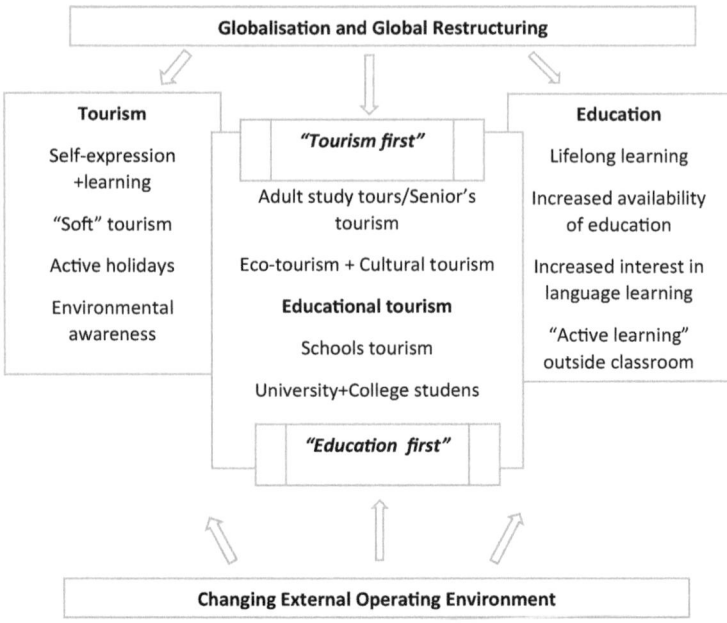

Fig. 2: Model of educational tourism (McGladdery & Lubbe, 2017, p. 321)

McGladdery and Lubbe (2017) emphasise the significance of educational tourism and discuss the duality of motivation to act, which is manifested through interaction "tourism first" and through "education first" and provide the touch points of this interaction (see Fig. 2).

In discussing the provided figure, it is possible to claim that tourism and education are closely interrelated when one plans to learn and explore while travelling, or when one intends to travel striving for new knowledge and experiences. For example, senior tourists are motivated to travel as the first component of an activity (Tourism First) which is associated with the desire to know, travel, and learn. Meanwhile, teachers that organise trips for students would consider education as the first component (Education First) although the students may associate such trips with travelling, pleasant experience, and learning outside school. According to Ritche et al. (2003), such tourism is related to rewarding, enriching, adventurous, and learning experience. Assessing the distinguished trends "Education First" and "Tourism First", it is assumed that in the first case the focus is placed on the curriculum and learning objectives at school, whereas the learning process itself is pre-planned and systemised. In the latter case, the

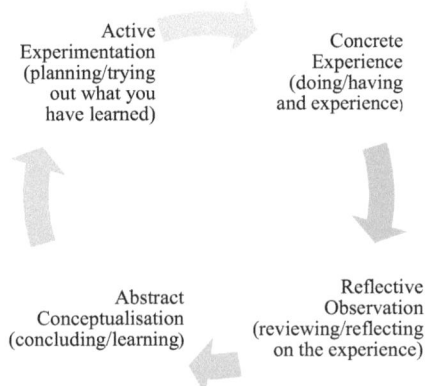

Fig. 3: Process of learning (McLeod, 2017)

focus is on visiting selected places, where the proposed non-formal learning programme is employed, yet its main motive is related to personal purposes rather than the intrinsic aspiration of a tourist/visitor to learn and discover. It shows that educational tourism is associated with learning goals, curriculum, and learning through experience and experimentation, including active engagement of students (Stone & Petrick, 2013; McGladdert & Lubbe, 2017; Harazneh et al., 2018, Nettleingham, 2018). In their study, Pitman et al. (2010, 223) highlighted three key ideas about the form of learning in educational tourism: "First, it was intentional, such as 'taking a trip specifically to broaden my horizons or enhance my knowledge.' Second, it was experiential, involving notions of 'immersion', 'hands-on', 'vivid' and 'evidence' and described as 'engaging with ideas in their original context.' Third, it was structured, such as one male academic's description of 'the combination of travel with a structured educational program."

When discussing the acquired experience in the process of learning via educational tourism, the model of experiential learning developed by Kolb (Pitman et al., 2010; McLeod, 2017; Dorfsman & Horenczyk, 2018) discloses the way how the participants of educational tourism learn. The four levels of learning (experience, reflection, concluding and practice) enable a tourist to create unique knowledge and reconstruct the acquired experience by supplementing it with new knowledge and skills (see Fig. 3). The main aspect of the Kolb's model of experiential learning is reflection which is an essential condition to achieve learning through experimentation and to acquire learning experience.

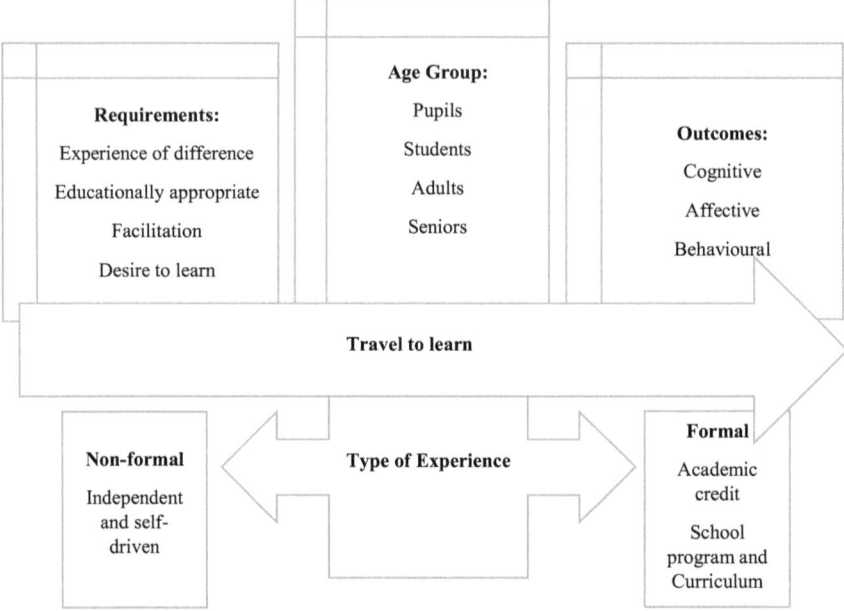

Fig. 4: A learning process model of educational tourism (adapted from McGladdery & Lubbe, 2017: 324)

In presenting the conception of experiential learning and integrating it into the process of educational tourism cognition, it is important to clearly identify the stages of learning and learning outcomes. In this case, the advantage of the outcomes is that they are measurable, and thus, can be verified and modified. McGladdery (2016: 84) distinguishes three categories within the presentation of the process of educational tourism (see Fig. 4): (1) Cognitive outcomes – measure what is to be learnt (knowledge acquired). (2) Affective outcomes – measure attitudes or ways of thinking that may change. (3) Behavioural outcomes – measure skills that will be developed.

Sharma (2015) notes that environmental factors are especially significant in creating an appropriate learning atmosphere via the consolidation of personal experience and knowledge, education examples, and a possibility to try out and get involved into the learning process. The environment with an abundance of direct distractors such as plants, animals, equipment and giant machinery, laboratory equipment, museum exhibits and artefacts activates thinking and encourages learning through observation, comparison, inquiry, classification, analysis, and experimentation. The model introduces the potential age groups of tourists, thus partly predefining the intended results of a selected travel or programme (Ritche et al., 2003; Vangas-Sánchez et al., 2009). For higher efficiency, it is important to make the tourist see and understand the difference between the experience of everyday reality and the newly created experience during the trip that is largely associated with motivation; a comprehensive discussion about the learning object; drawing conclusions, explaining the process, and creating rules/theories; and trying out and personally outliving the newly constructed experience.

Speaking about children and students, it is understandable that motivation and willingness to learn are not always a given priority; therefore, particular attention is allotted to the teacher or trip/education guide or leader. In this case, children and students broaden their knowledge in different fields important for learning at school or university/college studies, such as project preparation or group work; planning and implementation of experiments; nature preservation and sustainable consumption; interest in culture, history, and nature; and finally, interest in science and research. Moreover, it is important to remember that the goals of formal and non-formal education, as well as the elements of the learning process, are not identical.

As far as the goals of non-formal education are concerned, educational tourism should focus on constructing personal knowledge and experience while travelling and getting familiarised with objects and places. Meanwhile, formal education is more associated with the credit system, curriculum, and education content goals. Learning methods largely differ in the process of formal and non-formal education.

Different authors emphasise that it is highly important to combine nature and experience of the activity. Since education is largely associated with constructivism, it should embrace both formal and non-formal education, thus promoting continuous learning in an educational establishment, learning outside school, and using technologies in the learning process. In the process of educational tourism, organisation, and cognition, the most frequently applied learning methods are the ones that encourage engagement through active participation and application of transformative learning principles (Pitman et al., 2010; Dobrila, Sladjana & Maja, 2018; Dorfsman & Horenczyk, 2018):

- In formal education, the proposed methods include: regular lessons (preschools, schools and universities), thematic days, integrated thematic days, projects, ecological fieldwork, extra-curricular activities, study visits, research camps, summer and winter holidays research laboratory, visiting scholars, academic research programmes.
- In non-formal education, the proposed methods are: workshops (art, science, education, sport, psychology), conferences, study tours, sports and recreation activities, environment and humanitarian action, projects by NGOs, local community projects, education projects.

It shows that the content of educational tourism is clearly defined by focusing on the age group, the set goals, measurable results, or on the planned route/education topic, and the industry branch/objects of industrial heritage and the landscape are means for implementing the planned curriculum (McGladdery & Lubbe, 2017; Dobrila et al., 2018).

Alongside with the demand for learning and engagement, primary and secondary elements of the engagement into the learning process and experiencing are identified: the primary ones include culture and industrialisation objects, events, festivals, physical, social and cultural meeting points (heritage objects, museums, monuments, parks, zoos, laboratories, reservations, objects of archaeological excavations), and the secondary ones cover places meant for recreation (film spots, theatre, outdoor workshops, exhibitions, festivals, concerts, play like historical narratives). In this case, a tourist/visitor generally takes interest in the historical narrative of the place, the "reconstructed" industrial cities, waterfront developments, arts expression of the place, shopping and nightlife possibilities, convenient destination, and safe car parking (Murphy and Boyle, 2005). It should be taken into consideration that today the tourist is educated, has developed skills of analytical thinking, gets higher income, demonstrates exceptional interest in the visited site or object, and has a stronger connection with the visited location (McGladdery & Lubbe, 2017; Szromek & Herman, 2019).

Therefore, it is possible to claim that when trying to analyse the elements of educational tourism associated with heritage objects, it is difficult to define a clearly unified set of criteria encouraging the cognition and engagement in the learning process. Therefore, some authors (Murphy and Boyle, 2005; Vangas-Sánchez et al., 2009; Storm, 2014; Sharma, 2015; Loures, 2016) propose choosing an appropriate set from the available evaluation elements that would comply with the general conception of researchers and process organisers (see Fig. 5).

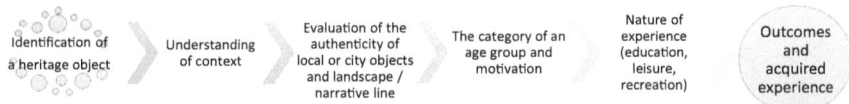

Fig. 5: Elements of analysing sustainable and long-term heritage including educational tourism (adapted from Murphy & Boyle, 2005; Vangas-Sánchez et al., 2009; Storm, 2014; Sharma, 2015; Loures, 2016; Sutestad & Mosler, 2016)

In summing up this section, it should be noted that the main goal of educational tourism is to direct people towards self-dependent learning, which depending on the age group, encourages the selection of a deliberate object or location to be explored and to seek not only knowledge, but also new experience based on personal motivation and personal experience. The new experience and new knowledge are constructed on the basis of the educational programme or the activity proposed by an educator/expert. Educational tourism gives a possibility to acquire new experience interacting with the real world outside school, or to create a space for unpredictability, experimentation, and real-time problem solving while travelling and exploring a place, object, or landscape.

Methodology Part 3: Presentation of the Research Method of Case Study

In order to understand and evaluate the changing locations and industrial cities affected by industrialisation, this study aims to answer the following questions based on the case of Ruhr area in Germany: how have the objects changed? What are the reasons for the change? What processes indicate that the transformation is taking place? In addition, the research aims to highlight the importance of educational tourism in newly revived industrial objects. The research data are presented using the descriptive type of a case study (Yin, 2003; Baxter & Jack, 2008), and focuses on the description of the area and objects, as well as identification of the changes (see Fig. 6).

The research begins with literature analysis that highlights the key aspects relevant to the review and analysis of places affected by industrialisation in relation to educational tourism. Authentic and unique objects in the Ruhr area in Germany were selected and visited by the author of the study from late January to early February 2019, namely, the Landschaftspark DuisburgNord in Duisburg, Zollern Colliery in Dortmund; World Heritage Site Zollverein Coal Mine Industrial Complex and Ruhr Museum in Essen; German Mining

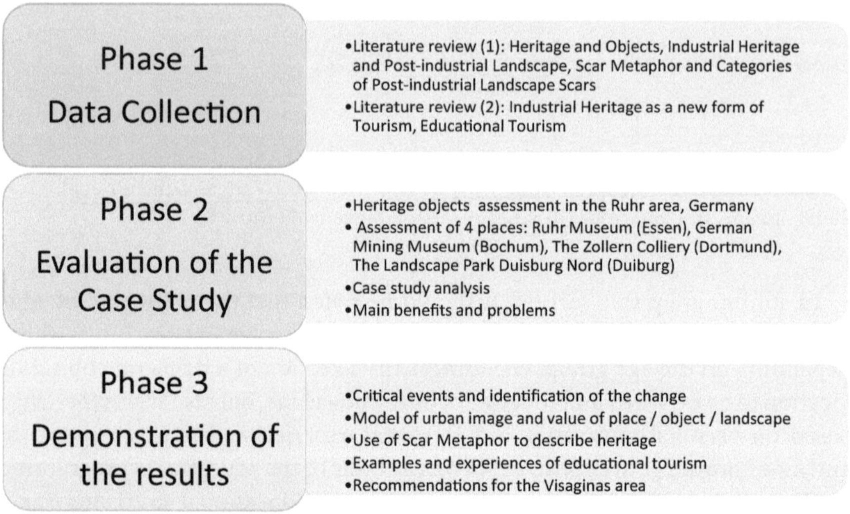

Phase 1 Data Collection	• Literature review (1): Heritage and Objects, Industrial Heritage and Post-industrial Landscape, Scar Metaphor and Categories of Post-industrial Landscape Scars • Literature review (2): Industrial Heritage as a new form of Tourism, Educational Tourism
Phase 2 Evaluation of the Case Study	• Heritage objects assessment in the Ruhr area, Germany • Assessment of 4 places: Ruhr Museum (Essen), German Mining Museum (Bochum), The Zollern Colliery (Dortmund), The Landscape Park Duisburg Nord (Duiburg) • Case study analysis • Main benefits and problems
Phase 3 Demonstration of the results	• Critical events and identification of the change • Model of urban image and value of place / object / landscape • Use of Scar Metaphor to describe heritage • Examples and experiences of educational tourism • Recommendations for the Visaginas area

Fig. 6: Methodological diagram (Capelo et al., 2011; Loures et al., 2017)

Museum in Bochum. Relevant information about the objects and the city, the transformation process, and changes were transcribed according to the above-mentioned aspects during the trip. Furthermore, the case study focuses on critical events that inspired the need for change, the nature of development, the distinctive features of the city and objects that were to be replaced, and emotions and experiences of educational tourism.

The Ruhr area in Germany is one of the most representative projects concerning industrial objects adapted to cultural spaces and recreational areas. This is a good example of sustainable development since tourism in this region serves as a tool for the revitalisation of the industrial heritage and landscape conservation. The areas of coal mines and factories have been transformed into galleries, museums, music spaces, recreational areas, and green sustainable spaces. The former industrial region has changed its image and become an important tourist destination with an emphasis on culture and educational tourism. The transformation of the Ruhr area in Germany through industrial heritage and educational tourism is presented as an ERIH project (European Route of Industrial Heritage). The aim of this project is to create a brand of industrial heritage tourism that could be used across Europe. Certain aspects of the German experience could be applied in similar zones or regions, such as the INPP and Visaginas Nuclear City.

Part 4. Presentation of the Research: The Case of the Ruhr Area Transformation in Germany

The Ruhr area is famous for its huge coal deposits and industrial history of coal mining and steel production. However, by the end of 1950, the coal mining industry was hit by the crisis and due to the increasing competition from the gas and oil industries, a wave of closures of coal mines and coal refineries broke out. Shortly afterwards (early 1970s), there was a crisis in oil industry which also hit the region hard. These events can be referred to as the critical events that led to a recession and a wave of closures. Of the nearly 200 mines in operation at the beginning of the century, this number fell to 125 in 1960 and plummeted to 29 in 1980. The number of workers in the mines decreased 50 percent from 1960 to 1980. The Ruhr area was referred to as a "giant" dying due to the unattractive physical space, social problems, negative image, soil contamination, mass unemployment, and limited opportunities to be reemployed in another sector of the labour market. In general, this region symbolised an economic recession.

Consequently, both politicians and business representatives understood the complexity of the situation and agreed that the change was needed. In this case, it is pertinent to mention that the re-structuring programme started (1989–1990) upon the call from the Ministry of Urban Development, Housing and Transport in the North Rhine Westphalia and targeted all sectors encouraging the submission of various restructuring project proposals. It is noteworthy that the decision to develop a new joint project for the revitalisation of the region consisting of many individual projects was advantageous in that different participants could express their ideas and present their visions. In general, the programme was initially based on the 80-kilometre landscape park along the Emscher River. At the time of the German reunification in 1990, this region was not the weakest part, but the programme was not fully operational, and there were over eight thousand hectares of abandoned industrial land in the region. In fact, this was a large area that could not easily be transformed into new types of spaces or parks as it obviously required huge funds and a well-thought conception (Storm, 2014).

Today, the Ruhr area consists of a 400-kilometre stretch of "Industrial and Cultural Heritage Route". The main goal of this immense decision was to connect the abandoned places by implementing three ideas that led to change and problem solving, namely, industrial nature, industrial heritage, and industrial art (Storm, 2014). Bogulim, Strohmeier and Lehner (2012) referred to the transformation of the region as an ascent of the "phoenix from the ashes". Hence, after a decade, the image changed, and the region became a custodian of cultural

Fig. 7: Route of industrial heritage (author of the photographs I. Tandzegolskienė)

and industrial heritage, which is something that the community members perceived as a positive transformation (Knoll, 2014). The route of industrial heritage in Ruhr now includes 10 cities with rebuilt authentic and historically important places and museums (see Fig. 7) that encourage exploration and experimentation. They include **Bochum** (Railway museum, Century Hall, German Mining Museum); **Dortmund** (Zollern colliery, Hansa Coking Plant, Westphalia Stadium); **Duisburg** (Inner Harbour, German Inland Waterways Museum, Landschaftspark DuisburgNord); **Essen** (World Heritage site Zollverein Coal Mine Industrial Complex, Villa Hügel mansion, Ruhr Museum; Gelsenkirchen (North Star Park (Nordsternpark); **Veltins-Arena** (Football stadium); **Hagen** (Hagen Open-air Museum); **Hattingen** (Henrichshütte Industrial Museum); **Oberhausen** (Gasometer exhibition centre, Rhineland Industrial Museum); **Waltrop** (Old ship lift, Henrichenburg boat lift); and **Witten** (Nachtigall Coal Mine, Muttental valley).

Authentic and unique locations showing the city's great changes and transformations of industrial places and landscapes have been selected for the case study. They include Landschaftspark DuisburgNord in Duisburg, Zollern colliery in Dortmund; World Heritage site Zollverein Coal Mine Industrial Complex and Ruhr Museum in Essen; and German Mining Museum in Bochum (Internet resource: https://alchetron.com/Rhine-Ruhr).

Presentation of Landschaftspark DuisburgNord in Duisburg

Critical events and identification of the change: internationally, the city of Duisburg is associated with heavy industry and an inland port. Today, it is still a city strong in steel production with impressive logistics centres. In order to identify the change factors, it is important to note the recognised and popular project of the Landschaftspark DuisburgNord (see Fig. 8) in the Duisburg Ruhr region. The Duisburg Iron Plant was closed in 1985, yet it was one of the few cases where the closure was not related to the economic change and the crisis. This closure was not emotionally difficult as all the workers retained their jobs, and the entire production process "moved" into another modernised factory. The closed factory was likely to be demolished, but a local initiative group started a discussion about preserving the site and eventually a 230-hectare site was sold by the factory owner for a symbolic price to a public body acting on behalf of Duisburg. The former industrial space with abandoned blast-furnaces, moulding machines, railway lines, tall cranes, tanks, warehouses, and administrative buildings has become a part of a large-scale revitalisation and transformation. Today, it is a landscape park which amazes with its industrial history, the "domestication" of the conception of industrial nature, and the establishment of recreational areas.

The model of the city image and the value of the place/object/landscape: Duisburg City is famous for its Landschaftspark DuisburgNord, which is considered a cultural and leisure park with several museums, such as German Inland Waterways Museum, Inner Harbour, and Museum of Modern Art and cycling trails such as Tiger & Turtle Magic Mountain. Like all the cities in the Ruhr area, Duisburg understood that the change was imminent and thereby became famous for cherishing its culture. Nowadays, the city can surprise by the creative reconstruction of closed objects. Apart from the former metallurgy giant which hosts one of the most beautiful industrial parks offering free movement and cultural and historical experience, the Duisburg harbour and storage facilities should also be mentioned. They have been converted into pedestrian zones and an office area. This immense project is spread over 89 ha and has the main goal: work, life, and leisure by the water. The city is also known for its cultural events taking place at the Theatre Duisburg, the Deutsche Oper

Fig. 8: The closed Duisburg plant is dominated by the concept of industrial nature (author of the photograph I. Tandzegolskienė)

am Rhein, and the Duisburg Philharmonic. Moreover, the city is known for its festivals such as the Duisburg Akzente and the Traumzeitfestival. Another cultural focus of the city is the Duisburg Film Week, which features German documentaries.

Using the scar metaphor to describe heritage: The Landschaftspark DuisburgNord has been recognised as an "abandoned" and devastated industrial site and can be described as its scar. Large and immobile buildings and the destroyed landscape are revitalised through the artificial creation of an eco-system. Travelling across the plant site or observing the landscape from the top of the observation platforms, we can see how the idea of turning gigantic installations and buildings into industrial sculptures was implemented by deliberate additional planting of vegetation around them. This is evidenced by clean water channels and agglomeration clusters, concrete partitions overgrown with green climbing plants, built-in recreation spaces, separate children's playgrounds and elements of an eco-garden.

Examples of educational tourism and experience: The newly revitalised Landschaftspark DuisburgNord is open to all visitors and invites them to explore, discover, experience, and capture it with their cameras. The park administration offers to visit observation platforms, walk along the park paths and capture stagnant moments in the old factory, climb the blast-furnace and observe installations from above. This conception is interesting because nothing has been changed or renovated since the factory closed. Such an idea encourages learning about the past, experience changes evoked by time and nature, and provides each visitor with an opportunity to create their own story and "gather a personal bouquet of emotions". Furthermore, this place is also known for sport and cultural events: a former gasometer has been turned into a diving facility which is open to the public and hosted by the Park Diving Club. Former ore bunkers provide walls for a number of climbing paths that are also open to the public and much used by the local division of the German Alpine Club. Visitors can participate in various educational activities: team building activities, canoeing and polo playing on an old gasometer, and activities in various workshops (blacksmithing, car making, woodworking, graffiti drawings, and photo workshops). Activities are offered to both adults and children. Cinema connoisseurs can visit the Stadtwerke-Sommerkino event, which annually attracts over 30,000 viewers of summer cinema novelties at Landschaftspark Nord, whereas the illuminations created by the world-famous lighting designer Jonathan Park immerse the night factory into an unusual sea of colours at the end of each show.

Presentation of Zollern Colliery in Dortmund

Critical events and identification of the change: The global crisis of the early 1960s and changes in the market economy affected both the region as a whole and the region of Dortmund. In 1987, the last coal mine in Dortmund was shut down, marking the end of centuries-old traditions. Another painful blow was the steel crisis due to the lowered demand for it. Despite various measures and efforts, such as structural reorganisation and merging of corporations, unemployment increased, and the region faced economic and social challenges to a stable life. During this period, the focus was on the reconstruction of the city centre and the preservation of the former industrial prosperity age. The aim was to create a unique, exciting, and travel-worthy concept of Dortmund. An important object of interest in Dortmund from the aesthetic perspective is the Zollern Colliery, the region's model coal mine built in 1989–1904. Some buildings of the mine were interesting architectural monuments built in Art Nouveau style and designed by the renowned architect Paul Knobbe.

Following the collapse of the coal mining industry in 1960s, the Zollern Colliery was closed in 1966. The preservation of the area was influenced by the emerging new trend. Namely, the concern to preserve the memory of the working-class history and the emergence of a new type of museums: decentralised industrial museums with a social history agenda and monuments left in situ, in combination with museum exhibitions located in the former industrial buildings. The site, buildings, and equipment were preserved and included in the Westphalian Industrial Museum in 1981. The Westphalia Museum of Industry is the first and the largest industrial museum in Germany, founded in 1979. The museum includes eight industrial sites and locations in different cities in the Ruhr area: Zollern Coal Mine in Dortmund, Hannover Coal Mine in Bochum, Nightingale Mine in Witten, Henrichshutte Metallurgy in Hattingen, Henrichenburg Ship Elevator in Waltrope, the Bocholt Textile Factory, the Lage Brick Museum, and the Gernheim Glass Factory in Petershagen. The mission of the museum is to communicate, research, and preserve the culture of the industrial period. The most important exhibits in this museum are the industrial buildings turned into monuments, and the focus is on people whose lives and work have been associated with factories and mills.

The model of the city image and the value of the place/object/landscape: In the past, Dortmund was a famous city with its iron and steel plants, yet the last plant was closed in 2001. As a traditional site for mining and heavy industry in the Ruhr area, Dortmund is today a venue for culture, exhibitions, various events, and museum presentations. The industrial city became the "carrier" of history via its museums. Besides, the city is also seen as a venue for Ruhr area "domicile" jazz, world-style, and avant-garde music events. The city has also taken a turn towards cherishing culture, which is evidenced by the opera house, drama theatre, children's and youth theatre, the newly founded children's opera theatre, the Oswal and Adlertum museums, and the Dortmund Philharmonic. The area is also famous for other industrial heritage objects, such as the Kokerei Hansa, a museum for the presentation of the world of work. The entire history

Fig. 9: Aesthetic value of Zollern Colliery architectural monument (author of the photograph I. Tandzegolskienė)

of labour and labour protection in a chronological order is presented considering technological advances, the environment, and its protection.

Using the scar metaphor to describe heritage: The object under analysis, the Zollern Colliery in Dortmund, is presented as an architectural monument and as an aesthetic place where the history of daily work is displayed and showcased through beautiful and exhilarating memories of work. Although much attention is paid to the experience and history of labour and the people who worked there, the scar metaphor is hidden under the facades of gorgeous buildings. The exhausting human labour and hard daily life is retouched by the exaltation of architectural buildings (see Fig. 9).

Examples of educational tourism and experience gained: In 1999, the Zollern Colliery started functioning as a museum of the social and cultural history of coal mining in the Ruhr area. The museum presents itself as a living industrial heritage site and a regional cultural forum. Although the branches of the museum are different, they are united by live presentation of the history of industry: history can be experienced there by weaving on the looms, laying bricks or blowing glass. The thematic division of the museum exposition is in line with its mission ("to communicate, explore and preserve the culture of the industrial era") and its title (Ruhr Museum of the Social and Cultural History). It emphasises not only the political-economic history of industrialisation, but also the daily life of miners and their families, their work, lifestyle, fashion, and leisure; as well as the political safety, hygiene, and other realities of miners. Alongside traditional museum exhibits (authentic objects, texts, photographs, videos, hands-on installations), there is a wealth of experiential activities: the visitors can go down into the cellars and experience the darkness of a coal mine, or they can climb onto the top of the production building and admire the greatness of the surroundings. They can also experience the scope and magnificence of production in the engine building because the original equipment was left in a giant empty space or can sit in a tiny coal-miners' wagon. The exhibition combines the presentation of the official historical information with a personalised narrative. An example is the "student Franz" who is a guide for children. The cardboard student Franz takes part in a special storytelling for children and appears in both production buildings and throughout exhibitions. Another example could be biographical objects and the life stories of concrete people. History is brought closer to visitors' lives through participation in shared human experiences that lead to involvement in learning.

Furthermore, science is represented in the Ruhr Museum of Social and Cultural History at the Zollern Colliery and reflects the features of a contemporary science museum. The narrative lines of the museum's exposition do not tell the immanent history of industrialisation (historical circumstances, economics, owners, development of the construction, coal mining industry). Rather, the exposition is clearly linked to the socio-cultural, economic, and historical context of the region. The museum raises questions about the dark sides of industrialisation (labour, exploitation, living and working conditions, disasters). The displays present industrialisation in an open, non-historical narrative, pointing to the different components of the phenomenon – work, everyday life, leisure, machinery, technical process, and architecture. Industrialisation is closely linked to the history of labour but avoids displaying "labour without workers" (where only the machinery and the production process are presented) and does vice versa – people are at the centre of the narrative.

Presentation of the World Heritage Site Zollverein Coal Mine Industrial Complex in Essen

Critical events and identification of the change: Essen has been known since the 19th century for its iron mines and industrial structures which occupied an area of 35 hectares. Although this period of industry and industrialisation is an issue of the past, industrial heritage still shapes the city. Closed mines and factories give the city its uniqueness and tell an enticing story. The Ruhr Museum in Essen was founded in 1904 as a museum of nature, history, and art. It was moved to a transformed coal mine building in 2010 (the authors of the transformation: Office for Metropolitan Architecture/Heinrich Boll and Hans Krabel). What makes this museum unique is that it presents not only the history of industrialisation, but also the natural and cultural history of the entire region from the geological period to the present day. Since 2001, World Heritage Site Zollverein Coal Mine Industrial Complex has been listed in the UNESCO World Heritage Sites. The mine is considered to be the centrepiece of the Route of Industrial Heritage and attracts more than 1.5 million visitors each year. The complex is unique as it dates back to the 19th century and reminds us of the building's magnificent existence (in 1970 it was the largest and the most modern industrial site in Europe). The year 1986 marked the end of this giant's life. The continuity of the complex was foreseen immediately after the closure, and the reconstruction work started in 1990. This is how the Ruhr Museum was opened in the coal-washing building, the Red Dot design centre was established in the boiler and a casino in the low-pressure compressor building. The World Heritage Site at the Zollverein Coal Mine Industrial Complex has spaces for artists, local salespersons, an ice arena, a swimming pool, a conference room, and a restaurant complex.

The model of the city image and the value of the place/object/landscape: Iron and steel used to be important in this city, but today it is admired for unconventional artistic solutions, museums, and art spaces. As an example, the Kulturpfad cultural trail is covered with blue stones. The trail stretches for 4 kilometres with 372 blue signs pointing to the 82 Stadtzeichen symbols of the city. They highlight the extraordinary architecture and public spaces that are adapted for arts and invite visitors to wonder, engage, and discuss. The city attracts by its cultural objects, such as the Poster Museum hosting a collection of more than 350,000 posters from the political, business, and cultural spheres, while largely focusing on documenting German poster design in the European context. Moreover, it attracts visitors by the Lichtburg Cinema, which is the largest cinema hall in Germany and also the oldest cinema still showing films.

Every year at the beginning of the summer, the Kulturpfad Fest is held in Essen with performances by music, theatre, and light professionals.

Using the scar metaphor to describe heritage: When reflecting on the Zollverein Coal Mine Industrial Complex, it is possible to talk about large structures, gigantic installations, and a fairly large area which was dedicated to industrial development. The size of the site itself and the visible devastation of nature can be described as a scar here. The inefficient buildings were "revived" and identified as necessary. They were resurrected as spaces that serve art, entertainment, and historical memory. The museum focuses on the activities which are very diverse and dedicated to exploring the region and preserving memory. The conception of the museum is focused on aesthetics and presentation of culture and education, based on various means (photography, short film screenings, artefacts, virtual maps, installations, visualisations).

Examples of educational tourism and experience: The Ruhr museum is founded in Zollvere in Coal Mine Industrial Complex. The main mission of the museum is to serve as a place of remembrance for the region. The Ruhr Museum is considered one of the first museums in Germany to combine the history of industrialisation with the theme of everyday life and social history. The museum's exposition is divided into three main themes: "the Present", "the Memory", and "the History." These themes, in turn, fall into subtler narratives. "The Present" introduces myths and stereotypes about the Ruhr region through such topics as work, solidarity, or homeland. In the section "Phenomena", the visitors can get acquainted with the ethnic and linguistic composition of the region's inhabitants, their hobbies, pets, small architecture (kiosks), leisure, culture, "industrial" nature, and the sounds and smells that are typical to the region. The narrative of the exhibition is complemented by a personalised way of presenting the story: talking to three "heroes" about what they value in the region. The second part of "the Present" is entitled "Time Stamps". In this section, natural objects (fossils, minerals) and "biographical objects" brought by people are displayed next to each other along special stories told about them. They present a map of modern collective memory.

"The Memory" exposition tells the geological, biological, archaeological, and cultural history of the region from ancient times to the present. It presents historical periods (Bronze Age, Stone Age, etc.); cultural movements (Reformation, Enlightenment, etc.); and various geological, archaeological, and cultural collections. "The History" exposition displays the history of industrialisation in the Ruhr region from the 19th century to the present day. This narrative is constructed on the basis of dramaturgical principles: prologue (history of the geological strata of the region), beginnings (1750–1830), fossils, technical innovations (1830–1870), the rise (politics and business 1870–1914), urbanisation (1914–1957), and transformations (1957–2010).

In addition, temporary exhibitions are held alongside the permanent exposition of the museum. The museum also offers "Coal Road" tours, where visitors have an opportunity to visit a coal processing plant with a guide and walk the coal mining and processing path in an industrial building that has been preserved in the same condition as it was left by the last miners in 1986. The thematic division of the museum exposition is in line with the aim of the museum's mission – to become the site of regional memory as it encompasses the whole of the region's history.

The museum's exposition combines learning and entertainment. The exposition presents a variety of exhibits: authentic artefacts (old industrial installations, biological, geological, cultural objects), art works, photographs, stories, video material, explanatory texts, sound and scent installations, and information terminals. It is important that exhibits from the museum's collections are displayed alongside with "biographical objects" that tell the story of everyday life. This way they combine natural and cultural history as well as historical facts and personal narratives of people. Leisure learning is illustrated by experiential activities, such as the "coal path" excursion and the opportunity to climb to the 45-metre height to admire the surroundings. The representation of science at the Ruhr Museum reflects the features of a contemporary science museum. The narrative lines of the museum's exposition do not emphasise the history of the industrialisation development but relate it to the natural, economic, social, and cultural processes of the region. The exposition presents industrialisation through its impact on nature and human life. The exhibition links the processes of industrialisation to the experiences of visitors through biographical and everyday stories. The Ruhr Museum focuses on audiences of different ages. Experiential activities are more focused on older high school children and young adults.

The authentic environment and the preserved equipment may interest science and technology admirers. Additional activities offered by the museum include a café and restaurant, an entertainment complex, and a variety of cultural events aimed at families who come here for longer periods of time. The Ruhr Museum uses various models of active learning: from interactive facilities, authentic artefacts, atmosphere of authentic environment to experiential activities, a factory tour, and information terminals. Emphasis is placed on aesthetic experience: smells, sounds, and theatrical atmosphere that is designed with the help of installations, colours, lighting, etc.

Presentation of German Mining Museum in Bochum

Critical events and identification of the change: The "golden age" of Bochum refers to the 19th century, when the Association for Mining and Steel Casting was established. A hundred years later, unfortunately, the first mines and pits were closed. One of the vital decisions of the post-war period was the establishment of the Ruhr University, which is currently one of the largest universities in Germany. This decision allowed the region to acquire a new role of an innovator in economic services, logistics, and healthcare. The Eisenbahnmuseum Bochum Railway Museum and the Deutsche Bergbau Museum, the German Mining Museum, are the reminders of coal mining and production facilities today. The German Mining Museum is one of the eight museums in Germany associated with archiving coal mining documents and recording historical industrial developments. The museum was founded and opened in 1930 in a former meat processing plant. In 2009, new buildings were erected by the industrialist architect Fritz Schupp, who converted the Essen coal plant into a museum.

The model of the city image and the value of the place/object/landscape: Nowadays, no one denies Bochum's past – it is presented as part of heritage and an example of great strategic planning that shows how an industrial city can become a cultural centre. Over 400,000 visitors annually descend to the museum grounds first and then ascend to a 63-metre-high supply tower overlooking Bochum and providing magnificent views. The beginning of the change was difficult because coal mining became unpopular in the 1960s, and the entire region had to be reborn immediately, literally overnight. With the establishment of the Ruhr University, which was run by different leaders at different stages (Peter Zadek, Claus Peymann and Leander Haußmann), and which became one of the best higher education institutions in the region, the Bochumer Schauspielhaus (Bochum Academic Theatre), the most innovative and radical theatre in Germany emerged alongside. Major international events, such as the Ruhr Triennial, the Klavier-Festival Ruhr, and the Ruhr Festivals have made the region one of the most picturesque cultural landscapes in the whole continent. With Bochum becoming the European Capital of Culture (RUHR, 2010), this new identity has become the symbol of the city: change through culture and culture through change where industrial installations have become new scenes for interesting and exciting art. The idyll of the Ruhr Region includes Dahlhauser Settlement in Heid and Miners Settlement in Bochum. The Dahlhauser in Heid was founded in 1906–1915 as an exemplary colony of workers in the adjacent Hanover pit. Also known as the Kappeskolonie, the settlement is among the most beautiful in the Ruhr region. It was designed and built as a green settlement by Robert Schmoll, the architect of the industrialist Krup family.

Using the scar metaphor to describe heritage: The analysis of this object reveals that the city itself can be distinguished as a large cultural centre which created its own traditions and new experiences. The history of industrial heritage and the changes that have taken place throughout this period are preserved and recorded in the city Museum. The suspension of the activities in the region, and the closure of mines and pits, can be considered as a scar of this region. The museum collects all the information about the period of stagnation in the region and depicts the nature of labour in the simulator, yet only the beautiful side of the story is visible. The other side of the story is presented with caution. This implies that the history of mining and coal mining is a thing of the past, and the focus is on the search for and creation of a new identity.

Examples of educational tourism and experience: The German Mining Museum invites the visitors to experience the subtleties of the mining world by visiting a 20-metre-deep demonstration mine with a guide underneath the museum. The route extends for about 2.5 km, where one can see the working routine of the miner as well as the tools and impressive machinery. The simulated experiential journey begins in the elevator descending at high speeds, thus allowing visitors to experience the vibrations and the noise. The narrator plays a very important role because the tour guide immerses visitors in the chronological events, which show the historical changes of tools and the work process, the use of tools and principles of machinery work, the everyday life and labour of workers, and the advantages of the latest industrial machinery. Thus, from the very beginning of the journey, the visitor is introduced to the first anchorage elements, the need for a horse as a vehicle, the train path, various drilling machines, and modern equipment. The route offers a powerful experience with the possibility of touching, testing, and discussing with the tour guide.

Since July 2019, the museum has been offering visitors a revitalised exposition that covers four themes: coal, mining, natural resources, and art. The first theme focuses on the history of the German coal industry. The second theme deals with the history of materials' science, mining, and industrialisation. The third theme presents natural resources, which are of immense benefit to humans today. The exhibits are presented using artefacts, laboratories, and installations. The exposition of the fourth theme presents mining peculiarities, where the scope of presentation ranges from artists' works to sculptures, paintings, and craft objects. The museum focuses on all age groups, creating a variety of options for experiential and new knowledge construction. In addition, students are offered various workshops like Coal Mining, Stone – a Window into the Past, Mining Yesterday and Today, How to Get Coal, and others.

After analysing industrial heritage and post-industrial landscape preservation, urban transformation, and heritage/landscape adaptation opportunities for new functioning, it is important to discuss a complex transformation at the cultural, historical, social, aesthetic, and ecological level, where planning and publicising cultural events as well as raising awareness of educational tourism at the local, regional and international levels are considered the key elements for change.

The presented cities focus on the socio-cultural context: theatre and opera performances, music festivals, various concerts, and cinema events; the historical context covers the use of museums and heritage spaces for meetings, concerts, and creative artistic solutions; whereas the ecological context embraces revitalisation of the landscape through creation of leisure areas, experiential paths, elements of eco-gardening, the use of water bodies and reservoirs for walking and recreation areas, and education. The city government, business, various public organisations, and community members could be involved in the project developing a new concept and restoring industrial heritage and the post-industrial landscape, thus ensuring creative and innovative ideas and public dialogue. According to Strom (2014) both Swedish and German strategies are similar in that the transformation process focuses on cultural and economic development. It is important that industrial heritage sites in this region are often associated with the preservation of heritage, values, and remembrance.

The Industrial Heritage in the Ignalina Nuclear Power Plant Region and Visaginas City

The case of the transformation of Ruhr area in Germany discussed above helps to acquire ideas and experience for the principles of possible identification of the value of industrial heritage in Lithuania. It would also be possible to use the metaphor of scarring to identify, discuss, and describe the visible, felt, or forgotten scars. The presented stages of industrial heritage construction development could also be used to transform and give meaning to objects, sites, or the landscape.

The city of Visaginas and the INPP, which is currently closed, are a distinctive example of industrialisation and "lost utopia" (Storm, 2014). Atomic

Visaginas city is a relatively young urban settlement founded in 1975 as a satellite to the INPP. Visaginas is an example of a city, the dominant sections of which were constructed to form habitats under the Soviet social paradigm. Most of the inhabitants were Russian speakers from other Soviet Socialist Republics. During the construction of the INPP, non-Lithuanian speaking highly educated persons moved to the area for work reasons, and the town of Visaginas was built to accommodate them. INPP is a closed nuclear plant consisting of two modules RBMK-1500 built close to the Visaginas City (Lithuania). Ignalina NPP decommissioning project includes decommissioning of Unit 1 and 2 and auxiliary facilities. The process is divided into two phases. The first phase started in 2004 and continued until 2013. The second phase was scheduled for 2014–2029. By 2030, the site of the two reactors should have been ready for re-use. All the decommissioning activities are planned to be finished by 2038.

The city of Visaginas, which is referred to as "the city of Lithuania, the avant-garde of high technologies" (Šliavaitė, 2010:99) in the Visaginas City Development Programme for 2001–2030, evolved from scratch: people, construction, and building materials for the city and the INPP. The older population of Visaginas remembers the construction period as very socially and economically beneficial to the country (the Soviet Union is meant), and even to a certain extent as a heroic period of their life (Šliavaitė, 2010). The future residents of the Visaginas City came from various parts of the large country and were promised an urban lifestyle – new dwellings in blocks of flats, cinemas, schools, a medical centre, and stable jobs at the INPP. The artificially formed sense of identity in this type of a city is difficult to replicate in other cities as it evolved and changed. People there had to learn to live together and to adapt to one another. The mother tongue of many Visaginas residents is Russian. The population consists of builders, engineers and physicists, and high-quality specialists who came to work at the nuclear power plant and who stayed in Lithuania after 1991. This again demonstrates the authenticity of the city because its people developed a collective attitude where the material reality is less important than the pursuit and promotion of the common goal. Unlike in the West, however, the vision of the collective spirit of a Soviet industrial city was carefully designed and controlled by the state and responsible institutions and in many cases was dependent on the state's political and economic strategy (Pusca, 2010). The collapse of this system, and the change in the economic situation led to a change in attitudes towards collective identity largely attributable to the exaltation of the work process, heroism, and five-year plans. Thus, political and economic changes can be described as a scar in the sense of energy security and in relation to the INPP, which are closely related to the city of Visaginas. The German scholar Ackermann (2017:280) aptly describes cardinal changes and context changes: "The most important decisions regarding the nuclear power plant were made in Moscow up to 1990. Since 2004, the bureaucrats in Brussels have been answering the most important questions about the power plant, as the closure of both reactors costing billions of Euros is only possible with EU support."

The critical discourse analysis performed by the authors Mažeikienė, Kasperiūnienė and Tandzegolskienė (2018) highlights the main themes discussed in the media describing the work of INPP decommissioning and dismantling. Namely, the most frequently discussed issues are decommissioning management problems, terms of waste repository installation, decommissioning complex and complicated processes such as the provision of a funding mechanism, cases of the lack of transparency at the INPP, and the persistent lack of information on the ongoing work and its safety. Meanwhile, Visaginas, which was built for the continuous and stable operation of the INPP, is mentioned much less in the media and is not associated with the INPP as an object adjacent to the city. The main themes of media coverage on Visaginas include the search for the urban identity and various cultural events, new professional discoveries of the city residents, and strategic considerations of attracting external investors. The city is often referred to as a dead city that is always seeking for its own identity and a new face.

The suspension of the reactors marked the decline of this industrial city and prompted the search for a new identity. Now it is important for this city to develop a strategy for some space and something new that can attract tourists/visitors while preserving actual formal spaces, restoring new similar spaces, and creating a lasting memory of the city and its industrial heritage and landscape. Meanwhile, the closure of the INPP creates a new narrative that is associated to "gate closure", job losses, destruction of the structures, destruction of the inventory, "scattering" of the local community identity, and more dramatically, INPP is not currently engaged in a mutual dialogue on a coherent and sustainable search for and creation of a common conception of the city and the region. Considering the changes in the city, it is possible to maintain that the history of the city attracts artists and scholars. Annual events are currently organised in the city: Visaginas city festival "Celebrate Visaginas", Country Music Festival, International Festival of National Cultures "Rudeninė". Scientists work on various projects in Visaginas and search for answers and changes in the identity of the industrial city. One of the most famous is the Laboratory of Critical Urbanism by Ackerman, Cope and Liubimau (2016) that receives support from the German Embassy. Researchers of Vytautas Magnus University also carry out a project in this region; its title is linked with the search for the opportunities to develop nuclear educational tourism in the region (EDUTOM project).

An interesting photographic project "Babochka" (Photographer Jonty Tacon) aims to present the idea of a Soviet nuclear power plant by capturing the exotics of the past idea, the realities of the present moments during the closure of the plant, the need to separate the city from the "atomgrad" conception, and the disintegration of the community identity. The mood of the city and the community during the closure of the INPP are reflected in the documentary "Butterfly City" (Olga Černovaitė). It raises discussion on the problems, changes, and visions of the Visaginas residents in the context of the INPP closure. Additionally, the play "Green Grass" was created based on the stories of the INPP and Visaginas and told by the residents and INPP workers. It was successfully shown in Visaginas and other cities. The directors Jon Tertel and Kristina Werner represent the idea of INPP workers and residents of Visaginas who ultimately dream about dismantling of the power plant until all the turbines are cut, the reactors dismantled and the radioactive fuel buried, and nothing is left in the place of the plant except a green lawn.

The INPP dismantling can be observed on the monitor in the information centre of the Power Plant. It is also possible to see the dismantling project and obtain information on the work progress from the professional staff of the communications department. Interestingly, after the release of the film about Chernobyl, the flow of tourists to the power plant increased as the previous model of the Reaktor bolshoy moshchnosty kanalny (RBMK) reactor was assembled in Chernobyl, and some scenes of the film were filmed on the premises of that power plant.

During the presentation, one can discover that the city is looking for a new face and is on the right path in doing so in the cultural context as illustrated by the examples described. Besides, it is possible to distinguish the authenticity of the city in that it is multicultural and because it is located on the shore of the lake. Although this is typical of industrial structures, the pine forest gives the city a green face. Hence, another potential element of the transformation is the ecological context which is highly welcomed at the political and community level in the city. However, the historical context as an integral part of the transformation plays an important role in constructing the new identity of the industrial city and the post-industrial landscape. The exhibits or heritage objects of the city and the INPP should have common spaces and a clear narrative line. Moreover, it is advisable to plan educational tourism programmes at both formal and non-formal education levels. The main themes that can be introduced in educational tourism could include:

- Interesting and inspiring stories from the early days of industry to the present day;
- The emergence and development of creative and attractive places;
- History of local identity formation;
- Themes on the authenticity of the area developed on the basis of cultural events;
- Educational lectures, field workshops, laboratory work, and simulations as an opportunity for different age groups to know the history of the atom;
- The city in the form of a butterfly, distinguishing authenticity;
- Tourism development based on the film industry;
- Gastronomy and heritage.

The provided example of the Ruhr area in Germany demonstrates that critical moments of change encourage the search for new strategies through collaboration between different organisations and stakeholders, bringing together certain elements of the transformation with clear objects and processes. An example is the "Kulturpfad" blue stones of the Essen City Cultural Trail, which loudly announces that the city focuses on cultural symbols and the opportunity to explore the city through culture. The INPP secrets and the past history if rebuilt and revived could be associated with the goals of educational tourism. Furthermore, in collaboration with urban activists, politicians, and the community, a common and uniform conception of urban and regional industrial heritage and post-industrial landscape could be formed. Murphy and Boyle (2005) emphasise community engagement. In the case of the city of Visaginas, this initiative is gaining strength and is a significant condition for the city to change its identity. It would be desirable to look for the uniqueness of the city and the area linking it to specific places, projects, and people's experiences.

Generalisation

The industrial changes have led to radical processes of the revitalisation of industrial objects, as well as urban and post-industrial landscapes that are associated with identity rethinking, transformation, denial, forgetfulness, or decline. This is not an easy process because on one hand, it involves searching for a new narrative seeking to preserve the past experiences and scars and on the other demonstrating that forgetting or reviving are not easy processes. In the analysed regions and areas, the transformations have been largely affected by the social identity formation. This revival is observed through nostalgia, retreat and default, which inspire new value and refocus. Therefore, Lithuania does not intend to attach itself to industrial heritage objects as they are associated with negative and painful experiences that serve as reminiscences of utopian goals and large constructions, hard and pointless work, and typical and unified architecture. Industrial heritage objects and distorted landscapes also presuppose discussion on the scars of industrial revolution that exist and manifest themselves through landscape devastation, urbanisation, and the gigantic nature and inefficiency of factories and plants. Such scars are most acceptable because they identify the cause and aim of revitalising the landscape, the city, and the objects themselves. Regarding these scars, the revitalisation of the landscape or object is attempted to be embedded in buildings, stories, or interesting structures by slightly covering or retouching the pain. It is a bit harder for objects to be labelled as industrial heritage scars that resemble a certain historical period, when a place, a particular object, or landscape is romanticised, yet at the same time frighten us with the remaining heavy ruins. The third form of the scar includes objects, locations, or landscapes that are not perceived and identified as part of a cultural object. Consequently, there is a danger of collapse, forgetfulness, and difficult search for own identity. By reconstructing the identity of a place, city, or object and creating the memory of a place of industrial heritage, industrial culture telling about community traditions, values, and customs is also created; it is associated with an epoch of industrial prosperity created by regional identities and symbols; give local residents and newcomers an opportunity to participate in the region's memory preservation and in the process of search for identity.

Industrial heritage objects are part of the cultural heritage, which implies that it is important to preserve industrial heritage to society. The constructivist philosophical approach encourages discussion on the construction of a relationship through the narratives of history and everyday life, the demonstration of objects through archival data, and authentic stories and installations that link the experience of the past to the present. During an experiential trip to these objects, tourists/visitors can learn from the scar metaphor either consciously or subconsciously and seek an emotional connection to the location or object. Industrial heritage objects or post-industrial landscapes attract tourists/visitors as they symbolise the recent past. They bring them back to remembrance and allow them to gain experience using the available tools of experiential learning. These places and objects offer experiential learning opportunities and experiences that help analyse unconventional solutions by viewing artefacts, listening to stories, or studying constructions of the atypical architectural decisions through innovative educational, leisure, and experiential programmes.

References

Abukabur, M., A., Shneikat, B., H., T., & Oday, A. (2014). Motivational factors for educational tourism: A Case study in Northern Cyprus. *Tourism Management Perspectives, 11*, 58–62.

Ackermann, F. (2017). *Labai blogai arba liuks: Aštuonios pamokos apie Lietuvą*. Vilnius: Lapas.

Alanen, A., & Melnick, R. (2000). *Preserving Cultural Landscapes in America*. Baltimore: The Johns Hopkins University Press.

Alexander, N. (2010). *Kolmanskop: An industrial heritage resource or only a tourist attraction: the assessment of value with regard to Kolmanskop Ghost Town and the industrial landscape of the Sperrgebiet National Park, Namibia*. University of Cape Town [Master thesis]. University of Cape Town.

Ankomah, P. K., & Larson, R. T. (2000). Education Tourism: A Strategy to Strategy to Sustainable Tourism Development in Sub-Sahara Africa. *DPMN Bulletin, 7(1)*, 19-24.

Bairašauskienė, L. (2018). Tapatybes ugdymas idntiteto ir socialinio identiteto teoriju kontekste. *Scientific Research in Education, 2*, 5–22.

Baxter, P., & Jack, S. (2008). Qualitative Case Study Methodology: Study Design and Implementation for Novice Researchers. *The Qualitative Report, 13*(4), 544–559.

Bogulim, J., Strohmeier, P., & Lehner, F. (2012). *Viel erreichen – wenig gewonnen: Ein Realistisches Blick auf das Ruhrgebiet*. Essen: Klartext Verlag.

Capelo, S., Barata, F. T., & de Mascarenhas, J. M. (2011). Why are Cultural Landscapes of various Values? Thinking about Heritage Landscape Evaluation and Monitoring Tools. *Journal of Landscape Ecology,* 4(1), 5-17.

Copic, S. & Tumaric, A. (2015). Possibilities of Industrial Heritage Reuse as Tourist Attractions – a Case Study of City of Zrenjanin (Vojvodina, Serbia). *Geographica Pannonica, 19*(2), 44–49.

Čepaitienė, R., & Mikailienė, Z. (2017). *Pasaulis prasideda čia. Paveldo ugdymo principai mokyklinio amžiaus vaikams*. Metodinė priemonė. Vilnius: Didakta.

De la Torre, M., & Mason, R. (2002). *Assessing the Values of Cultural Heritage*. The Getty Conservation Institute, Los Angeles, 3–4.

Dobrila, L., Sladjana, A., & Maja, V. (2018). Sustainable Educational Tourism Potential of Djerdap National Park. Forum geografic. *Studii și cercetări de geografie și protecția mediului, 17*(2), 159–169.

Dorfsman, K. I., & Horenczyk, G. (2018). Educational Approaches and Contexts in the Development of a Heritage Museum. *Journal of Experiential Education, 41(2),* 170–186.

Drėmaitė, M. (2002). Pramonė kaip paveldo objektas. *Kultūros paminklai, 9,* 110–118.

Harazneh, I., Al-Tall, R. M., Al-Zyoud, M.,F., & Abubakar, A. M. (2018) Motivational factors for educational tourism: marketing insights. *Management & Marketing, 13*(1), 796–811.

Ismailova, G., Safiullin, L., & Gafurov, I. (2015). Using historical heritage as factor in tourism development. *Social and Behavioral Siences,* 157–162.

Jensen, U. J. (2000). Cultural Heritage: Liberal Education, and Human Flourishing. In E. Avrami & R. Mason (Eds.), *Values and Heritage Conservation*. Research Report (pp. 38–43). Los Angeles: Getty Conservation Instutute,.

Knoll, GM. (2014). *Roud der Industriekultur: Bewahrtes Erbe des Ruhrgebiets.* Grebennikov Verlag GmbH.Krinke, R. (2001). *Overview: design practice and manufactured sites.* In: Kirkwood, N. (Ed.), Manufactured Sites: Rethinking the Post-Industrial Landscape (pp. 125–149). Taylor & Francis, New York.

Loures, L. (2008). *Post-Industrial Landscapes: dereliction or heritage?* 1st WSEAS International Conference on landscape architecture (la '08), Algarve, Portugal,

Loures, L. (2016). Using Post-industrial Landscape Transformation as an Urban Development Strategy – Tales from Portugal. *Miesto zemelapiu formavimas, 13,* 217–227.

Loures, L. C., Rodrigues, F. M., Nunes, J. R., & Loures, A. (2017). Post-Industrial Landscapes: are they threats or opportunities? *International Journal of Environmental Sience, 2,* 295–301.

Maga, A. A., & Nicolau, P. E. (2018). Conceptualizing Educational Tourism and the Educational Tourism Potential (evidence from ASEAN countries). *Advances in Economics, Business and Management Research (AEBMR), 39,* 343–348.

Mason, R., & Avrami, E. (2000). *Heritage Values and Challenges of Conservation Planning.* In J. M. Teutonico & G. Paumbo (Eds.), *Management Planning for Archaelogical Sites:* An International Workshop Organized by the Getty Conservation Institute and Loyola Marymount University (pp. 13–16). Los Angeles: The Getty Conservation Institute.

Mažeikiene, N., Kasperiūnienė, J., & Tandzegolskienė, I. (2019). Nuclear media discourses after the slousure of the Ignalina Nuclear Power Plant: Is the game over? *Central European Journal of Communication, 12*(3), 335–360.

McGladdery, C. A. (2016). *The relationship between international educational tourism and global learning in South African high school learners*. [Unpublished doctoral thesis]. Pretoria: University of Pretoria.

McGladdery, C. A., & Lubbe, B., A. (2017). Rethinking educational tourism: proposing a new model and future directions. *Tourism Review, 72*(3), 319–329.

McLeod, S. A. (2017). *Kolb's Learning Styles and Experiential Learning Cycle. Kolb - learning styles*. Accessed 29-05-2019: https://www.simplypsychology.org/learning-kolb.html.

Meethan, K. (1996). Consuming (in) the Civilized City. *Annals of Tourism Research, 23*(2), 322–340.

Meier, L., & Aytekin, E. A. (2019). Transformed landscapes and a transnational identity of class: Narratives on (post)industrial landscapes in Europe. *International Sociology, 34* (1), 99–116.

Metropole Ruhr. Accessed 25-07-2019 http://www.route-industriekultur.ruhr/

Murphy, C., & Bole, E. (2005). Testing a concept al model o cultural tourism development in the post-industrial city: A case study of Glasgow. *Tourism and Hospitality Research, 6*(2), 111–128.

Nettleingham, D. (2018). Heritage Work: Preservations and Performances of Thames Sailing Barges. *Cultural Sociology, 12*(3), 384–399.

Pitman, T., Broomhall, S., McEwan, J., & Majocha, E. (2010). Adult learning in educational tourism. *Australien Journal of Adult Learning, 50*(2), 219–238.

Pusca, A. (2010). Industrial and Human Ruin of Postcommunist Europe. *Space and Culture, 13*(3), 239–255.

Regionalverband Ruhr: http://www.route-industriekultur.ruhr/, Accessed 02-08-2019.

Rehberg, W. (2005). Altruistic individualists: motivations for international volunteering among young adults in Switzerland. Voluntas. *International Journal of Voluntary and Nonprofit Organizations, 16*(2), 109–122.

Richards, GW. (2011). Cultural tourism trends in Europe: a context for the development of Cultural Routes. In Khovanova-Rubicondo, K. (Ed.). *Impact of European Cultural Routes on SMEs' Innovation and Competitiveness* (pp. 19–36). Strasbourg: Council of Europe Publishing..

Ritchie, B. W., Carr, N., & Cooper, Ch. (Eds.) (2003). *Managing Education Tourism* Clevedon: Channel View Publications.,

Rodríguez-Ferrándiz, R. (2014). Culture Industries in a Postindustrial Age: Entertainment, Leisure, Creativity, Design. *Critical Studies in Media Communication, 31*(4), 327–341. DOI: 10.1080/15295036.2013.840388.

Ruhr museum. Internet resource: www.ruhrmuseum.de Accessed 02-08-2019.

Sharma, A. (2015). Educational Tourism: Strategy for Sustainable Tourism Development with Reference of Hadauti and Shekhawati Regions of Rajasthan, India. *Journal of Knowledge Management, Economics, and Information Technology, 5*(4), 1–17.

Sie, L., Patterson, I., & Pegg, Sh. (2016). Towards an understanding of older adult educational tourism through the development of a three-phase integrated framework. *Current Issues in Tourism, 19*(2), 100–136. DOI: 10.1080/13683500.2015.1021303.

Šliavaitė, K. (2010). Deindustrializacija, socialinis nesaugumas ir išgyvenimo strategijos. *Lietuvos etnologija: socialinės antropologijos ir etnologijos studijos, 10*(9), 93–116.

Smith, L. (2006). *Uses of Heritage*. London and New York: Routledge.

Stone, M. J., & Petrick, J. F. (2013). Th Educational Benefits of Travel Experiences: A Literature Review. *Journal of Travel research, 52*(6), 731–744.

Storm, A. (2014). *Post-industrial Landscape Scars*. Macmillan, 5–98.

Sutestad, S., & Mosler, S. (2016). Industrial Heritage and their Legacies: "Memento non mori: Remember you shall not die". *Procedia-Social and Behavioral Sciences, 225*, 321–336.

Szromek, A. R., & Herman, K. (2019). A Business Creation in Post-industrial Tourism Objects: Case of the Industrial Monuments Route. *Sustainability, 11*(5), 2–17.

Urry, J. (1995). *Consuming Places*.

London and New York: Routledge.

Vangas-Sánchez, A., Plaza-Mejía, M. A., & Porras-Bueno, N. (2009). Understanding Residents'Attitudes toward the Development of Industrial Tourism in a Former Mining Community. *Journal of Travel Research, 47*(3), 373–387.

Yin, R. K. (2003). *Study Research. Design and Methods*. SAGE Publications. pp. 19–82.

Linara Dovydaitytė

The Pedagogy of Dissonant Heritage: Soviet Industry in Museums and Textbooks

Abstract: This chapter examines how Soviet industry is remembered and studied in both formal and informal education in post-Soviet Lithuania. Industrialization of the country coincided with and was forced by Soviet occupation (1940–1990), and the legacy of this industrialization is explored using a concept of "dissonant heritage". The central focus of this chapter is the in-depth interrogation of five museum displays and twenty history textbooks covering the period of high industrialization in Soviet Lithuania. Content analysis and ethnographical study of these pedagogical sources reveals that there are quite different and even competing narratives around Soviet industry. These range from celebratory stories of technological inventions and rapid growth of new industries to negative narratives about the Russification of the country and contamination of the land. While the image of industrialization as a Soviet colonial project prevails, the work and life of industrial communities remains untold. The absence of working-class experiences not only creates a double dissonance in the heritagization of Soviet industry but also creates a gap between educational narratives and the live memory students encounter in the reality. In addition, the chapter reveals that nuclear energy plays a significant role in the narratives of Soviet industry as the nuclear theme introduces the importance of sensibilities such as nostalgia and fear, into the discourse of "dissonant heritage".

Keywords: Soviet industry, nuclear energy, heritage, museum, textbook, Lithuania

Introduction

The initial idea for the chapter emanated from an observation of growing academic, cultural and artistic interest in the Ignalina Nuclear Power Plant (INPP) and its satellite town Visaginas (formely Sniečkus), situated in North Eastern Lithuania.[1] The INPP was built in the territory of Soviet Lithuania between 1975 and 1987, and housed two largest Reaktor bolshoy moshchnosty

---

1   Since the early 2000s the INPP and its satellite town has been an object of numerous international research projects not only in nuclear sciences but also humanities and social sciences, including history, architecture and social anthropology. The scholars are mosly interested in the INPP as an example of Soviet nuclear industry while Visaginas has drawn attention to it as a (post)socialist mono-industrial town. For references see: Freimane, 2016b; Stepanov, 2018. Among many other artistic and

kanalny (RBMK) reactors in the world. Engineered as an ambitious nuclear energy project of the Soviet empire, the INPP was attacked by environmental and national activists following the Chernobyl disaster in 1986.² After gaining independence in 1990, Lithuania decided to close its only nuclear power plant in order to meet the conditions of entry to the European Union, beginning the decommissioning in 2004³. Together with the closure of the plant, the thirty-year history of Visaginas as an atomic town came to an end. It seems that the end of the nuclear era in Lithuania motivated the need to begin the heritagization of this quintessential technology and industry of modernity. Yet the process of turning the past into heritage only starts here, and it is complicated.

It is complicated for at least two reasons: firstly, various groups of society relate differently to the INPP and the history of Visaginas. For the nuclear engineers and the majority of the population of Visaginas, predominantly Russian-speaking immigrants from the entire former Soviet Union, that place, represent both the progressive project of modernity and its collapse. For the environmental activists and the majority of the population of Lithuania, the INPP is associated with Soviet occupation – for most Westerners, with an insecure nuclear power station in the East. Yet Storm claims that none of these groups consider the INPP as "their own" heritage.⁴ After having investigated the INPP in 2010 and included it in an international comparative study, Storm states that "still the Ignalina is activated in neither local nor international context as a means for memory work and future orientation" (2014, p. 95).

Secondly, in post-Soviet Lithuania, the complicated relationship with the past is conditioned not only by the particularities of the INPP as a Soviet nuclear project, but also by a wider range of issues related to the Soviet industrial heritage in general. Until World War Two, Lithuania was essentially an agricultural country. Industrialization began here at the end of the 1950s and became the most important economic and social phenomenon (Misiunas & Taagepera, 1993, p. 183). The Soviet Union, which prized the geopolitical situation of Lithuania, invested huge resources in the country's infrastructure. In

---

cultural events the INPP was presented in the Baltic Pavilion at the 15th Venice Biennale of Architecture in 2016.
2   The INPP has the same reactor type as used in Chernobyl.
3   Two reactors were put into operation in 1983 and 1987 and were closed in 2004 and 2009, respectively.
4   According to Storm the nuclear community is the most obvious group who could claim Ignalina as their heritage. But she suggests that there is a grief rather than a conscious heritage process that takes place here (2014, pp. 94–97).

Soviet Lithuania, the development of new branches of industry was initiated; new factories and industrial complexes were built resulting in an escalation of rapid urban growth;[5] and industry, and, later, services surpassed agricultural employment.[6] Industrialization altered the social structure of the society and thus introduced the modern lifestyle and urban culture.[7] Therefore, the period of high industry in Lithuania means both Soviet occupation of, and radical change in the society, which subsequently turned away from rural and agricultural towards urbanization and industrialization.

Yet when Lithuania regained independence in 1990, the legacy of Soviet industry did not become heritage. On the one hand, with the change in the political system and the transition from planned socialism to "wild" capitalism, many branches of industry collapsed, factories went bankrupt, and industrial workers lost jobs and had to modify their qualification. Under the circumstances of the swift historical and political changes, nobody was interested in the collapsed industry. On the other hand, there was no place for narratives, experiences and memories of Soviet modernization, urbanization and economic welfare in the post-Soviet politics of memory. The dominant discourse presented the Soviet period as one of loss and trauma (Nikžentaitis, 2013); a negative assessment of the entire period (Safronovas, 2009), the narratives of the perpetrator and the victim, repression and opposition prevailed. According to Drėmaitė, popular consciousness, which identifies national identity with the rural past, associates the heritage of Soviet industry, primarily understood as "alien", with pollution, a Russian immigrant workforce and poor quality of products: "Industrialization is a Soviet legacy, thus it is not ours" (2012, p. 72).

Since the beginning of the 2000s, the concept of Soviet heritage began expanding and became more varied in Lithuania. Historiography started covering topics related to the life and culture of the society during the late Soviet period alongside the previously dominating political themes of Soviet

---

5   In 1950, 28 % of the Lithuanian population lived in the cities; in 1970, they made up 50 %; and in 1989, 68 % (Meškauskas, 1994, p. 253).
6   In 1970 industrial and construction employment surpassed agricultural employment. In 1980–85 industrial and construction employment peaked with around 40 % of total population, and services became the major sector of growing employment (Meškauskas, 1994, p. 260).
7   A modern family consisting of two people became commonplace, an urban youth culture formed, the forms of spending leisure time changed with the spread of television and cars, with the rise of living standards, consumerism became a dominant force, etc. (Misiunas & Taagepera, 1993, pp. 218–227).

repressions and opposition to the regime; diverse interpretations of the heroic period – the anti-Soviet guerrilla war – emerged. Museums and exhibitions began displaying fragments of Soviet industry, modern lifestyle and urban everyday life.[8] The history of the everyday, including the Soviet period, was incorporated into school textbooks dominated previously by the narrative about politics and economy (Bitautas, 2018). Such a change allows us to ask – what is considered to be the Soviet industrial heritage in contemporary Lithuania, how is it interpreted and for what purpose is it created?

The goal of this chapter is to provide a wider context for the heritagization of the nuclear industry in Ignalina by investigating how Soviet industry, including the nuclear industry, is remembered and how people can learn about it in Lithuania today. Two types of pedagogical sources have been chosen for analysis: museum displays and history textbooks. Pedagogy in this text is broadly conceived as a learning process taking place both at the institutions of formal educational and in the realm of broader culture. The latter is defined as "public pedagogy", which covers various forms and sites of education, including informal educational institutions (i.e. museums), popular culture (i.e. movies), dominant discourses (i.e. public policy) and social activism (i.e. grassroots movements) (Sandlin et al., 2011).

In accordance with critical pedagogy and culture studies, pedagogy is understood both as a practical and a political activity. It is associated not with the transfer of knowledge, but with the formation of experience and subjectivity: "When one practices pedagogy one acts with the intent of creating experiences that will organize and disorganize a variety of understandings of our natural and social world in particular ways" (Giroux & Simon, 1988, p. 12). In this sense, pedagogy is understood as a set of practices, inside or outside of schools, that "organizes a view of, and specifies particular versions of what knowledge is of most worth, in what direction we should desire, what it means to know something, and how we might construct representations of ourselves, others, and our physical and social environment" (Giroux & Simon, 1988, p. 12). Thus, both school textbooks and museum displays may act pedagogically

---

8 In the open-air museum in Grūtas Park, which was established in 2001 and soon became very popular, not only the remnants of Soviet culture are exhibited, but also the "banal socialism" is turned into heritage. For example, the experiences and memories of Soviet everyday life are revived through Soviet dishes in the Grūtas café (Lankauskas, 2006). The Energy and Technology Museum opened in the defunct Vilnius Electric Plant in 2003 presents the history of Lithuanian technology allocating a very important place to Soviet industry.

through the production of narratives and the creation of experiences that make us think about the past in some ways rather than others.

This chapter analyses five museum displays and temporary exhibitions, and twenty history textbooks recount Soviet industry. Among several science and technology museums open in Lithuania today, I have selected those that are clearly concerned with the industry of the Soviet period, including the nuclear industry. All currently approved[9] 5th, 10th and 12th grades' history textbooks covering the history of Lithuania of the second half of the 20th century have been analysed.[10] Two interrelated layers of museum displays and textbooks have been examined by using the method of content analysis: the semantic layer that comprises the thematic choices of the museum and textbook discourses; and the layer of material, visual and linguistic realization of the semantics. In other words, according to the general theory of semiology, the question is not only what is being said, but also how it is said.[11] Analysis of the museum exhibitions is accompanied by specially conducted interviews with museum workers and study of secondary sources (museum documents, websites, etc.). The main purpose of the research is to find out how Soviet industry is narrated in contemporary museums and school textbooks. The main focus of this study is not only the associations with the past whether negative or positive, condemning or sympathetic, created by the narrative, but also a more complex question inseparable from the study of heritage: what and whose histories are told by the

---

9 The textbooks approved by the Ministry of Education, Science and Sport of the Republic of Lithuania, valid for the academic year 2018–2019, are published in the database of textbooks and other school supplies supervised by the Education Supply Centre of the Ministry of Education, Science and Sport of the Republic of Lithuania: https://www.emokykla.lt/bendrasis/vadoveliai The database also includes several editions of the same textbook. If the new edition of the textbook completely repeats the contents of the old edition, only the new version of the textbook is used in this study. If the new edition contains significant changes, not necessarily in the text, but also in visual material, both editions are analyzed.
10 According to The General Curriculum Framework for Basic Education (2008) and The General Curriculum Framework for Secondary Education (2011), history is taught at Lithuanian schools in three concentric circles. The material of the first circle is taught in the 5th and 6th grade, the second, in the 7th–10th grades, the third, in the 11th and 12th grade history programme. This way the 50 years Soviet period (1940–1990) is studied in the 5th, 10th and 12th grade in Lithuania, starting with general introductory and going over to more complex questions.
11 For more on the semiological interpretation of museums see Ravelli (2006), on analysis of textbooks see Klerides (2010).

heritage of Soviet industry in post-Soviet Lithuania? What role is played here by scientific discoveries, technological objects, industrial management and the life of industrial society? What narratives, memories and experiences of Soviet industry are available for the practices of pedagogy today?

A (Double) Dissonance in Industrial Heritage

It is worth discussing the issue of Soviet industrial heritage (including the nuclear industry in Ignalina) using the concept of dissonant heritage. In their acclaimed book on heritage management, Tunbridge and Ashworth argue that all heritage is inherently dissonant: discrepancy and incongruity is characteristic of any product of heritage (1996). This is related to the very definition of heritage. Heritage is a product not of the past, but of the present, consciously produced with regard to the needs of the present. Heritage is produced by selecting and interpreting the past, thus it always contains a certain message that, implicitly and explicitly, conveys certain values. Due to this selection and interpretation process, dissonance is not a simple byproduct of the production of heritage, but a constitutive part of that product. To illustrate this thought, Tunbridge and Ashworth point out that "all heritage is someone's heritage and therefore logically not someone else's: the original meaning of an inheritance implies the existence of disinheritance and by extension any creation of heritage from the past disinherits someone completely or partially, actively or potentially" (1996, p. 21). The heritage produced through selection and interpretation always represents historical experiences of specific social, ethnic, or religious groups and disinherits other groups whose "distinctive historical experiences may be discounted, marginalized, distorted or ignored" (Tunbridge & Ashworth, 1996, p. 29).

In the 1990s, soon after the restoration of Lithuania's independence, persons and their families who had suffered from the occupation regime became the principal group which inherited the Soviet past. The Lithuanian Union of Political Prisoners and Deportees representing them initiated the main museum in Lithuania dedicated to the Soviet past: the Museum of Occupations and Freedom Fights (founded in 1992, it was titled the Museum of Genocide Victims until 2018). The community of former prisoners and deportees have founded in total around 40 museums and expositions in Lithuania, which, like the main museum, also present the Soviet period through the narrative of terror and resistance (Rindzevičiūtė, 2015). The social group that was clearly dispossessed of the Soviet past was the former working class. During the late Soviet period industrial and construction workers constituted a considerable portion of the society making up around 40 % of the total population in the early 1980s. Yet, considering the fact that the Soviet period is a period of occupation, one should not oppose those groups to each other. On the contrary, the membership of both groups – former prisoners/deportees and former workers – partially overlaps. Thus, in this case, we can state that in post-Soviet Lithuania, the former working class turned part of Soviet legacy into heritage (traumatic and heroic historical experiences), and the other part of legacy (historical experiences of work and everyday life) have remained the past, partially history, but have not become heritage. And a portion of the former working class, which consists of Russian-speaking immigrants[12] or persons that have not suffered from the regime,[13] have become disinherited from the Soviet past.

Considering the heritagization of Soviet industry in post-Soviet Lithuania, one could speak about a double dissonance by which I mean that former industrial communities have been doubly disinherited from both the Soviet past and the Soviet industry.[14] On the one hand, after the historical turning point in 1990, the (former) industrial communities did not claim the heritage of Soviet industry as their own (Drėmaitė, 2012) and were disinherited. On the other hand, the issue of social class alone causes dissonance in production of industrial heritage (Tunbridge & Ashworth, 1996). Having referenced industrial museums of Western countries, especially Britain, as an example, Tunbridge and Ashworth state that the main disagreement lies in the question: which and whose stories are being told by former factories and industrial sites. Class dissonance is the most important element when one decides which narrative about the industrial past to choose: that of the technical progress or of the history of work, that of free enterprise or of capitalist exploitation, that of the elite or of the working class (Tunbridge & Ashworth, 1996, p. 78). Class dissonance defining capitalist industry may be relevant also to Soviet industry, which was formed in reality by the ruling class of the Communist Party (or the nomenclature) and

---

12 Immigration to Soviet Lithuania where Lithuanians constituted about 80 % of the population was negligible.
13 For example, those who owned little land in pre-war Lithuania and for whom Soviet occupation brought greater economic welfare and social stability.
14 By double dissonance I mean that in the case of Soviet industrial heritage former working class is doubly disinherited: 1. The main inheritors of the Soviet past are those who suffered but not benefited from Soviet regime (workers got jobs, flats and social insurance so we can say they benefited from the system) 2. The main inheritors of Soviet industry tend to be Soviet ruling class and central government but not industrial workers which I try to show in the text afterwards.

the lower working class.¹⁵ Thus, the question which and whose stories should be told by Soviet factories and industrial sites that are being converted into museums or texts about former industries is open and problematic.

Industrial Heritage: What and Whose Stories?

In Western countries, the heritagization of modern industry started in the 1960s and 1970s as a consequence of and reaction to the end of the industrial era that lasted for 200 years.¹⁶ After the collapse of various industries, many factories and industrial sites were closed, demolished or left to decay. At the same time, academics and history amateurs started collecting, preserving and exhibiting industrial heritage, at first in Great Britain and later in other countries. Today industrial heritage is considered to be part of cultural heritage: it is defined by the various legal documents of international organizations, it is included in the lists of values to be protected, some countries treat industry as a symbol of their national identity (Vargas-Sánchez, 2015), and industrial tourism is considered to be an important factor in local regeneration processes (Hospers, 2002).

The most important international guidance document passed by The International Committee for the Conservation of the Industrial Heritage (TICCIH)¹⁷ gives the following definition:

> Industrial heritage consists of the remains of industrial culture which are of historical, technological, social, architectural or scientific value. These remains consist of buildings and machinery, workshops, mills and factories, mines and sites for processing and refining, warehouses and stores, places where energy is generated, transmitted and used, transport and all its infrastructure, as well as places used for social activities related to industry such as housing, religious worship or education (2003).

---

15 The opposition of Soviet colonialism could match another division characteristic to the imperialist capitalist industry of "provincially/colonially ruling classes or the exploitation of peripheral/subject populations": the central power in Moscow or the workforce of the occupied countries.
16 Although the notion of "technical heritage" was used in Soviet Lithuania the heritagization of modern industry did not take a place here. In the late 1960s and 1970s, when Western countries already faced industrial crisis, the Soviets experienced the culmination of industrialization and felt no need to preserve it (Drėmaitė, 2012).
17 TICCIH has been ICOMOS's specialist adviser on industrial heritage since 2000 and assesses industrial sites for the World Heritage List.

This universally accepted document relates industry to all areas of life – from science and economics to everyday work and life. From the beginning, however, in the practice of industrial heritage there is an ongoing debate around its principal subject: the history of technology and business or social history of work (Storm, 2014).

In the 1990s, while critically analysing the growing number of industrial museums, Fitzgerald distinguished three widespread ways of showing industrial heritage: the "internalist style", the "celebratory style" and the "social and cultural historical style" exhibitions (1996). The internalist style is characteristic to the oldest displays of science and technologies that present industry through showcasing technical objects. Machines – steam turbines or locomotives – are exhibited as works of art: intrinsically grand, beautiful and valuable. This style of exhibiting separates industry from the history of hard work, from the wider social and cultural context. The celebratory style exhibitions relate technological objects to people's stories, but only those of a certain group – industrialists and engineers, and not the stories of workers. Furthermore, they present industry uncritically as a constant technological progress, thus "the worst celebratory style exhibitions are guilty of propaganda" (Fitzgerald, 1996, p. 119). Internalist and celebratory exhibitions conceal complexities, ambiguities and controversies, which are characteristic to any industrial heritage (Cossons, 2012). Those displays remind rather of temples where science and technology are worshipped (Fitzgerald, 1996).

The third, the newest, type of exhibitions links modern technologies to their use and impact on people's lives. The social and cultural historical style exhibitions focus on the way technologies have affected different social groups, including workers. Those exhibitions use more varied sources; alongside technological objects and documents, they utilize personal testimonies and memories. Yet Fitzgerald points out that focus on the social and cultural impact of technologies does not necessarily make this type of exhibition more analytical and critical. He quotes *The Great Railway Show* at the National Railway Museum as an example which paid much attention to the travelling habits of the British royal family but told nothing about railway accidents or railway workers' strikes (1996, p. 124).

Davies adds to this debate by discussing the presentation of work and communities related to that work at museums and heritage sites. She claims that the debate over what is more important – technological equipment or workers' activities – does not exhaust the subject of work. Any work, industrial activity in this case, creates an entire social and cultural phenomenon: "Work has many socio-political by-products and is not simply a process of performing a particular task" (1996, p. 114). Industrial work creates industrial communities, which are characterized by a social and cultural lifestyle defined by their work activities, but not limited to them (for instance, football as an important part of the life and identity of an industrial community). Today the social and cultural historical style exhibitions seek to present an exhaustive image of industrial community life by recounting narratives of work, education, leisure, the everyday and the lifestyle in and around factories and industrial sites.

The aforementioned typology of industrial exhibitions helps us to understand the diversity of ways to show industrial heritage, but in practice those three different types often coexist in one exhibition or heritage project. A good example of the latter are industrial museums and heritage sites making up the Route of Industrial Heritage in the Ruhr area in Germany. The history of local industry is told here in very different ways: from the Engine House in the Zollern colliery, which is presented as a modern temple of industry, to the half-celebratory, half critical display of industrialist life and activities in the former administrative building of the same colliery, to simple, mundane objects such as the half-full bowl of soup displayed as a rare specimen in the Ruhr museum which informs us about the peculiarities of daily life of an industrial worker's family.[18]

Industrial Heritage and Nostalgia

Like other kinds of heritage, industrial heritage is strongly related to the question of identity. Heritage, similarly to language or religion, functions as an identity marker. It is used "to construct narratives of inclusion and exclusion", which usually define a community by outlining its specific characteristics through the difference from another community (Graham & Howard, 2008, p. 5). Heritage is also used in order to support the continuity of identity. The narratives and representations of the past provide the feeling of continuity for the present society, relate the present to the past and simultaneously present the past as a finished stage and opens opportunities for the future. Thus, people's lives are localized in a "safe" linear narrative that links the past, the present

---

18 Through an ordinary bowl of soup Mrs Ritter from Kirchhellen tells a story that she could only half-fill the soup bowls as otherwise the soup would spill out over the side. This is only one of the effects on daily family life of the subsidence caused by mining in the building zone. For years one side of the house has been sinking by a little more than millimetre a year. This information is presented in accompanying label in the Ruhr Museum.

and the future. The past, "once translated into heritage", appears as the basis of identity, which provides "familiarity and guidance, enrichment and escape" (Graham & Howard, 2008, pp. 5–6).

Macdonald relates the movement of industrial heritage to the efforts to preserve and rethink the identity of a community in the changing world. As the need to preserve rural heritage emerged under the circumstances of industrialization, attention to the industrial past emerged during the period of de-industrialization, i.e., in the face of change and loss. By referring to the museums of everyday life, Macdonald argues that such means of heritage help to articulate local and communal identity, to salvage ways of life that have been vanishing and preserve a certain difference and independence while encountering changes (2013). While understanding that the efforts of communities to musealize everyday objects and everyday lives could mean simply a defensive or compensatory reaction upon encountering menacing social changes, she claims that these efforts are still worthy of academic analysis. From the anthropological point of view, "the emphasis on everyday things (and lives) is an ultimate extension of this idea – everything can be salvaged [. . .], and all lives given recognition (in an appropriate identity-displaying agency such as a museum)" (2013, p. 160).

A certain product of heritage is inseparable from the efforts to preserve the past, especially its ordinary, mundane aspects: nostalgia.[19] Nostalgia as a longing for a (lost) home is an especially popular and ambiguous concept in post-Soviet memory studies. It is possible to distinguish two ways of using the concept of nostalgia in Lithuanian memory studies: a negative/evaluative one and the neutral/analytical one. The first is used in studies where nostalgia is understood as any remembrance of the Soviet era not marked by clear negative attitudes. In her analysis of the public discourse (such as public holidays, monuments and museums), Čepaitienė notes that post-Soviet memory is framed by two opposite attitudes: the denunciation of the Soviet period and its nostalgic remembrances (2007). Both the interest in Soviet culture (films, art exhibitions, heritage tourism) and positive memories of Soviet everyday life are treated as nostalgia here (Čepaitienė, 2009). Thus, nostalgia not only

---

19 "If it is the interpretation that is traded, not its various physical resources, then at one level a heritage product is a particular service, such as a visit to a museum, theme park or historic city, but at a deeper level it is an intangible experience – whether it is nostalgia, pleasure, pride or something else – that is the product." (Tunbridge & Ashworth, 1996, p. 27).

starts mistakenly referring to various memories of the Soviet period,[20] but also becomes a negative concept: "Nostalgia for Soviet times is related to anti-democratic, pro-communist, and populist sentiments."[21]

Social anthropologists use the concept of nostalgia more productively by treating it as an analytical tool for understanding social memory. While examining biographical narratives of ordinary people, including workers, Šutinienė observes that the interpretations of the Soviet period recorded in autobiographies are different from the public discourse; they include fewer ideological appraisals, stereotypes and "amnesias" (2003). Anthropologists emphasize the ambivalence of memories and evaluations. The analysis of biographical narratives indicates that today people long for everyday life but not for the Soviet system (Klumbytė, 2004), for sociability rather than socialism (Lankauskas, 2006), for the past "not as a good life but as a life lived well – with dignity, pride, a sense of purpose, with social savvy and skill" (Lankauskas, 2015, p. 53). Soviet nostalgia is studied as a social practice which conveys more about the present than about the past. Klumbytė understands nostalgia as a force structuring post-Soviet social life:

> "[I]n nostalgic reminiscences of Soviet times villagers and marginalized urban residents reclaim visibility, voice their concerns, and appeal for respect, recognition, and inclusive citizenship. By accusing the nostalgics of having a false consciousness and remaking them into social others, the mainstream public [...] repeatedly deny their right to a respectful citizenship and exclude them from the post-Soviet modernity project" (2009, p. 110).

Thus, nostalgia here is not a longing for the past, but a comment on current social injustices.

---

20 "[...] memorial discourses and practices are constituted through a multiplicity of competing genres. Nostalgia is just one of them. Genres and subgenres of memory may be helpful heuristic devices, but they can (and do) quickly distract us from the complexity and complications of 'really existing' memory in social life. They tidy up or model memory into bounded units of analysis. It is not, however, the tidiness but messiness of memory that we need to describe and interrogate" (Lankauskas, 2014, p. 41).

21 "[...] Negative labelling of nostalgic people circulates in stories about village and small-town residents drowning in alcoholism, women giving birth to children solely in order to get government benefits, and people either avoiding work or relying on a questionable work ethic and dishonesty" (Klumbytė, 2009, p. 100). The author notes that nostalgia in the post-Soviet public memory is associated with a disease, a virus and similar negative phenomena.

Nostalgia could also be understood as an important tool for creating personal identity. In their study of the autobiographies of the first generation born during the Soviet period, Žilinskienė et al. note that "ordinary" people remember the Soviet period either neutrally or with ambivalence or nostalgically. Nostalgia appears when the Soviet period is remembered as an important time for forming identity (2016). The biographies of industrial communities allocate a particularly important place to nostalgia. Since the working class and industrial work played a very important role in Soviet ideology, a radical "devaluation" of this social class happened after 1990.[22] In the changing political and economic situation, industrial communities encountered not only unemployment and social insecurity, but also an absence of meaning in their life and work. Šliavaitė who has analysed the reactions of the INPP workers to the closure of the plant notes that at least part of this nuclear community grounded their worldview on the ideas of modernity, such as progress and growth. Thus, in the face of de-industrialization, they feel that they are losing not only the economical, but also the ideological foundation (2010a).

Moreover, in the works of anthropologists, post-Soviet nostalgia is "legalized" by comparing the experiences of the Lithuanian population to the experiences of Western societies. For example, one of the most important things that people still hanker after is a standardized life marked by a clear scenario, characteristic of modern societies. Not only post-Soviet, but also post-industrial Western societies long for the stability and security associated with such a life, while trying to deal with the unpredictability and uncertainty of post-modern life (Žilinskienė et al., 2016). The collapse of industry elicits a similar response from local people who can nostalgically remember former industry as the core of a local community, as related both to local history and to the inhabitants' identities and ways of life (Šliavaitė, 2010b).

Although nostalgia is a longing not only for a lost home, but also for a home that has never existed and has been idealized, it is an important part of individual and collective memory. Anthropological memory studies ascribe a formative function to nostalgia, both in the work of creating identity and in dealing with social challenges. On the other hand, nostalgia rarely appears in its pure form and not all positive or neutral, non-evaluative narratives about

---

22 Kideckel defines the situation of the working class in post-communist Eastern and Central Europe in the following way: "The meaning of the workers' lives and concerns are dismissed and the very category 'worker' or 'industrial worker' is made almost invisible in public discourse" (2002, as cited in Šliavaitė, 2010a, p. 68).

the past are nostalgic (Lankauskas, 2015). Most often, nostalgia gets mixed with recollections of a different type that make up a complex, contradictory and ambiguous memory.

What's Industrial in Industrial Museums?

The central museum presenting industrial heritage in Lithuania is the Energy and Technology Museum opened in 2003 in Vilnius.[23] This is a site-specific museum founded in a defunct electric power plant,[24] which is in a way symbolical, bearing in mind that the basis of modern industrialization is electrification. The museum serves a dual purpose: the preservation of the authentic building of the plant and its equipment as well as the presentation of the history of Lithuanian energy and technology.[25] The museum display consists of four themes. The exhibition "Energetics" tells the general history of electricity and the development of the energy industry in the country, while highlighting the Vilnius power plant. The exhibition "Made in Vilnius" presents technologies and various industries developed in Vilnius. The exhibition "Transport" showcases the historical collection of automobiles and motorcycles. The historical part of the museum display is supplemented by the interactive exhibition called "Science and Technology for Children". This study analyses exhibitions presenting the first two topics.

These exhibitions use several ways of presenting industrial heritage. The major part of the "Energetics" display is based on the object-centred principle; it

---

23 The museum was opened in 2003, celebrating the 100th anniversary of the first public power station in Vilnius. In the beginning, the museum used only 13 % of the floor space of the former power station. At the end of 2008, the whole building was reconstructed preserving the authentic equipment of the power plant and installing the permanent exhibitions in the 5000m^2 area. In spring 2019, a part of the museum was closed for renovation, which is planned to be finished in 2020. The exhibitions discussed in this text can be seen on a virtual tour: https://etm.lt/virtualus-3d-turas/
24 This was the first public electric plant in Vilnius, which operated from 1903 to 1998.
25 On the museum's website, the main purpose is defined in the following way: "to collect, preserve, research, exhibit and promote the history of Lithuanian energetics and technology and events related to it, to design a display and organise other exhibitions and events". The mission of the museum found in the 2014–2018 strategic plan says that this is a museum "cherishing the object of industrial heritage – the building and equipment of the old electric plant, exploiting the unique cultural educational space dynamically and harmoniously, environment (not only nature) friendly, creating a unique educational, cultural and scientific added value".

Fig. 1: Model of Ignalina NPP at the Energy and Technology Museum, Vilnius, 2018. Photo by the author.

mostly focuses on displaying technological equipment: steam-boilers, turbines and the plant's control panel. Models of other Lithuanian electric plants, such as the INPP (Fig. 1) and the hydroelectric plant in Kruonis, are exhibited adjacent to it. Technological machines, accompanied by the labels with limited information, are supposed to fascinate visitors with their shapes, magnitude and authenticity. Here, similarly to art museums, visual and textual didactics providing context, in this case presenting the history of the electric plant, are displayed next to objects valuable in themselves. The internalist style of the display might be illustrated by an excerpt from the annotation of the Turbine hall, which is clearly dominated by technological aesthetics at the expense of human work: "The Power plant's control panel was installed on a special platform with marble plates framed by oak fillet and decorated by a clock on the top. The control panel was manned permanently by a person."[26]

---

26 This is the only mention of work in the annotation of around 150 words length.

Fig. 2: Fragment of the exhibition "Made in Vilnius" at the Energy and Technology Museum, Vilnius, 2018. Photo by the author.

The exhibition "Made in Vilnius" says more about the social and cultural history of local industry, from pre-modern crafts to contemporary laser industry. Naturally, the major part of the exhibition is dedicated to the high industry of the Soviet period. The centre of the exhibition features industrially produced objects, such as sewing machines, vacuum cleaners, television and radio sets, or shoes. Objects perform a dual function here: they represent the products of modern industry (made in Soviet Vilnius) as well as modern domestic appliances, everyday life utensils. The impact of industrialization on the way of life is further emphasized by exhibiting a reconstruction of the apartment typical of the late Soviet period (Fig. 2). The easily recognizable furniture and domestic utensils, the "scenography" of the room and typical interior details (such as a popular domestic plant) easily act as memory triggers for visitors of a certain age, which in turn can invoke nostalgic or other memories. The panels displaying texts and documentary photographs, on the contrary, inform about industrialization as a part of the Soviet regime (by emphasizing such themes as planned economy or Soviet customs, e.g., product thefts) and as a force organizing the social life (by portraying the factory as the main source of material and social life for the workers).

Interestingly, the museum presents various industries from Soviet period in a positive or neutral light – whether those are the models of electric plants displayed for the viewer's pleasure, or a panel presenting local laser industry, which has been a source of pride since the Soviet period. The celebratory style of displaying technologies does not raise problematic questions such as the safety of nuclear power or the manner in which industry damages the environment. In other words, the museum does not participate in the debate on controversies surrounding practically all modern technological inventions and their use. Ambivalence rather than celebration accompanies the theme of industry in relation to a modern and urban life. The museum presents modernization as a Soviet project. Although the exhibition "Made in Vilnius" shows modern objects and the living environment as important markers of modern lifestyle, the visual-textual panels contextualizing the exhibition emphasize Sovietization rather than modernization. This might be illustrated by the visual-textual panel in which the facts of the growth of industry are presented ("In 1985, there were 84 industrial enterprises employing 114,000 people in Vilnius."). The text with facts is accompanied by a photograph of the interior of a modern flat, its owner hovering dust, and the caption under the photograph declares: "The modernization of domestic environment hardly concealed the abnormality of Soviet life."

Fitzgerald notes that, despite the fact that technological progressivism "in the post-Chernobyl era" is no longer valid, yet for industrial museums, celebratory style interpretations are still the "common-sense" approach. The reasons for that are the following: the museum's natural wish to celebrate its collections; popular sources suffused with technological progressivism (books, television, films), which often serve as reference for museum curators, and also corporate sponsorship of a museum or an exhibition (Fitzgerald, 1996). The fact that the Energy and Technology Museum displays (Soviet) industry in a celebratory rather than critical way, thus presenting a partial view of science and technology, could be somewhat determined by its stakeholders. Besides the Vilnius city municipality, other stakeholders of the museum include organizations directly related to the energy and district heating business: National Association of Lithuanian Energy, Lithuanian Electricity Association, AB Vilnius Heating Networks, and Lithuanian District Heating Association.

A further similar example of a museum, only this time funded by a single corporation, is the INPP Visitor Centre (named as The Communication

Division in the structure of INPP, see the website of the plant). It was founded in 1995 as a department of communication with the society in the then still operating nuclear power plant. Today it functions as the corporation's visitor centre and is the main arena presenting the nuclear energy industry in Lithuania. The Visitor Centre is established in the administration building; its exhibition modestly occupies one room and the adjacent corridor. The main objects in the exhibition space are the model of the nuclear power plant and models of two repositories for radioactive waste (Fig. 3). The rest of the space is occupied by the free-standing panels featuring short texts and one or several photographs accompanying them. The exhibition also displays an original nuclear fuel channel and the spent fuel cask CASTOR, a couple of dummies dressed as workers of the power plant, and also a collection of costumes for actors who participated in the series *Chernobyl*, abandoned after the filming in the nuclear power plant.[27] Historical photographs documenting the construction of the plant are displayed on the walls in the corridor. A television screen presents a live broadcast of the works at the Spent Fuel Storage Pool.

It goes without saying that the main focus of the exhibition is on the INPP itself. First, it presents technical features of the RBMK reactor as well as the history of building and operating the INPP. The second important theme is the decommissioning of the nuclear power plant and the handling of the radioactive waste as well as related technological challenges and solutions. The third topic discussed is the natural and man-made radiation; the use of radiation in various industries, including nuclear energy. One could say that the display, despite one panel presenting Visaginas as the satellite town of the INPP, is produced in the internalist style and concentrates solely on the technological aspect of the nuclear industry and the decommissioning of its objects. Yet the INPP Visitor Centre does not function as a traditional museum where the visitors are invited to explore its contents independently. Almost all visits to the Visitor Centre take place in the form of guided tours. Thus, in order to learn how nuclear industry is narrated here, one needs to join the guided tour.[28] The

---

27 In spring 2018 episodes of the historical drama TV miniseries *Chernobyl* for HBO were filmed at the INPP. INPP was chosen for filming due to its visual similarity to the Chernobyl nuclear power plant and also because both plants used the same RBMK type reactors.

28 This research is based on the interviews with two employers of the INPP Visitor Centre (26 September 2018), participation in the guided tour around the display of the Visitor Centre and an interview with the guide (27 September 2018) and participation in the guided tour inside the plant (23 September 2019).

Fig. 3: The models of two repositories for radioactive waste at the INPP Visitor Centre, 2018. Photo by the author.

contents of the tours are mostly determined by the group of visitors, their background knowledge and interests.[29]

The INPP Visitor Centre organizes two types of visits. The free guided tours take place around the display in the Visitor Centre as described above. Paid tours are organized to the premises of the nuclear power plant, including the reactor hall, the control room and the turbine hall, in which the dismantling of works is being carried out. Thus, the visitors can experience the nuclear power plant in both direct and mediated ways. In both cases the plant is presented or experienced as an exceptional *oeuvre* of science and technology, surprising in its scale, complexity and (former) capacity, in accordance with what Hecht

---

29 Visitors to the INPP might be roughly divided into two unequal groups: nuclear energy professionals, i.e., those who have specialist knowledge in this area, and other visitors, i.e., the so-called lay public. In this study I focus on the second group of visitors and the narrative about nuclear industry offered to them.

calls "nuclear exceptionalism" (2012). For instance, the story in the Visitor Centre begins with the highlights such as the fact that in 1993, the INPP was included among the Guinness World Records because it produced the largest percentage of the general electric energy needed by the state throughout the entire history of nuclear energetics. The visitors walking around the power plant are astonished by the size and complexity of its premises and equipment, the embodiment of technological expertise and mastery, an experience similar to the admiration of complex and great works of art. Yet a visit to the power plant is also about the fascination with grandiosity and insecurity at once, since we are dealing with highly dangerous technological outcomes such as radioactive waste here. During the tours, questions are often asked about the radiation level; sometimes a strong fear of radiation is expressed.

The participants of tours into the so-called controlled zone are most impressed by the very procedure of entering the power plant. The visitors need to book in advance, their identities are checked, all personal items should be left behind, one has to change all clothes down to underwear, wear helmets, gloves and, in some places, a respirator. After the tour, one not only gets changed, but also radioactivity levels are checked and if they have increased, one might have to take shower. In the power plant that is being dismantled, the same requirements of radiation safety and nuclear security are applied as were in place while it was operating, thus producing some sort of performance involving the participants of guided tours. Therefore, visitors can perceive the magnitude and danger of the power plant as a specific technological object live, through bodily and sensory experience.

The theme of safety is central both to the Visitor Centre and while walking in-situ, yet it is not questioned critically. One of the main references without which no tour can take place is the question of Chernobyl: "The negative shadow of Chernobyl is always near us." Tour guides discuss this topic with the visitors not as a wider and still unsolved (or unsolvable) problem of nuclear technology and industry, but as a particular case caused by the faulty management of Soviet nuclear industry. Chernobyl is presented not as a disastrous consequence of the nuclear industry, but as an accident that has prompted positive changes such as technological safety improvements in operating nuclear power plants, including the INPP.

A further important topic is the decommissioning of the INPP, i.e., the process that is taking place now, which can be experienced while walking around the power plant or watching a live broadcasting at the Visitor Centre. The decommissioning is presented as a process at least as complex as the construction and management of the most powerful nuclear reactors some time ago, because this is the first project of immediate dismantling of RBMK type reactors in the world.[30] While briefing the visitors about the challenges of decommissioning, the tour guide mentions that it is still not clear how to destroy radioactive graphite, and the global community of nuclear industry is waiting for the results of an experiment that will be carried out at the INPP. Radioactive waste, its longevity and means of managing it are also discussed during the tours. One of the main facts stimulating the visitors' (lack of) understanding and imagination is information that approximately 500,000 years will be needed for the used fuel to become equal to natural uranium in terms of radio toxicity. The guide presents another fact as a scientific (and political) challenge: nobody knows yet how the long-lasting radioactive waste will be managed after the 50 years during which it will be preserved in constantly monitored containers.

Depending on the visitor group interests, topics related to the social and cultural history of the INPP are also considered. While visiting the power plant, one has an opportunity to talk to the employees, the majority of whom have been working here since the very opening of the plant. From the people working in the plant one could hear stories about their job and life story, learn why they came to the INPP and how they settled here. The guides also present curiosities of living in a closed city known as Soviet atomgrad. The visitors are interested in the urban development of the city built from scratch, the multinational and bilingual community still living here, the health and diseases of the plant personnel or even fishing in the waters of the Drūkšiai Lake which was used to cool the reactors. The effect of nuclear industry on the environment, similarly to the theme of safety, is not discussed critically. For example, the guide, while talking about fishing, mentions that due to the impact of the INPP, the temperature of water in the lake has risen several degrees, in the wake of which the flora has changed and some kinds of fish have disappeared. Yet she also points to a positive change: more carps have emerged in the lake, and next to the water pumps, one could catch fish with bare hands.

As a place of nuclear industry heritage, the INPP is a paradoxical place. The defunct nuclear power plant, thanks to the complicated and partially

---

30 There are three possible options for decommissioning: immediate dismantling, deferred dismantling, and safe conservation and entombment. Ignalina NPP uses immediate dismantling when the equipment is dismantled practically immediately after the closure of reactor's operation. https://www.iae.lt/en/activity/decommissioning/153

experimental decommissioning process, is still an actively functioning nuclear enterprise[31] and will remain so at least until 2038. Here, the history of nuclear industry is both demonstrated and simultaneously destroyed; although it is practically impossible to destroy it, bearing in mind the longevity of radioactive waste. Perhaps, the planned nuclear waste repositories will at some time become places of nuclear industry heritage. Today the nuclear industry heritage is celebrated via emphasizing the uniqueness of the INPP by pointing out that this was the most powerful nuclear power plant in the past and is an exceptional decommissioning project in the present. Compared to the Energy and Technology Museum, it is interesting that the Soviet story-line is almost eliminated in the narrative of the INPP, although Soviet legacy is vividly experienced during the visit, e.g., through the domination of the Russian language inside the power plant.[32] The Visitor Centre presents the INPP as a scientific and technological oeuvre rather than a product of the complex sociotechnical system. Museum narratives emphasize the safety of nuclear industry,[33] although physical experiences during the visit may evoke insecurity and nuclear fear.

If both museums with permanent displays under discussion here present Soviet industry as more or less a history of technological innovation, various temporary museum projects rather focus on people's life around the industries. The first such project was a travelling exhibition "Dream Factories. Industry and Modernism in the Baltic Sea Region 1945–1990" hosted by the Energy and Technology Museum in autumn 2009. Organized by The Workers' Museum in Copenhagen and the National Cultural Heritage Agency under the Danish Ministry of Culture, the exhibition stemmed from a three-year collaborative scholarly research project carried out at Northern and Baltic universities with the aim "to examine the connections between industry and modernism and explore how technology, industry and modernism have affected the everyday life and culture of the North European people" (Drėmaitė, 2009, p. 142).[34]

---

31 According to the data of the 1 January 2019, 1,901 employees were involved in the dismantling of the power plant. Around 5,000 people were employed when the power plant was operating.
32 The aforementioned possibility to talk to an employee of the power plant means to speak Russian.
33 Emphasis on security is a common practice in visitor centres of nuclear power plants in order to counter anti-nuclear arguments and sentiments.
34 The historian of architecture, Marija Drėmaitė, who participated in the research project, was responsible for the Lithuanian part of the exhibition, in collaboration with the Energy and Technology Museum and Vytautas Suslavičius, an engineer of the Elektrėnai power plant.

The exhibition was based on the first-person approach and told stories of the life and work of seven workers at seven different industrial enterprises in Soviet Lithuania, Latvia and Estonia as well as social democratic Denmark, Finland, Norway and Sweden. The Lithuanian case was presented through the story of Algis Mišinis who worked on the construction of the largest industrial enterprise of the 1960s – the Lithuania Power Plant and its satellite city, Elektrėnai. The story of every protagonist of the exhibition consists of six themes: dreams about the factory, work at the factory, factory and home, factory and leisure, factory and the society, and dreams that have been changed. Thus the focus is not on the state or politics but on the role of workers in the process of modern industrialization and how abstract ideas of modernization were implemented in their daily life (Drėmaitė, 2009).

This exhibition not only returned the industrial past to its legal heirs, the workers, but also introduced other important revisions in the discourse around Soviet industry. While comparing the stories of workers in Soviet and capitalist states, common features of industrial societies emerged, such as universal education, healthcare and homes, modern consumption and leisure, mechanized and rationally organized work. In both political systems, according to Drėmaitė, "the directions and ideology of the processes of modernization were identical. We must admit that, despite the Soviet system, there were economic achievements in the Lithuanian SSR as well"[35] (2009, p. 157). Yet more importantly, this project focused not only on the historical factual material, but also on subjective memories and emotional experiences. These are the members of the industrial society who share their life stories, dreams, attachments and disappointments. The beginning of high industry is remembered as a period of dreams and faith in a better future; the collapse of industry is perceived as a personal loss: "At the beginning, when we were still young, we believed in a better future. We lived very well then. Yet later disappointment came: it seemed that everything stopped and would not change, whether we do anything or not. After independence, everything started to collapse, many people did not adapt to the new situation" (quoted from Drėmaitė, 2009, pp. 154–155). The exhibition makes it clear that, despite the differences in political systems, people's dreams about the modernization of life were related to the goals of personal welfare, including material wellbeing.

---

35 Despite the similarity among industrial societies, Drėmaitė emphasizes also the particularities of Soviet industrialization, which resulted from, for instance, Soviet planned economy and the limitations of consumer good and trade (2009).

The first-person approach is characteristic also of another temporary exhibition organized in quite an unexpected place: at an art museum. In 2017, The M. K. Čiurlionis National Museum of Art opened a community gallery in one of its venues together with the inaugural exhibition, titled "The Great Industry". The exhibition presented the history of Kaunas as an industrial city as well as the life in two textile factories in the late Soviet period. The exhibition was co-produced together with two former industrial communities which live next to already defunct factories. People's memories and curious objects from private collections were collected as a basis for the exhibition narrative. The narrative was created around two central figures – two female artists who worked at Soviet textile factories. Both of them were the first artists who went to work in industry; both of them, now in their 90s, are extraordinary personalities. Their personal lives and work stories, together with mundane but curious objects (such as a jacket made of award-winning fabric, examples of special elastic used for female underwear or even an artificial vein created by a textile factory technologist), suggested different nuances in the history of Soviet industrialization and modernization. The exhibition shed some light on industrial production as a place for creativity and discoveries, on gender issues in Soviet industry, and on the participation of Soviet Lithuanian industry in the Cold War competition.

Like "Dream Factories", the exhibition "The Great Industry" was not so much about industry as technology, but about people who worked in the industry. As such, based on subjective memories and emotional experiences, "The Great Industry" exhibition provoked the question of nostalgia for the Soviet past. The production of a community exhibition is a complex curatorial work involving forging contacts, earning trust, long communication and collaboration with various people. It emerged, however, that one of the main problems in the process of creating this exhibition was the fear of nostalgia. The curator of the gallery, Auksė Petrulienė, herself an artist experienced in community art projects, admitted in an interview: "While inviting people to join the project, I was afraid to encounter nostalgia for Soviet times but luckily, nobody expressed any nostalgic feelings." Thus, Soviet nostalgia is perceived as a bad thing, which has no place in the museum, even if a part of the society has this feeling. This story can be understood not as the museum's intention to present a partial view of the past, but as a fear of Soviet nostalgia provoked by the earlier discussed negative concept of nostalgia which prevails in current memory culture.

The third and last example of attempts to turn Soviet industry into a museum narrative through social and cultural history is the Museum of Visaginas. This

Fig. 4: A view of the permanent display at Visaginas Museum, Visaginas, 2019. Photo by the author.

museum is still in the process of being created,[36] thus it is only possible to analyse its realized temporary projects to date. It was founded in 2014 as a department of the Visaginas Culture Centre. The museum consists of a modest display in one hall, several rooms allocated for the office and the collection and one museum worker. Being so small and dependant on the Culture Centre, also still lacking a strategy approved by the municipality, it is still an interesting example of work with the history of industrial community. What is now visible and accessible at the Museum of Visaginas is an artistic installation "Valley of Butterflies", which contains celebratory wishes to Visaginas written by both the former and current residents of the town (a result of an interactive internet project "Celebrate Visaginas") (Fig. 4). Together with a small photography

---

36 This research is based on the interview with the employer of the museum, 6 March 2019.

exhibition documenting the life in the nuclear town, this artistic installation marks symbolically the very orientation of the museum collecting practice.

The rare visitors who see the museum's collection, stored in several rooms, are surprised when they find archaeological and folk art items here, which refer to the history of the place before the emergence of the town in 1975. Sometimes this is explained as an attempt by Lithuanians working at the Culture Centre "to narrate the history of Visaginas as having pre-socialist layers, and thus reinstall their symbolic authority over the place" (Freimane, 2016a, p. 43). However, ethnographic objects make up only a part of the museum's collection. It also preserves a large collection of photographs by the main photographer of the town who recorded Visaginas from 1978 to 2003 (they are in the process of being digitalized). The collection also consists of albums, medals, badges and other "souvenirs" that had been owned by various organizations, including the workers of the INPP, as well as items of Soviet daily life and technology (from telephones and photo cameras to collections of cosmetics) and even a dried Christmas tree decorated with Soviet toys. Yet, according to the museum's curator, they are interested not in old things, but in people's stories. Mass produced stuffs from the Soviet period usually do not have a clear artistic or historical value, thus only narratives attesting to the experiences, memories and identities of the community can afford them value and turn them into heritage.

While focusing on the testimonies of the community, the museum has several projects in-process. A good example of such an activity is a project about the history of the Visaginas Acrobatic Sports School. It is interesting because it was the only professional school of acrobatics in Soviet Lithuania and also because it was initiated not by the city authorities, but by two sportsmen. The museum volunteers collect interviews with the school's coaches and their pupils; digitalize their photographs, albums, medals and other things; and create a virtual exhibition on this basis. According to the museum curator, this will be an exhibition not only about some of the city's past, but also an impulse to rethink the present by demonstrating how even in a Soviet-planned town, unplanned grassroots initiatives could be born "from below".

Another important aspect is that the museum still does not attempt to expand its material collection and rather concentrates on oral history and the digitalization of private archives. Collecting is perceived here not as collecting exhibits, but as a way to start contacts with the local community and encourage it to create the museum together. The Museum of Visaginas sees itself not as storage of things, but as a tool for people to speak about themselves to others. It is a question for the future what role nuclear industry will play in this story. According to the museum curator, local residents want to talk about the nuclear power plant as an important part of the identity of the town; they appreciate the exceptional urbanism and architecture of the atomic town, the particularities of life in it. Yet the life in the town comprises increasingly more varied aspects because most of its inhabitants did not even work at the nuclear power plant. Their personal memories and stories reveal the diversity of the city, which the museum would like to represent. Thus, so far, the narrative about the modern living environment and collective forms of leisure determined by industrial production, whether it was Soviet or nuclear industry, prevail in the Museum of Visaginas. As the museum curator points out, "political systems differ but people's goals and needs in life are similar".

Interestingly, museum projects oriented towards (former) industrial communities emphasize not the past, but the present or even the future. Both the community gallery in Kaunas and the Museum of Visaginas stress the significance of the communal and partially operate as places of community activism. While creating the exhibition "The Great Industry", one of the collaborating communities lost their premises because the municipality cut their funding, thus the museum offered its premises for the meetings of the community members. Moreover, the museum publicized this problem both in the exhibition itself and in the media. The Museum of Visaginas also centres its projects on current urgent issues. For example, the lack of community initiatives from below has served as an inspiration to revive the history of the Acrobatic Sports School. Thus, we could say that social and cultural narratives about the industrial past seek to recover values characteristic to an industrial society, such as social equality, collectivity and optimistic hope for a better future. It leaves aside its negative features, such as unification, the absolutism of rationality or technocratic fundamentalism (Drėmaitė, 2009).

Thus, what is the meaning of industrial in post-Soviet industrial museums and museum projects in Lithuania? One could say that it is both industrial technology and the life of industrial communities. One tendency is to celebrate the achievements of science and technology through the preservation of authentic buildings and equipment, presenting machines as *objets d'art* and providing the experience of the sublime of technology. Despite the fact that those technologies were created during the Soviet period, they are perceived positively, in the spirit of technological progressivism. A critical evaluation of the Soviet aspect emerges when the impact of industrialization on the society is presented. The modernization of life is understood here first as subjugation of the society to the Soviet regime. The other, opposing, tendency is only emerging via temporary and work-in-progress type projects. They seek to musealize the social and cultural history of Soviet industry through collecting and displaying subjective stories, experiences and memories of (former) industrial communities. Narratives of work and leisure rather than that of "work without workers" (Davies, 1996) prevail here. These projects contribute to the historical legitimacy and agency of (former) industrial communities, since they try both to articulate local and communal identities, and to re-affirm the values of industrial society. Both tendencies lack a critical assessment of the impact of modern industries on the environment and on people's lives.

The Narration of Soviet Industry in History Textbooks

A similar question "what's industrial" can be asked also while reading Lithuanian history textbooks, which address the Soviet past. Soviet industrialization as well as the related urbanization and modernization are discussed in a more or less detailed manner in all 5th, 10th and 12th grades textbooks.[37] These topics are included also in normative documents regulating history teaching in Lithuanian general education schools: "The General Curriculum Framework for Basic Education" (2008) and "The General Curriculum Framework for SecondaryEducation" (2011). According to the curriculum, students of the 5th grade should study the following topics of the Soviet period: Soviet occupation (through the annihilation of the population and resistance against the occupants) and everyday life of the people in Soviet Lithuania. History should also be taught through references to both the past and the present of the family and native place. Students of the 10th grade analyse both the history of Lithuania and global history. The history of the Soviet period should emphasize the following topics: occupation and resistance, the impact of science and technology on the economy and the development of the society, the development of culture and everyday life. In the 12th grade the greatest attention should be paid to the history of the society by referencing politics, economics and culture to reveal the changes in public life. The studied topic "The Society during the Times of Cold War and Collapse of Communism" should cover guerrilla war and dissident resistance to the Soviet regime, collaboration with the regime and adaptation to it, the "Sovietization" of Lithuanian economy and society as well as liberation from the regime.

---

37 Among the 20 analysed textbooks, the only 11th–12th grade textbook written by Makauskas says almost nothing about Soviet industry and its impact on people's lives. It narrows this down to three sentences claiming that heavy industry started to be developed in Lithuania from 1950 and industrialization was a means to Russify Lithuania (2006).

Tab. 1: History textbooks covering the period of Soviet Lithuania and approved by the Ministry of Education, Science and Sport of the Republic of Lithuania as valid for the academic year 2018–2019. Compiled by the author.

5th Grade History Textbooks

No.	Author(s)	Title	Publisher	Year
1.	Jakimavičius Viktoras	Gimtoji šalis Lietuva. Lietuvos istorijos skaitiniai. 5 kl.	Alma littera	1998
2.	Zakarauskienė Izolda	Lietuvos istorija. Skaitiniai. 5 kl.	Agora	2000
3.	Brazauskas Juozas	Lietuvos istorija. Skaitiniai. 5 kl.	Šviesa	2000
4.	Stašaitis Stanislovas, Šačkutė Jūratė	Tėvynės istorijos puslapiai. Lietuvos istorijos vadovėlis. 5 kl.	Margi raštai	2000
5.	Litvinaitė Jūratė	Palikimas. Istorijos vadovėlis. 1/2-oji kn. 5 kl. (serija „Šok")	Šviesa	2007
6.	Laužikas Rimvydas, Mickevičius Karolis, Tamkutonytė-Mikailienė Živilė, Kapleris Ignas	Kelias. Istorijos vadovėlis. 5 kl. 2 d.	Briedis	2008
7.	Petreikis Darius, Litvinaitė Jūratė, Meškuotis Faustas, Ramoškaitė-Stongvilienė Rūta, Bitautas Algis, Stankutė Simona	Istorija. 5 kl. (serija „Atrask")	Šviesa	2014

10th Grade History Textbooks

No.	Author(s)	Title	Publisher	Year
1.	Kasperavičius, Algis, Jokimaitis, Rimantas	Naujausiųjų laikų istorija. 10 kl.	Kronta	2003
2.	Brazauskas, Juozas, Makauskas, Bronius	Lietuvos praeities puslapiai. Istorijos vadovėlis. 3-ioji kn. 10 kl.	Šviesa	2004
3.	Kapleris, Ignas, Meištas, Antanas, Mickevičius, Karolis, Laužikienė, Andželika, Tamkutonytė-Mikailienė, Živilė	Laikas. Istorijos vadovėlis. 2 d. 10 kl.	Briedis	2017 (2007)
4.	Bakonis, Evaldas	Tėvynėje ir pasaulyje. Istorijos vadovėlis. 10 kl.	Šviesa	2009
5.	Kraujelis, Ramojus, Streikus, Arūnas, Tamošaitis, Mindaugas	Istorijos vadovėlis. 2 d. 10 kl. (serija „Raktas")	Baltos lankos	2010

(continued on next page)

Tab. 1: Continued

12th Grade History Textbooks			
1. Mäesalu, Ain, Kiaupa, Zigmantas, Straube, Gvido, Pajur, Ago	Baltijos šalių istorija. 10–12 kl.	Kronta	2000
2. Civinskas, Remigijus, Antanaitis, Kastytis	Lietuvos istorija. 12 kl.	Vaga	2001 (2000)
3. Makauskas, Bronius	Lietuvos istorija. 11–12 kl., 2-oji kn.	Šviesa	2006 (2000)
4. Gečas, Algirdas, Jurkynas, Juozas, Jurkynienė, Genia, Visockis, Albinas	Lietuva ir pasaulis. Istorijos vadovėlis. 12 kl.	Šviesa	2001
5. Kaselis, Gintaras, Kraujelis, Ramojus, Lukšys, Stasys, Streikus, Arūnas, Tamošaitis, Mindaugas	Istorijos vadovėlis. 2 d. 12 kl.	Baltos lankos	2008
6. Kapleris, Ignas, Laužikas, Rimvydas, Meištas, Antanas, Mickevičius, Karolis	Laikas 12. Istorijos vadovėlis. 1/2 d. 12 kl.	Briedis	2016 (2011)
7. Anušauskas, Arvydas, Kaselis, Gintaras, Kraujelis, Ramojus, Lukšys, Stasys, Streikus, Arūnas, Tamošaitis, Mindaugas	Istorijos vadovėlis. 2 d. 12 kl.	Baltos lankos	2012
8. Navickas, Virginijus, Svarauskas, Artūras	Istorijos vadovėlis 12 kl. (IV gimnazijos kl.)	Ugda	2015

The topics of industrial, urban and modern life are presented differently in textbooks written by different authors and published at different times (Tab. 1).[38] A separate chapter could be dedicated to industry and everyday life, for example: "Huge Factories Arose" (Jakimavičius, 1998), "Everyday life of People during the Soviet Period" (Laužikas et al., 2008) or "The Sovietization of Lithuanian Economy and Society" (Anušauskas et al., 2012). These topics may also form an integral part of the narrative about Soviet politics (Civinskas & Antanaitis, 2001) or public life during different periods of Khrushchev's "thaw" and Brezhnev's "stagnation" (Kaselis et al., 2008; Kapleris et al., 2016; Navickas

---

38 Today textbooks published earlier than the aforementioned normative documents are also approved.

& Svarauskas, 2015). In the 5th grade textbooks, industry and modern life are sometimes described as part of the history of the entire Soviet period (Petreikis et al., 2014; Stašaitis & Šačkutė, 2000) or even of all general history of Lithuania (Litvinaitė, 2007). Despite these differences, an analysis of Lithuanian history textbooks could trace prevailing narratives about the Soviet industrialization, urbanization and modernization.

In most cases industrialization is presented as an important feature of Soviet period life. This is immediately noticeable upon looking through various visual highlights which appear in textbooks. The beginning of high industry appears among the dates of the most important events of the Soviet period. For example, in the chronological table highlighted in the 5th grade textbook, the 1960s as the period during which large factories were constructed is presented next to the dates of the Nazi and Soviet occupations, deportations, formation of collective farms and the Sajūdis movement for independence (Stašaitis & Šačkutė, 2000). Industry appears even in more telling examples of infographics. In the 5th grade textbook, the chapter "Life in Soviet Lithuania" opens with a chronological table with only two dates: 1950 as the year when the Lithuanian Anthem was banned and 1980, the year when the Mažeikiai Oil Refinery began operating (Laužikas et al., 2008) (Fig. 5). And in the 12th grade textbook, the chapter "The Lithuanian Society: from the Soviet Period to the Restoration of Independence" opens with the map of Lithuania where only the large industrial enterprises are marked (Mažeikiai Oil Refinery, Lithuania Power Plant in Elektrėnai, Jonava Nitrogen Fertilizer Factory and INPP) together with strategic Soviet military objects, such as airports and the nuclear weapons storage[39] (Kapleris, 2016).

The industrial in textbooks most often denotes various industries themselves as well as factories and industrial enterprises. New industries, such as energy, metal, chemical, and electronics industries, that emerged during the Soviet period are proudly listed in the textbooks. In almost all textbooks, in both texts and photographs, the largest Soviet industrial companies that operated in Lithuania are presented.[40] There are often attempts to demonstrate the rapidity

---

39 The map has to illustrate a clearly exaggerated statement by the textbook's authors that "Occupation government turned the LSSR into a giant military base" (Kapleris et al., 2016, p. 215).
40 Such as the Kaunas Hydroelectric Power Plant, Elektrėnai Power Plant, Ignalina Nuclear Power Plant, Mažeikiai Oil Refinery, Jonava Nitrogen Fertilizer Factory, Kėdainiai Chemical Plant, Akmenė Cement Plant, Kinescope Plant in Panevėžys and Alytus Cotton Textile Factory.

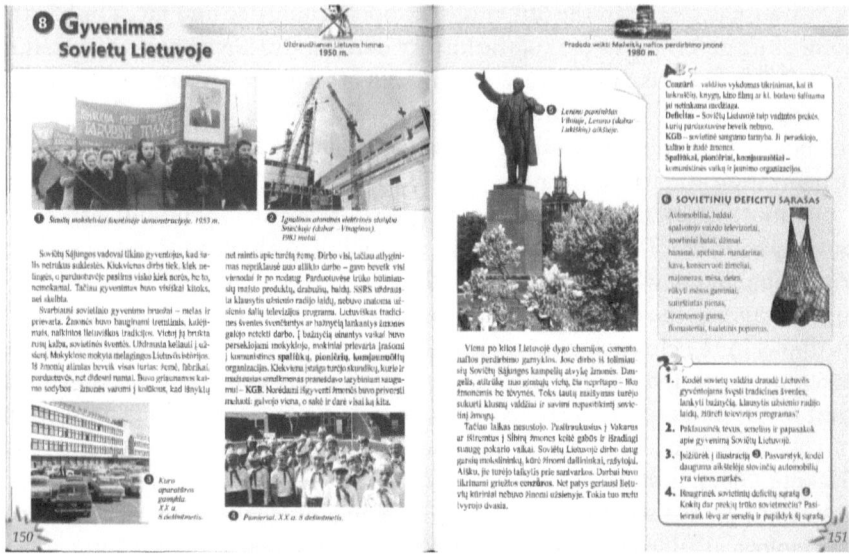

Fig. 5: Fragment of the history textbook by Laužikas, R. et al. (2008).

and the vast scale of industrialization through linguistic means. Industrial enterprises are called "industrial colossi" (Brazauskas, 2000; Stašaitis & Šačkutė, 2000), "industrial giants" (Kraujelis et al., 2010; Bakonis, 2009; Kaselis et al., 2008). Rapid industrialization is described with such phrases: "Factories [...] sprouted in Lithuania one after another" (Laužikas et al., 2008, p. 151), "Soviet government decided to turn our land of agriculture into the land of factories and plants" (Stašaitis & Šačkutė, 2000, p. 168). Sometimes the power and modernity of Lithuanian industrial enterprises is emphasized: "In 1965 [started operating] the Kaunas Artificial Fibre Factory, the most modern factory of chemical industry. This was one of the largest factories in the Soviet Union and Europe. The largest factory of this profile in the Soviet Union was the Vilnius Drill Factory" (Brazauskas & Makauskas, 2004, 150).

Energetics is often singled out among other industries as the basis of industrialization in history textbooks. Photographs of electric power plants in Kaunas, Elektrėnai and Ignalina dominate amongst images of other factories with the INPP playing a major role. The texts not only single out the date (1983) when the INPP was launched, but also emphasize its exceptional status by mentioning that this was the only nuclear power plant in the Baltic States, that its reactor was the most powerful in the world, included among Guinness World Records (Navickas & Svarauskas, 2015; Kapleris et al., 2017; Kapleris et al., 2016). The 5th grade textbook explains how a nuclear power plant operates and in what way it is different from other electric plants by pointing out the low cost of nuclear energy (Jakimavičius, 1998), although history textbooks usually do not explain how various industries operate and focus on their historical development instead. One could say that the INPP is presented as a symbol of modern industrialization although, as we shall see later, not necessarily in a positive sense.

The formation of the urban and modern society in Lithuania is widely discussed as a consequence of industrialization in history textbooks. The motif of rapid growth and radical changes also prevails here. The textbooks emphasize the fact that rapid industrialization resulted in extensive construction: multi-storey blocks of flats, schools, kindergartens were built, new residential districts and even new cities (Elektrėnai, Sniečkus) were designed, new roads were built and infrastructure improved on a mass scale. The chapters on Soviet social life in textbooks are most often illustrated with the photographs of new residential districts. The scale of urbanization that took place in the 1960s and 1970s, i.e., migration from countryside to the city, is emphasized. Sometimes this is illustrated by providing numbers: in 1960, 40 % of the Lithuanian population lived in the cities, and in 1980, already around 60 % (Navickas & Svarauskas, 2015; Kraujelis et al., 2010). The authors of most textbooks point out that during the Soviet period, Lithuania was transformed from an agricultural country into an industrial one, and Lithuanian society became urban for the first time in history. Yet next to the quite neutral and even celebratory listing of the facts about growth and change, the textbooks provide a somewhat negative image of Soviet modern life. It is possible to distinguish several dominating narratives about Soviet industrialization and modernization in history textbooks.

First, the narrative of a centralized/colonial control of industry prevails. The textbooks allocate the largest amount of space to the discussion of how Lithuanian industry functioned in Soviet economy based on centralized administrative planning. The authors criticize the flawed process of Soviet industry when imported materials were used for local production, and the produced goods were distributed around the entire Soviet Union in a centralized manner. The negative attitude is also expressed in the titles of the chapters such as "Producing for Everyone, Destroying Our Nature" (Stašaitis & Šačkutė, 2000) or "In the Grip of Planned Economy" (Kapleris et al., 2016). Large industrial companies and factories are called not only the giants of industry, but also the "monsters of industry" (Kapleris et al., 2017). Because of the centralized distribution of production, Soviet industry is discussed as an obvious form of colonial exploitation (Kraujelis et al., 2010; Anušauskas et al., 2012). Moreover, Lithuanian industry is linked not to local, but to colonial needs. On the one hand, industry dependent on imported materials had to guarantee the irreversible integration of Lithuania into the Soviet Union (Kaselis et al., 2008). On the other hand, local industry was used to produce for the military needs of the Soviet Union (Kapleris et al., 2017; Anušauskas et al., 2012). Several authors present Soviet industry as "alien" by pointing out that it was created without taking into account local conditions and traditions (Kapleris et al., 2016).

The stories about the "alien" Soviet industry are enhanced by the second narrative linking industrialization to the Russification of the country. Although Russian immigration to Lithuania was relatively insignificant, in comparison with Latvia and Estonia,[41] Russification performs an important role in the narratives about Soviet industry. Most authors of textbooks state that industrialization itself was a means to Russify Lithuania, since immigration from all over the Soviet Union was necessary to both build and man large factories.[42] The satellite city of the INPP, Visaginas, is often quoted as a typical example of "Soviet colonialism", since around 30,000 Russian-speaking immigrants moved there to live (Kaselis et al., 2008; Anušauskas et al., 2012). Next to the prevailing narrative of industrialization as Russification, an alternative story can be found, mostly in the 12th grade textbooks. Some authors present the reform of Soviet economy that took place from 1957 to 1962 (the so-called *sovnarkhoz* system), which gave more power to local authorities for a short while and encouraged regional planning. The consequence of this reform for Lithuanian industry was an even distribution of factories across the entire territory (Civinskas & Antanaitis, 2001; Kaselis et al., 2008; Anušauskas et al., 2012; Navickas & Svarauskas, 2015; Mäesalu et al., 2000). This allowed for the hiring of local people in industry and "saved [us] from the influx of colonists" (Gečas et al., 2001, p. 335). Interestingly, the procurement of work for local inhabitants and the preserved certain "Lithuanianness" of industry is generally presented

---

41 During the entire Soviet period Lithuanians constituted around 80 % of the Lithuanian population.
42 "In order to develop industry many working hands were needed. This was a pretext to send workers from other republics of the USSR. In particular, many arrivals came to build such giants of industry as the Mažeikiai Oil Refinery and Ignalina Nuclear Power Plant." (Bakonis, 2009, p. 137). Similar stories are presented in other textbooks; see Laužikas at al., 2008; Brazauskas & Makauskas, 2004; Kaselis et al, 2008; Anušauskas et al., 2012; Makauskas, 2006; Mäesalu et al., 2000.

in textbooks not as a positive aspect of industrialization, but as a deliverance from yet another of its negative aspects.[43]

The third narrative, which strongly supplements the negative image of Soviet industry, presents pollution as a consequence of industry. The theme of ecology is discussed in almost all textbooks. Soviet production is condemned for polluting the environment, for not taking care of environment protection and hiding accidents from the society. Some textbook authors dedicate separate chapters to the topic of ecology, with graphic titles, e.g., "The Blackened Sky and Dead Trees" (Stašaitis & Šačkutė, 2000) or "The Threat of Ecological Catastrophe" (Kapleris et al., 2017). The factories are referred to as "horrors" of nature (Stašaitis & Šačkutė, 2000), facts about the impact of pollution on the number of oncological diseases are used (Kapleris et al., 2017), and witnessed accounts of accidents in industrial plants are presented. The texts are accompanied by photographs depicting views of contaminated soil or withered trees. The INPP has a symbolical place in this narrative. The textbooks present the INPP not only as an especially powerful, but also a very dangerous industrial plant. They even call it "a nuclear bomb of delayed action" (Kapleris et al., 2017). The authors mostly associate the threat of the INPP with the accident at Chernobyl by pointing out that both electric power plants used the same type of unsafe reactors (Kapleris et al., 2017; Bakonis, 2009; Kapleris et al., 2016, Gečas et al., 2001). Texts are illustrated with memories of contemporaries and images from the Chernobyl nuclear power plant and photographs from the meetings of anti-nuclear activists in Lithuania and abroad (Kapleris et al., 2017, Gečas et al., 2001) (Fig. 6).

The fourth narrative presents the urban and modern life which stemmed from industrialization as the Sovietization of Lithuanian society. The authors of textbooks argue differently in order to demonstrate social processes and everyday life as a part of the Soviet regime. One way is to associate the events and processes in the past with negatives. We can find such a simplistic attitude in the 5th grade textbooks, in which Soviet everyday life is presented as "grey"[44]

---

43 A sole exception is the textbook by Navickas and Svarauskas which argues that the territorial distribution of industry helped to solve the occupational problem in the province (2015, p. 157).

44 "The Greyness of Soviet Everyday Life" is the title of one of three topics in the textbook's chapter "Life in the Soviet Countryside and in the City" (Petreikis et al., 2014). Greyness refers here to the deficit of goods and uniformity of fashion.

Fig. 6: Fragment of the history textbook by Gečas, A., et al. (2001).

(Petreikis et al., 2014) and "lacking freedom"⁴⁵ (Litvinaitė, 2007). The textbook chapter about social life in Soviet Lithuania begins with the statement that is repeated in the summary of the chapter: "The most important features of Soviet life were falsehood and violence" (Laužikas et al., 2008, p. 150). Another way, mostly used in the 10th grade textbooks, is to construct an argument based on the principle of thesis and antithesis. Thus, all achievements of modern life are diminished and acquire a different, almost opposite, meaning. The telling examples are excerpts from different textbooks: "Healthcare was free in the Soviet period, but one had to wait for hours queuing up in the clinics ..." (Kapleris et al., 2017, p. 232); "There were also minimal social guarantees: pensions, unemployment benefits, exemptions for families with many children etc. Yet people were very oppressed by universal deficit" (Bakonis, 2009, p. 138); "Living conditions in the flats were improved for most families created during the post-war years. The price of that was that a part of the society started not only to come to terms, but also to *identify with the imposed system*" (Brazauskas & Makauskas, 2004, p. 155). Finally, one more way is to treat the forms of modern life as indirect tools of Sovietization. Some authors of the 12th grade textbooks interpret such features of a modern industrial society as education and healthcare accessible to all, social guarantees and the provision of jobs and homes to everybody as only a means by which the Soviet regime attempted to "assume control of the Lithuanian society"⁴⁶ (Anušauskas et al., 2012, p. 131). The modern way of life which encompasses leisure entertainment and consumerism is seen as encouraging a conformity with the Soviet system and therefore is represented negatively in the textbooks (Anušauskas et al., 2012; Gečas et al., 2001; Kapleris et al., 2016). Modernization is paradoxically presented not as an improvement in living conditions, but as the "coming to terms with the occupation authorities"⁴⁷ (Gečas et al., 2001, p. 335).

---

45 A short paragraph about Soviet everyday life states: "Life became easier. But there was no freedom of action. After work people were forced to attend mandatory meetings and various events" (Litvinaitė, 2007, p. 19).
46 "Soviet authorities created good opportunities for social mobility for those layers of society who did not have necessary conditions to seek higher education or a career before. The Soviet system identified education with ideological education, therefore the number of pupils of general and professional secondary schools, of students of higher education increased rapidly" (Anušauskas et al., 2012, p. 130).
47 "Already during the times of Khrushchev, most inhabitants came to terms with the occupation authorities, their national feelings were overshadowed by the

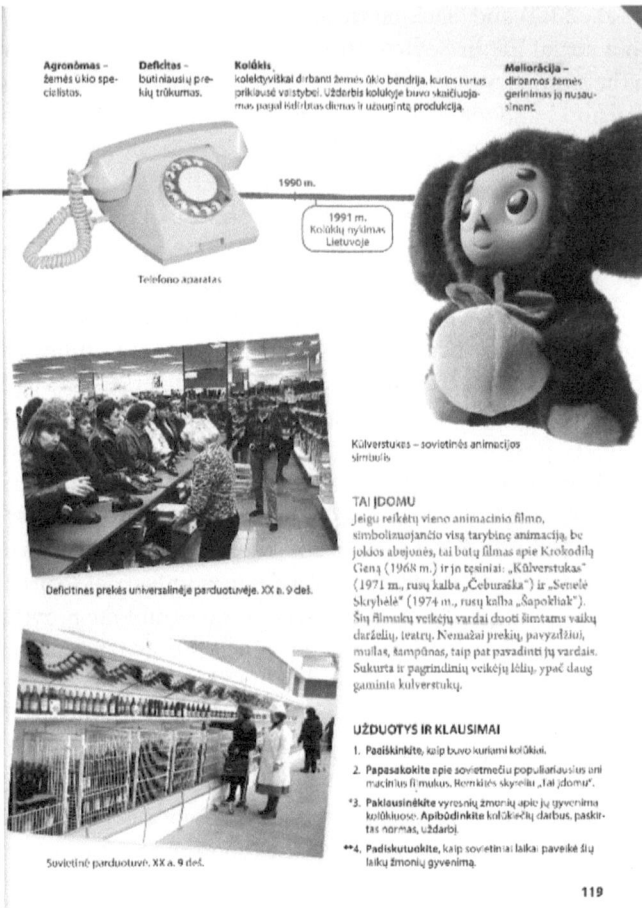

Fig. 7: Fragment of the history textbook by Petreikis, D., et al. (2014).

Images illustrating the chapters in textbooks about Soviet life tell more varied stories. They are particularly abundant in lower grade and more recent textbooks. The photographs most often showcase domestic appliances (telephones, television sets, and radios), means of transport (Zhiguli cars,

---

accumulation of material values" (Gečas et al., 2001, p. 335). It is necessary to point out that several textbooks for the 12th grade (Civinskas & Antanaitis, 2001; Navickas & Svarauskas, 2015) present the urban and modern life in more neutral terms.

motorbikes), new residential districts and interiors of flats, the currency (roubles), youth fashion and sports festivals. These telling details of daily life are presented together with images from Soviet demonstrations, working bees and party meetings, Soviet youth organizations (pioneers) or queues in half-empty shops. A telling example of the complexity of visual storytelling is four photographs illustrating the chapter in the 5th grade textbook on life in Soviet cities (Fig. 7). They depict a telephone, Cheburashka, a character from a Soviet animation movie (Kūlverstukas in Lithuanian, known as Topple in English translations), and two interiors of shops in one of which people crowd around the goods that were largely unavailable (Petreikis et al., 2014). Yet if visual narratives sometimes present a many-sided image of the Soviet life, the texts usually emphasize its negative aspects, such as the shortage of consumer goods, the poor quality of products, a lagging behind the West, mass alcoholism and plunder of state property. In higher grade textbooks students are sometimes asked, e.g., what are positive and negative changes that happened in Lithuania during the Soviet period. Yet often it is impossible to find an answer to the first part of the question in the text.[48]

Thus, in history textbooks, the industrial signifies first of all the development of industry and its Soviet control. The rapid growth of new industries and the construction of industrial enterprises are partially presented in a celebratory way, but a negative evaluation of industrialization as a colonial project controlled from the centre prevails. Much like the industrial museums, to industrialize and to modernize indicate here to Sovietize the society, to subjugate Lithuania ideologically to the occupation regime. In this way, the history textbooks present the story of Soviet authorities as owners of industry sidelining the role of industrial communities. The topic of work and workers is not presented at all in the stories of history textbooks about the Soviet industrial society. Although the authors discuss the question of social class, they focus mostly on the presentation of a specific Soviet class – the nomenclature.[49] Only two textbooks briefly mention the working class, which increased to around

---

48 As if by foreseeing this problem, the question is sometimes formulated differently: "Assess the impact of the Soviet period on Lithuanian society. Which negative aspects inherited from that period are reflected also in our time?" (Kapleris et al., 2016, p. 215).
49 Nomenclature and its privileges are discussed in chapters with such telling titles as "Some Are More Equal than Others" (Kapleris et al., 2017) or "Life Was Good Not for Everyone" (Brazauskas & Makauskas, 2004).

40 % of the Lithuanian population.[50] Some authors of textbooks encourage students to ask their parents and grandparents about life in Soviet Lithuania and also about work in industry.[51] In this way, a voice may be given to those who have lost the right to inherit their lived Soviet period. However, as Kohrs notes, the narratives of textbooks and live memory often have nothing in common (Kohrs, 2006). Talking to their family students often receive a completely different image of the Soviet life, which is more about the normal everyday life than about an active or passive resistance against the occupation, the topic that dominates the textbooks. For example, while reading textbooks, students could not understand in any way how and why nostalgia for the Soviet times appears in the stories of people from the older generation (Kohrs, 2006). The fact that the theme of industry is presented in textbooks not from the point of view of people who created it or felt its impact, but from the point of view of the authorities, only confirms this idea.

Conclusions

A widespread opinion is that the Soviet industrial heritage is a marginal area of heritage in Lithuania or is not regarded as heritageable at all (Drėmaitė, 2012; Storm, 2014). In fact, "the authorized heritage discourse" (Smith, 2006) in Lithuania does not treat Soviet industry as a valued legacy. For example, there is not a single industrial building from that period listed in the Register of Cultural Values. Despite that, there are relatively many diverse practices of memorializing and studying Soviet industry in contemporary Lithuania and one of the underlying goals of this chapter was to demonstrate this. Currently, one can see and experience Soviet industry, including the nuclear industry, in various museum projects and even assist with collecting for it. It also constitutes quite a large part of the narrative about the Soviet past in history textbooks.

Soviet industry is a clearly dissonant heritage in the sense that producers of this heritage reveal and try to maintain a binary relationship between "us" and

---

50 One textbook presents the particularities of the Soviet working class by claiming that "in reality workers as a group of the society were only a workforce in state factories" (Civinskas & Antanaitis, 2001, p. 201). The second textbook presents workers through the changes in the social composition of the society: "In 1960, there were around 490,000 workers in Lithuania, and in 1980, already over a million" (Navickas & Svarauskas, 2015, p. 157).
51 Mostly the authors of the 5th grade textbooks give such tasks to the students. See Litvinaitė (2007), Laužikas et al. (2008).

"them" as well as between Lithuanians and the Soviets. History textbooks tell, at the first glance, a celebratory story about high industrialization and rapid urbanization, but in the end they present this as a negative narrative about the Russification of the country and contamination of its land. Contrary to the textbooks, industrial museums tend rather to celebrate technological advances and invite their visitors to admire the greatness of the machines without any critical environmental assessment. Both museums and textbooks, however, share a negative image of modernization as a means to Sovietize Lithuanian society rather than a complex process of societal and economic change.

The heritagization of Soviet industry in Lithuania creates also an effect of a double dissonance, since it raises the issue of social class and hides it at the same time. The narratives in permanent expositions in the museums and in textbooks clearly "depopulate" the history of Soviet industry, to use Davies term (1996). The centralized development and control of industry from Moscow rather than the work and life of industrial communities prevails in the narratives in history textbooks. In the same way, turbines and reactors dominate the museum halls and industrial sites. In both cases, the heritage of industry owners is preserved and presented, as it comes, negatively or positively, once more – paradoxically – disinheriting those who worked in the industry and whose lives it formed.

The pedagogy of such a dissonant heritage is in no way a simple task. The analysis of museum displays and textbooks as educational sources has limitations because they constitute only a part of the learning process, which usually occurs through interactions during visits to museums, guided tours, teacher's work in classroom and other activities. On the other hand, the narratives they create can be understood as a foundation on which the experiences and subjectivities of learners are formed. As this research has shown, museums and history textbooks often emphasize different things in their narratives about Soviet industry. There are two important themes related to industry requiring exceptional attention: those are the environmental and social issues. The environmental issue is widely discussed in history textbooks, despite the fact that pollution is almost exclusively linked here with the Soviet type industry and not with modern industry in general, which would be more correct. Yet this theme is completely missing from museum displays and projects, thus presenting not only a partial, but also a distorted view of the industrial past and its impact on the current situation of climate change. Similarly, the social and cultural history of industry is mostly concealed in textbooks, while emerging museum practices attempt to present it. Perhaps, such different representations of these issues may be used productively in order to reveal a fuller image of the history of industry in education if it takes place both inside and outside of school.

As the chapter reveals, the nuclear energy industry performs an important role in the narratives of Soviet industry. Not only the INPP becomes a symbol of high industrialization, but, I argue, the nuclear theme introduces the importance of sensibilities into the discourse of Soviet industry. One of these, nuclear fear, may be felt both while reading the textbooks and through "museum" experiences while visiting the defunct nuclear power plant, now in the process of becoming a site of nuclear heritage. Another feeling, nostalgia, is produced by the emerging museum practices, which seek to give a voice to the former industrial communities as an important participant of the processes of industrialization and modernization. One of them is the Museum of Visaginas which collects and shows the lived experiences and memories of people whose lives were shaped by the industry. By seeking to re-actualize certain values of the industrial society, such as sociability, the museum clearly resists the aforementioned fear of post-Soviet nostalgia. Precisely by emphasizing feelings of nostalgia and fear, these narratives of the nuclear industry may become powerful tools of the dissonant heritage pedagogy, encouraging both empathy and critical thinking.

References

Bitautas, A. (2018). *Daugiaperspektyvio požiūrio ugdymo raiška Lietuvos istorijos vadovėliuose (Expression of Multi-perspectivity Approach Development in Textbooks on Lithuanian History)* [Doctoral dissertation], Vytautas Magnus University.

Čepaitienė, R. (2007). Sovietmečio atmintis – tarp atmetimo ir nostalgijos (Soviet Memory from Rejection to Nostalgia). *Lituanistica, 53*(4), 36–50.

Čepaitienė, R. (2009). Sovietinės kultūros šaltiniai tarp futurizmo ir paseizmo (Sources of Soviet Culture: Between Futurism and Pastism). *Darbai ir dienos, 52*, 85–104.

Cossons, N. (2012). Industrial Heritage: Treasure or Trash? *Journal of Cultural Property Conservation*, (22), 7–26.

Davies, K. (1996). Cleaning Up the Coal Face and Doing Out the Kitchen: The Interpretation of Work and Workers in Wales. In G. Kavanagh (Ed.), *Making Histories in Museums* (pp. 105–115). London: Leicester University Press.

Drėmaitė, M. (2009). Svajonių fabrikai? Industrializacijos palikimas Baltijos jūros regione (1945–1990 m.) kultūros istorijos požiūriu (Dream Factories and the Legacy of Industrialization in the Baltic Sea Region (1945–1990) from the Viewpoint of Cultural History). *Darbai ir dienos, 52*, 141–158.

Drėmaitė, M. (2012). Industrial Heritage in a Rural Country. Interpreting the Industrial Past in Lithuania. In M. Nisser, M. Isacson, A. Lundgren, & A. Cinis (Eds.), *Industrial Heritage around the Baltic Sea* (pp. 65–78). Uppsala: Uppsala Universitet.

Fitzgerald, L. (1996). Hard Men, Hard Facts and Heavy Metal: Making Histories of Technology. In G. Kavanagh (ed.), *Making Histories in Museums* (pp. 116–130). London: Leicester University Press.

Freimane, I. (2016a). The Centrality in and of Visaginas. In F. Ackermann, B. Cope, & S. Liubimau (Eds.), *Mapping Visaginas. Sources of Urbanity in a Former Mono-Functional Town* (pp. 41–47). Vilnius: Vilnius Academy of Arts Press.

Freimane, I. (2016b). Socialist Heritage? Mapping the Existing Academic Literature on Visaginas. In F. Ackermann, B. Cope, & S. Liubimau (Eds.), *Mapping Visaginas. Sources of Urbanity in a Former Mono-Functional Town* (pp. 21–27). Vilnius Academy of Arts Press.

Giroux, H. A., & Simon, R. I. (1988). Schooling, Popular Culture, and a Pedagogy of Possibility. *Journal of Education, 170*(1), 9–26.

Graham, B., & Howard, P. (2008). Heritage and Identity. In B. Graham, & P. Howard (Eds.), *The Ashgate Research Companion to Heritage and Identity* (pp. 1–15). London: Routledge.

Hecht, G. (2012). *Being Nuclear. Africans and Global Uranium Trade.* Cambridge, MA: MIT Press.

Hospers, G. J. (2002). Industrial Heritage Tourism and Regional Restructuring in the European Union. *European Planning Studies, 10*(3), 397–404.

Klerides, E. (2010). Imagining the Textbook: Textbooks as Discourse and Genre. *Journal of Educational Media, Memory, and Society, 2*(1), 31–54.

Klumbytė, N. (2004). Dabarties istorijos paraščių žmonės: atsiminimai apie sovietmetį ir kasdienybės patirtis kaimo bendruomenėse (People at the Margins of History: Remembering Soviet Times and Experiencing Present in Village Communities in Lithuania). *Politologija, 3*(35), 1–21.

Klumbytė, N. (2009). Post-Socialist Sensations: Nostalgia, the Self, and Alterity in Lithuania. *Lietuvos etnologija: socialinės antropologijos ir etnologijos studijos, 9*(18), 93–116.

Kohrs, M. (2006). The Communist Past in Lithuania and its Portrayal in the Latest History Textbooks. In B. Šetkus, & R. Šetkuvienė (Eds.), *Mokykliniai istorijos vadovėliai ir europinės visuomenės ugdymas Rytų ir Vidurio Europos šalyse (School Textbooks of History and the Development of European Society in the Countries of Eastern and Central Europe)* (pp. 64–72). Vilnius: Vilniaus pedagoginis universitetas.

Lankauskas, G. (2006). Sensuous (Re)Collections: The Sight and Taste of Socialism at Grūtas Statue Park, Lithuania. *The Senses and Society, 1*(1), 27–52.

Lankauskas, G. (2015). Missing Socialism Again? The Malaise of Nostalgia in Post-Soviet Lithuania. In O. Angé & D. Berliner (Eds.), *Anthropology and Nostalgia* (pp. 35–75). New York: Berghahn Books.

Macdonald, Sh. (2013). *Memorylands. Heritage and Identity in Europe Today.* London: Routledge.

Meškauskas, K. (1994). *Lietuvos ūkis 1940–1990 (Lithuanian Economy 1940–1990).* Vilnius: Lietuvos istorijos institutas.

Misiunas, R. J., & Taagepera, R. (1993). *The Baltic States. Years of Dependence 1940–1990.* Berkeley, Los Angeles: University of California Press.

Nikžentaitis, A. (2013). Atminties ir istorijos politika Lietuvoje (The Politics of Memory and History in Lithuania). In A. Nikžentaitis (Ed.), *Atminties daugiasluoksniškumas. Miestas, valstybė, regionas (The Multilayerness of Memory. The City, the State, and the Region)* (pp. 517–538). Vilnius: Lietuvos istorijos institutas.

Ravelli, L. J. (2006). *Museum Texts: Communication Frameworks.* London: Routledge.

Rindzevičiūtė, E. (2015). The Overflow of Secrets. The Disclosure of Soviet Repression in Museums as an Excess. *Current Anthropology, 56*(12), 276–285.

Safronovas, V. (2009). On Tendencies of the Politics of Remembrance in Contemporary Lithuania. *Ab Imperio. Studies of New Imperial History and Nationalism in the Post-Soviet Space,* (3), 424–458.

Sandlin, J. A., Wright, R. R., & Clark, C. (2011). Re-examining Theories of Adult Learning and Adult Development Through the Lenses of Public Pedagogy. *Adult Education Quarterly, 63*(1), 3–23.

Šliavaitė, K. (2010a). Social memory, Identity and Narratives of Decline in a Lithuanian Nuclear Plant Commmunity. *Acta Historica Universitatis Klaipedensis XX, Studia Anthropologica IV,* 52–71.

Šliavaitė, K. (2010b). Deindustrializacija, socialinis nesaugumas ir išgyvenimo strategijos posovietinėje erdvėje: Visagino atvejis (Deindustrialization, Social Insecurity and Strategies of Survival in the Post-Soviet Region: the Case of Visaginas). *Lietuvos etnologija: socialinės antropologijos ir etnologijos studijos, 10*(19), 93–116.

Smith, L. (2006). *Uses of Heritage.* London: Routledge.

Stepanov, A. (2018). *Lithuania – Short Country Report.* History of Nuclear Energy and Society (HoNESt, project Ref. 662268). http://www.honest2020.eu/sites/default/files/deliverables_24/LT.pdf, accessed 16 October 2019.

Storm, A. (2014). *Post-Industrial Landscape Scars.* New York: Palgrave Macmillan.

Šutinienė, I. (Ed.). (2003). *Socialinė atmintis: minėjimai ir užmarštys (Social Memory: Remembrance and Oblivion).* Vilnius: Socialinių tyrimų institutas.

The International Committee for the Conservation of the Industrial Heritage (TICCIH). (2003). *The Nizhny Tagil Charter For The Industrial Heritage.* http://ticcih.org/about/charter/, accessed 27 September 2019.

Tunbridge, J. E., & Ashworth, G. J. (1996). *Dissonant Heritage: The Management of the Past as a Resource in Conflict.* Chichester: John Wiley & Sons.

Vargas-Sánchez, A. (2015). Industrial Heritage and Tourism: A Review of the Literature. In E. Waterton, & S. Watson (Eds.), *The Palgrave Handbook of Contemporary Heritage Research* (pp. 219–233). New York: Palgrave Macmillan.

Žilinskienė, L., Kraniauskienė, S., & Šutinienė, I. (2016). *Gimę socializme: pirmoji sovietmečio karta (Born in the Socialism: The First Soviet Generation).* Vilnius: Vilniaus universiteto leidykla.

Textbooks

Anušauskas, A., Kaselis, G., Kraujelis, R., Lukšys, S., Streikus, A., & Tamošaitis, M. (2012). *Istorijos vadovėlis. 2 d. 12 kl. (History Texbook, 2nd part, 12th Grade).* Vilnius: Baltos lankos.

Bakonis, E. (2009). *Tėvynėje ir pasaulyje. Istorijos vadovėlis. 10 kl. (In the Homeland and the World. History Texbook. 10th Grade).* Kaunas: Šviesa.

Brazauskas, J. (2000). *Lietuvos istorija. Skaitiniai. 5 kl. (History of Lithuania. A Reader. 5th Grade).* Kaunas: Šviesa.

Brazauskas, J., & Makauskas, B. (2004). *Lietuvos praeities puslapiai. Istorijos vadovėlis. 3-ioji kn. 10 kl. (The Pages of Lithuanian Past. History Textbook. 3rd Book. 10th Grade).* Kaunas: Šviesa.

Civinskas, R., & Antanaitis, K. (2001). *Lietuvos istorija. 12 kl. (History of Lithuania. 12th Grade).* Vilnius: Vaga.

Gečas, A., Jurkynas, J., Jurkynienė, G., & Visockis, A. (2001). *Lietuva ir pasaulis. Istorijos vadovėlis. 12 kl. (Lithuania and the World. History Textbook. 12th Grade).* Kaunas: Šviesa.

Jakimavičius, V. (1998). *Gimtoji šalis Lietuva. Lietuvos istorijos skaitiniai. 5 kl. (Lithuania, The Native Country. A Reader of Lithuanian History. 5th Grade).* Vilnius: Alma littera.

Kapleris, I., Laužikas, R., Meištas, A., & Mickevičius, K. (2016). *Laikas 12. Istorijos vadovėlis. 2 d. 12 kl. (Time 12. History Textbook. 2nd part, 12th Grade)*. Vilnius: Briedis.

Kapleris, I., Meištas, A., Mickevičius, K., Laužikienė, A., & Tamkutonytė-Mikailienė, Ž. (2017). *Laikas. Istorijos vadovėlis. 2 d. 10 kl. (Time. History Textbook. 2nd part, 10th Grade)*. Vilnius: Briedis.

Kaselis, G., Kraujelis, R., Lukšys, S., Streikus, A., & Tamošaitis, M. (2008). *Istorijos vadovėlis. 2 d. 12 kl. (History Textbook. 2nd part, 12th Grade)*. Vilnius: Baltos lankos.

Kasperavičius, A., & Jokimaitis, R. (2003). *Naujausiųjų laikų istorija. 10 kl. (History of Modern Period. 10th Grade)*. Vilnius: Kronta.

Kraujelis, R., Streikus, A., & Tamošaitis, M. (2010). *Istorijos vadovėlis. 2 d. 10 kl. (serija „Raktas"). (History Textbook. 2nd part, 10th Grade (series 'Key'))*. Vilnius: Baltos lankos.

Laužikas, R., Mickevičius, K., Tamkutonytė-Mikailienė, Ž., & Kapleris, I. (2008). *Kelias. Istorijos vadovėlis. 5 kl. 2 dalis. (Road. History Textbook. 2nd part, 5th Grade)*. Vilnius: Briedis.

Litvinaitė, J. (2007). *Palikimas. Istorijos vadovėlis. 1/2-oji kn. 5 kl. (serija „Šok"). (Legacy. History Textbook. 1st and 2nd books, 5th Grade (series 'Dance'))*. Kaunas: Šviesa.

Mäesalu, A., Kiaupa, Z., Straube, G., & Pajur, A. (2000). *Baltijos šalių istorija. 10–12 kl. (History of Baltic Countries. 10th–12th Grades)*. Vilnius: Kronta.

Makauskas, B. (2006). *Lietuvos istorija. 11–12 kl., 2-oji kn. (History of Lithuania. 11–12th Grades, 2nd book)*. Kaunas: Šviesa.

Navickas, V., & Svarauskas, A. (2015). *Istorijos vadovėlis 12 kl. (IV gimnazijos kl.). (History Textbook for 12th Grade)*. Kaunas: Ugda.

Petreikis, D., Litvinaitė, J., Meškuotis, F., Ramoškaitė-Stongvilienė, R., Bitautas, A., & Stankutė, S. (2014). *Istorija. 5 kl. (serija „Atrask"). (History. 5th Grade. (series 'Discover'))*. Kaunas: Šviesa.

Stašaitis, S., & Šačkutė, J. (2000). *Tėvynės istorijos puslapiai. Lietuvos istorijos vadovėlis. 5 kl. (The Pages of Homeland's History. Lithuanian History Textbook. 5th Grade)*. Vilnius: Margi raštai.

Zakarauskienė, I. (2000). *Lietuvos istorija. Skaitiniai. 5 kl. (History of Lithuania. A Reader. 5th Grade)*. Vilnius: Agora.

Ineta Dabašinskienė

Place and Language Transformations in a Post-Soviet Landscape: A Case Study of the Atomic City Visaginas

Abstract: The chapter analyzes the case of a post-Soviet city Visaginas (Lithuania) due to its socialist and mono-industrial heritage at present experiencing an extremely complicated transition period. Today Visaginas provides a very special example of Lithuanian ethnic landscape and represents a geographically, culturally and ideologically isolated place. Its ethnic composition is very diverse, but mainly consists of Soviet-period immigrants with a strong pro-Soviet identity who arrived in the 1970s. The collapse of the regime and the local economy have brought anxiety, uncertainty and fear not only to the inhabitants, but also to the city itself, to its identity and its future. This study focuses on the multiple issues, relying on the concepts of place identity, language ideologies, policies and practices in the framework of the global new economy and commodification. The linguistic landscape and soundscape of Visaginas demonstrated varied linguistic resources of the city: from the dominance of Russian, efforts to use Lithuanian, and English in written signs. The attempts to manifest linguistic diversity as a social capital of the city are obvious and have good potential. Whatever changes in individual repertoires and group preferences will take place in the future, bilingualism at the level of society seems to be the most desirable outcome.

Keywords: linguistic landscape, multilingualism, Russian-speakers, language commodification, language attitudes, atomic city, Visaginas

Introduction

Mobility is a reality of human life. People have always moved around and changed places for different reasons – political, economic, social and cultural. Wars, changes of regimes, conflicts, lack of resources, climate disasters or just a desire for adventures have forced humans to dislocate. In the age of globalization and high technologies, mobility has become a typical feature of modern societies. Diverse forms of mobility bring about changes in neighborhoods and communities, making them more hybrid as different languages and cultures blend. This is a fact. However, the imminent question is always the same: do we consider this diversity as a strength of social and economic well-being of the society in question? Or do we fear that multiculturalism and multilingualism will bring instability to the homogeneity of the society and weaken its traditional values? It is obvious that multilingual and multicultural groups are more complex than "pure" and homogeneous societies. The complex issues of diversity and integration are closely tied with the manifestation of power – social, political, economic, and linguistic, which fosters opposition and possible conflicts between the groups of "we, ours" and "they, theirs" (Schieffelin, Woolard, and Kroskrity, 1998). Tensions between the majority of population and minorities, mostly immigrants, have become evident in many European countries where cultural and linguistic diversity has been promoted and served to advocate tolerance and openness. However, other cultures and ethnicities are greatly welcomed mainly in "authentic" domains, such as folklore, crafts, music and, especially, cuisine and are tolerated under one condition – they have to be well "integrated", which most often means "assimilated" (Bloomaert and Verschueren, 1998). Moreover, linguistic integration is considered a must to become a true member of the society.

Thirty years ago, linguistic integration of Russian speakers was (and still is) at the center of the integration policy in the Baltic States because of the Soviet regime, which was marked by the supremacy of the Russian language and asymmetrical bilingualism, with Russian dominating in high-level spheres and Russian-speaking minority groups (Marten, Lazdiņa, Pošeiko and Murinska, 2012, p. 290). The Baltic States had to take necessary steps to re-introduce all functions of national languages after many years of Russification policy, therefore they had implemented "thick", "control-oriented" policies (Spolsky, 2004 ; Siiner, 2006). All three countries (Estonia, Latvia and Lithuania) have introduced national language laws, supplemented by a number of amendments and normative acts that define the status, teaching and use of languages in the state. Language policies guarantee that the state language is used and promoted, and that the relevant institutions work properly to maintain and develop the standard language. On this view, bilingual practices, code switching and borrowing become a problem (Vihalemm and Siiner, 2011).

The Russian-speaking population in the Baltic countries has lost its "guaranteed" position; as a result, the three states have been faced with the challenges of adopting new language planning and integration policies regarding Russian nationals (Vihalemm, Siiner and Masso, 2011, p. 116). Today, national languages are dominant, but Russian keeps a strong position in many domains despite the fact that language acquisition policies for the Russian-speaking population have aimed at developing their competence in the state languages[1]. Lithuania due to certain historical and political circumstances had a "better" situation than other Baltic States regarding its ethnic composition, as ethnic minorities, including Russian speakers, compose less than 20 per cent of the population. In Lithuanian cities the monolingualism of the dominant language (an official state language) is a result of the national language law requirement. However, local practices show the use of minority languages as well as an increasing visibility and use of global English. The English language is now the first foreign language taught at schools and is used as the main lingua franca in international communication. Moreover, not only does it appear in public signage for symbolic effects, but is also used in many activities (professional, leisure), especially by young people.

As is known, the build-up of specific contexts and unique arrangements of demographic, social, political and attitudinal factors in different localities may play an important role. The aim of this study[2] is to analyze the case of a post-Soviet city with special focus on its community, to reflect on the socialist past, to discuss the complicated present, and to envisage perspectives for the future. The town of Visaginas, a "migrant island" (Baločkaitė, 2010) in Lithuania, due to its socialist and mono-industrial heritage at present experiences an extremely complicated transition period. In what follows, the multiple issues of Visaginas will be disputed relying on the concepts of place identity, language ideologies, policies and practices.

---

1   The linguistic policies of the Baltic countries and the integration issues of the Russian-speaking population have been reported and discussed in many publications (Hogan-Brun, Ozolins, Ramonienė and Rannut, 2009; Rannut, 2008; Hogan-Brun and Ramonienė, 2005; Muiznieks, 2010; Kasatkina, 2007; Potashenko, 2010, etc.). For the new directions and discussions on languages in the Baltic States see Lazdiņa and Martnen (2019).
2   This study is a part of the EDUATOM (The Didactical Technology for the Development of Nuclear Educational Tourism in the Ignalina Nuclear Power Plant (INPP) Region; No. 01.2.2-LMT-K-718-01-0084/232) project, which aims to develop an educational nuclear tourism route in the INPP region in Lithuania. It is funded by the European Regional Development Fund according to the supported activity "Research Projects Implemented by World-class Researcher Groups" under Measure No. 01.2.2-LMT-K-718.

Visaginas: From the Planned Soviet Past to the No-Where Future?

The integration of the soviet republics, including the Baltic states, via industrial projects was very important for the Soviet Union to establish and maintain fixed economic structures: "(...) construction of large-scale industrial structures and special industrial towns served as an important tool for integrating the Baltic States into the united network of Soviet space" (Cinis, Drėmaitė and Kalm, 2008, p. 227).

Visaginas presents the case of "the planned socialist towns". As argued by Baločkaitė, "the planned socialist towns emerged first as the workers' settlements for socialist industrial enterprises. They, alongside their industrial enterprises, served not only economic aims, but also ideological ones" (2012, p. 45–46). These planned towns were mostly mono-industrial, with specific industries, such as nuclear energy in Visaginas, or uranium production activities at the plant in Sillamäe, Estonia and the hydroelectric power plant in Aizkraukle (former Stučka) in Latvia.

These towns were constructed as exclusive sites of socialism where city infrastructures were planned by leading architects, the living standards were above the average of the country (Cinis et al., 2008), and "the socialist culture and way of life were openly celebrated" (Baločkaitė, 2012, p. 46). These towns, according to Baločkaitė, "were projects of social engineering designed to develop a new type of community and personality. As model communities for socialism, they were meant to legitimize their socialist regimes, draw a line with the past, and signify the beginnings of a new socialist era" (2012, p. 46). A very specific feature of these towns was the absence of history, as they were built as a new enterprise in scantily populated regions. The first inhabitants usually were migrant workers without a common past, only sharing a socialist present.

Visaginas (former Sniečkus), as a planned soviet town, has become a symbol of economic and industrial progress of the USSR. It was built in 1973 as a satellite settlement to the Ignalina Nuclear Power Plant (INPP), the most progressive nuclear power plant at that time. In 1975, the symbolic cornerstone of the town was laid (see Fig. 1), and workers from different areas of the Soviet Union came to build the atomic town. The town was named after the first secretary of the Central Committee of the Lithuanian communist party, Antanas Sniečkus (in 1992, Sniečkus was renamed Visaginas). In 1983, the nuclear power plant was launched. Visaginas was supposed to become "the town of nuclear energy". The main employee was the Nuclear Power Plant, which played a vital role in

Fig. 1: The bilingual sign (Lithuanian-Russian) on a stone marking the establishment of the town in 1975 (The town for people of a nuclear power plant will be built here, August, 1975).

establishing the town's identity. In 1999, there were 5108 jobs at INPP, making up 38 per cent of the town's employment (Kavaliauskas, 1999, p. 248).

However, in 1986 the tragic catastrophe in Chernobyl temporarily stopped construction works. After the re-establishment of Lithuanian independence, emergence of the Green movement and other protests, the construction of the third reactor was suspended and its demolition began in 1989 (Baločkaitė, 2010; Kavaliauskas, 1999). In negotiations with the EU, Lithuania had to fulfill the requirement to gradually close the atomic plant. Therefore, the first reactor was stopped in 2004, the second in 2009. The two reactors are currently undergoing a decommissioning process. By 2030, the site of the two reactors should be ready for reuse.

During the first two decades of a new city's life, Visaginas was the most rapidly growing city in Soviet Lithuania due to immigration and a high birth rate; more than 25,000 immigrants arrived (Kavaliauskas, 1999, p. 30). However, the number of Lithuanians grew slowly from 5.8 per cent in 1979 to 14.96 per cent in 2001 (Kavaliauskas, 1999, p. 59). At the power plant, the percentage of Lithuanian workers was also quite low. After 1990, with uncertainties about the future of the nuclear power plant and employment looming, the number of inhabitants stabilized at around 30,000.

Today Visaginas provides a very special case of Lithuanian ethnic landscape and represents a geographically, culturally and ideologically isolated place. Its ethnic composition is very diverse, but mainly consists of Soviet-period immigrants with a strong pro-Soviet identity who arrived in the 1970s. After the declaration of Lithuanian Independence, Visaginas went through different stages of difficult developments, tensions and uncertainties in a search of its new identity and forms of co-existence with the rest of Lithuania. Today, the ethnic composition of the town still reminds the former Soviet Union with 52 per cent of the population being ethnic Russians, Belarusians (9.89 per cent), Poles (9.32 per cent) and Ukrainians (5.16 per cent), the rest belong to almost 40 different nationalities that mainly speak Russian. Lithuanians are a minority group with only about 15 per cent. Due to emigration in 2001–2011, the number of inhabitants in Visaginas decreased by 25 per cent; today there are only 19,000 inhabitants, of which almost 80 per cent are Russian speakers.

The number of Russian-medium secondary schools and pupils attending them is decreasing, with some Russian-speaking families sending their children to Lithuanian primary schools and Lithuanian kindergartens. Therefore, the number of schoolchildren is also decreasing, and in general, quite a number of young people consider emigration as their future choice. Almost 60 per cent of school graduates choose to seek higher education; however, half of them study not in Lithuania, but in the UK, Finland, Latvia, Russia, Belarus or Ukraine. Research shows that mainly in mixed families, if one family member is Lithuanian, children continue their education in Lithuania (Šliavaitė, 2012).

Many inhabitants of Visaginas were first-generation immigrants, with their relatives and professional ties in the Soviet Union; they still maintain strong diasporic connections with Russia and other former Soviet republics. However, thirty years of living in a new reality force Visaginas community to develop its relations with Lithuania and Lithuanians and call them to balance their loyalties between the two states (Baločkaitė, 2012). In order to understand this locality better, we must keep in mind that Visaginas has experienced a twofold "tragedy": the collapse of the Soviet Union and declaration of Lithuanian independence along the decommissioning of the INPP. These realities have brought anxiety, uncertainty and fear not only to the inhabitants, but also to the city itself, to its identity and its future. As it looks now, the search of a new identity, including opportunities for the economic development, does not provide any sound solution for the city. It is important to note that the population of Visaginas was and still is (to a certain extent) highly educated and qualified, as the power plant needed experts for the complicated jobs. Therefore, small businesses or local industrial projects are not considered an attractive alternative for the city's future. Today, due to a long stagnation period, emigration and slow infrastructural developments, the city lost its economic and social status, prestige and became *"the dying city", "the Soviet city",* and *"the ghost town".*

Theoretical Approach and Methodological Remarks

Social scholarship features intensive debates on place and identity; *sociolinguistics of place, belonging and mobility*, on the other hand, are rather new concepts providing valuable knowledge on language and place relationship (Auer, 2013; Blommaert, 2010; Cornips and de Rooij, 2018; Pennycook, 2010). Usually these concepts include discussing actual resources of languages deployed in real sociocultural, historical and political contexts, which, according to Blommaert, focus "not on language-in-places but on language-in-motion, with various spatiotemporal frames interacting with one another" (2010, p. 5). We also see that "territorialized" patterns of language use are complemented by "translocal" or "deterritorialized" form of language use, and that the combination of both often accounts for unexpected sociolinguistic effects (2010, p. 4–5).

These approaches invite to rethink the locality from the perspective of authenticity and tradition co-existing with various forms of local and global realities: urbanization, the new economy, multilingualism, hybrid communications and new types of identities. This study intends to contribute to the scholarship of place identity related to the soviet and mono-industrial heritage as well as the Russian-speaking community and its future perspectives based on a qualitative form of analysis. The aim of this paper is multifold: to provide not only a sketch of historical development of the town, which is essential in order to understand the complexity of the place, but also to discuss and analyze the concepts of place, language and identity in the framework of the global new economy and commodification. The main approach of the study is a qualitative one. The Linguistic Landscape (LL) methodology was used for initial screening of Visaginas public and private written signage in order to recognize the patterns of language policies, status and use. The main task was to take photos of private and public signs where at least two languages were present. Up to forty photos were collected in different areas of Visaginas. The objects included private (companies, shops, services, hotels) and public (cultural center, municipality, school, etc.) institutions. However, it should be noted that in general not many signs were found compared to other Lithuanian cities (Muth, 2012; Ruzaitė, 2017). Our intention was not collecting as many pictures of a location as possible as usually is the case of the LL method, which often relies on a quantitative aspect. For this study, simple quantitative data are unlikely to help explain complex societal issues. We assume that the detailed explanation of facts by reflecting on the political, social or psychological context might provide a better understanding of the people, locality and its linguistic dynamics.

Additionally, interviews with local representatives of the most prevalent ethnic groups (Russians, Belarusians, Ukrainians, Kazakhs and Uzbeks) were conducted to critically reflect on their linguistic practices and attitudes. In total, nine representatives were recorded: 2 Kazakhs, 2 Russians, 2 Ukrainians, 1 Lithuanian, 1 Belarusian and 1 Uzbek. There were seven females and two males participating in this study, all of them were of senior age (half of them retired). The duration of interviews varied greatly, from two hours to half an hour. The data were gathered during several sessions at different times (in 2018 and 2019) and amount to eleven hours of recordings. We suppose that both public arena (signs) and the informants' personal reflections will provide a possibility to understand how the local community marks its identity and demonstrates linguistic practices.

Let us introduce the Linguistic Landscape approach as for the potential reader of this volume this sociolinguistic methodology might be unfamiliar. LL is a relatively new field of study that encompasses diverse approaches to sociolinguistics, language policy, semiotics, etc. (Backhaus, 2009; Cenoz and Gorter, 2006; Landry and Bourhis, 1997; R. Scollon and S. W. Scollon, 2003; Spolsky, 2004). LL usually performs two main functions: the first is to inform about the diversity of languages present on signs and to provide information about the sociolinguistic composition of an area; the second is a symbolic function, as the presence of one's own language on signs can be interpreted as the fact that this language has a special value and status in a certain sociolinguistic context (Ruzaitė, 2017).

Globalization and mobility have different impacts on cultural, social and political life of many places. The city of today reflects dynamic linguistic landscapes and mainly exhibits signs of multilingualism. Linguistic landscapes, according to Landry and Bourhis, usually refer to "the language of road signs, advertising billboards, street names, commercial shop signs, and public signs on government buildings that combines to form the linguistic landscape of a given territory, region, or urban agglomeration" (1997, p. 25). Moreover, linguistic landscape includes "any sign or announcement located outside or inside a public institution or a private business in a given geographical location" (Ben-Rafael, Shohamy, Hasan Amara and Trumper-Hecht, 2006, p. 14). Urban spaces with various written signs refer to different modes of linguistic diversity; help to discover attitudes and beliefs, ideologies and power relations; to understand private and public, global and local interactions; and to analyze the relationship between languages, people, communities and identities. Languages visible in a public space provide information about their use, status and spread; about possible differences between official language policies and real local practices. Official language policies and power relations between different groups can also be determined (Backhaus, 2007, p. 11). Linguistic signs could be helpful to understand the representations of minority languages as well as their coexistence with the state language. Thus, language use and visibility reflect the constant negotiation of various identities, ideologies, policies and practices.

In other words, linguistic landscape "refers to the visibility and salience of languages on public and commercial signs" (Landry and Bourhis, 1997, p. 23). LL is thus perceived as a public space that displays languages and consists of varied discourses and types of genre, which are typically characterized by multimodality and multilingualism. At the surface level, public space may seem to offer an open area for versatile language exposures, but in practice it often turns into an arena of ideological and political struggle for ownership of space, representation and control (Ruzaitė, 2017). Display of languages are often predetermined by a variety of linguistic, economic, political and other factors. It also has to be mentioned that linguistic landscaping is a dynamic process; therefore, it is interesting to observe linguistic landscapes of post-soviet places for identifying essential changes. As Du Plessis (2010, p. 74) states, "a change in regime can bring about a change in the linguistic landscape". LL focuses on urban sites since modern cities are mostly multilingual and reflect the global trends of the new economy and commodification of social and cultural phenomena. Moreover, cities exhibit competing powers of local and global, including language policies and practices. From this viewpoint, Visaginas has its own justification. The language situation there indicates the changes in the city, shows the representatives of diverse population, and might serve as a laboratory to focus on the issue of how different forces conflict or coordinate their attitudes (Shibliyev, 2014). The function of English, as a global player, is interesting to observe as well because the power and prestige of Russian is still dominating the locality.

Language Policies, Attitudes and a Sense of Belonging

During the soviet times Russian was dominant as a language of soviet ideology and a promoter of "brotherhood". Since the restoration of independence, the sociolinguistic situation has been gradually changing as a state language, Lithuanian, is compulsory in the public sector. However, despite serious

Fig. 2: Graffiti in the residential area of Visaginas (Stop NATO. No war)

attempts to rearrange the linguistic landscape of Visaginas, Russian is still actively used in a private sector as well as in everyday communication.

There are many reasons for slow changes in this city. Scholarship on Soviet period immigrants to the Baltic States stresses the challenges of integrating this population into the social and cultural life of local societies (Vihalemm and Siiner, 2011). Russia, as the "country of origin", pronounces its concerns for its compatriots abroad and encourages "the maintenance of ties and Russian speaker identification with Russia as homeland" (Birka, 2016, p. 219). Recent events in Ukraine have demonstrated Russia's plans to protect and support the Russian language and culture of its compatriots living abroad. The awareness of complex relationships between social integration and feelings of belonging could provide a better understanding of identity transformation in Visaginas (see Fig. 2).

In this context of geopolitical events, Visaginas becomes a platform for negotiations of diverse ideologies and identities. The challenging feelings of belonging and attachment to the country of origin vs. country of residence are sensitive for the population and require a subtle approach. Therefore, it is quite clear that Lithuanian government is concerned about the loyalty of the Russian-speaking population and its attachment to Russia. The interview data show that Russian speakers in Visaginas favor integration over assimilation (for similar results in Latvia see Pisarenko, 2006); therefore, they uncompromisingly continue to cultivate their culture and use Russian extensively in all domains. They also explore different media platforms, predominantly in Russian. However, they admit the need to know Lithuanian, and research has shown that the competence of Lithuanian today is higher than ten or so years ago, especially among the young population (Lichačova, 2014).

The sociolinguistic situation in this city is quite complex. Despite governmental attempts lasting for thirty years to financially support language teaching, organize Lithuanian classes for adults, and introduce more hours of Lithuanian at schools and kindergartens, residents of Visaginas, especially older ones, still show quite poor competence of Lithuanian and prefer to speak Russian privately and in public. The issue is not only of a linguistic nature – the overall societal disappointment and the feeling of helplessness is obvious:

> "*three times I was attending Lithuanian courses. But they teach us about grand dukes, what they dream, but not the language. People are tired, stumbled, do not see any sense to study it for the fourth or fifth time*" (74, female).
> "*if then someone has been told us that we need to study Lithuanian or even other languages, that languages are important for a person, for his or her development, we did not know that...*" (72, male).

Even though elderly people claim that the age is an obstacle to learn Lithuanian, they also admit that the situation has changed now and the pressure is not so high as it was right after 1990, when "*only in one night we had to start speaking Lithuanian*" (70, female). Overall, the older Russian-speaking generation has developed more positive attitudes toward Lithuanian but have difficulties speaking or even understanding Lithuanian. They have mentioned bad memories of the attended language courses, poor quality of teaching materials and teachers' competence as non-motivating factors for learning.

Additionally, the most important factor for learning the language – the environment – was mentioned by the majority of respondents as well: "no one to talk to in Lithuanian" – that is the expression often mentioned during the interviews. Most of the informants mentioned poor motivation to study Lithuanian, as Lithuanian linguistic environment for them is very limited:

> "*Visaginas is a very closed town. It is closed because of the inability to communicate in Lithuanian. It is like a cage*" (55, female)
> "*Motivation to study Lithuanian should come from the pressure of the environment; however, that pressure is experienced when people from Visaginas go to other places in Lithuania, but not here*" (46, female).

It is clear that social networks among people in Visaginas are mainly with Russians; therefore, communication takes place only in Russian. Those who cross the border of Visaginas have both the advantage and pressure to learn Lithuanian:

> "*I like Visaginas very much. I go to Vilnius very often and I come back here...... I go to Vilnius every week as I have to arrange many things related to the performances and other business. And I have very good friends there...*" (46, female).

The social networks approach (see Milroy, 1987) indicates that population groups having relatively many contacts with neighbors and only few contacts with outsiders use primarily the language their neighbors are using because they are less exposed to other languages. Clearly, the Visaginas case confirms this. Unfortunately, the town is still a very closed place and its residents rarely go to other places in Lithuania; therefore, they are not encouraged and motivated to advance their Lithuanian.

Despite the many issues, including linguistic, the positive attitudes toward Lithuania were observed in the data of older respondents as they seem to be reconciled with the present situation:

> "*my children and grandchildren do not live here. One daughter is in London now. They are speaking English very well (…) and Russian. My grandchildren know Russian and we speak in Russian with them (…). But we don't want to go anywhere. We like Visaginas*" (76, male).

Younger people's attitudes toward Lithuania and the EU are even more positive, mainly due to pragmatic reasons, such as favorable social welfare and political security. A Lithuanian passport is important for young people as it allows them to travel and work in many European countries (Labanauskas, 2014). This possibility enables young people to leave Visaginas and choose other places as their homes.

The respondents for the most part regard Russia or other countries of origin (the post-Soviet space) positively as well. Moreover, all survey participants are willing to rediscover, maintain and transmit their ethnic cultures, languages, histories and traditions to their children and grandchildren. The participants of diverse ethnicities have expressed their high motivation to discover their own roots. They engage in various cultural activities organized by their ethnic communities and organize Saturday schools for children to teach the heritage language. The motivation to re-separate from the Russian cultural space and rediscover their own ethnicity, culture and language was mainly expressed by the Kazakh, Uzbek and Ukrainian representatives; likewise, other ethnic minority groups are very active as well (see Visaginas cultural center/minorities). Their enthusiasm to discover and promote their own ethnicities, however, are limited to folklore, authentic cuisine, crafts and some language classes; but this is only a symbolic action because they all navigate in the Russian informational space, and the Russian language still remains the only language of communication:

> "*We all came here with the same aim, to build an atomic plant, to live and work here. We came from different parts of the Soviet Union and nobody cared what your*

nationality is or if you speak another language. We all spoke Russian, and we speak Russian now....[..]. Everyone was in good relations with the others, we lived in peace... [...]." (79, female).

The concept of "ethnicity" was not important in the Soviet period, and the leitmotif of "druzhba narodov" ("friendship of nations", "brotherhood") and unity still dominated the interview data:

"We live here, we spent here all our life, we raised our children here. We like Visaginas, its nature, people. …. It's our home" (60, female).

Despite mainly positive attitudes toward Lithuania and Visaginas as their home, the respondents indicated disappointment regarding many aspects of social, economic and political life in the country. It is important to note that most of the residents follow political processes in Russia and consider the Russian language as a global language worth knowing and learning.

The long-term, strictly state-promoted, mono-ethnolinguistic (Lithuanian) approach, where "integration" means "assimilation", seems to be hardly applicable in Visaginas. The language situation adds even more complexity to the place. Although Russian is dominating, the reality of the place is a linguistic triangle: Russian – as a language of the dominant Russian-speaking community, Lithuanian – as an official language (but rarely used), and English, both a symbolic sign and a possibility to be part of global exchanges. The linguistic landscape demonstrates that official written signs are mainly in Lithuanian, private signs are bilingual (Russian – Lithuanian or Russian – English), but the spoken language is predominantly Russian (see Fig. 3). The level of formality of social space defines the use of the Russian language: if the social environment is less formal, then Russian is used; formal environments require Russian speakers to use Lithuanian (Labanauskas, 2014). However, this does not happen often as Lithuanians change to Russian when communication in Lithuanian is impossible.

The urgent question today is how to deal with the Soviet legacy: should the teaching of Lithuanian as the main language of all domains be promoted and strengthened? Or should bilingual or just Russian dominating linguistic practices be maintained? There is no one clear-cut answer to this complex issue. However, offering a different perspective toward Russian and the soviet heritage of the place might be useful if we view language and culture in the time of late capitalism as commodity.

Fig. 3: Bilingual signs of Russian-Lithuanian or Russian-English in Visaginas in private domains

Language, Authenticity and Commodification

Today Visaginas is in the active process of developing its new identity. Different ideas and approaches are put forward by diverse stakeholders, from the de-ideologized scenario of the "city of happiness, youth" or "green city" to a very clearly articulated project as the city of "socialist atomic past". The idea "to sell" the soviet past is actively discussed and promoted not only by the residents of the city, but especially by outsiders. The concept of "heritage tourism" mainly refers to the consumption of historical and cultural heritage testimoning the past (Poria, Butler and Airey, 2003); it is also linked to the places, artefacts and activities represented by the narratives of locals and include cultural, historical and natural resources (Yale, 1997). Therefore, tourism in sociolinguistic peripheries urge local communities to rethink their cultural, political and economic conditions, and this involves the reconstruction of linguistic capital for the new and potential economic capital (Heller, Pujolar and Duchêne, 2014). The theme of identity is closely linked to the issues of heritage. In the context of global economy and tourism industry, Russian, as an identity marker for Visaginas, could be reexamined. Through museums and other heritage sites, tourists can be told the local story presented in such a way as to affirm and reinforce the place identity and self-image. The construction of identity is integrally bound to tourism discourses that seem to claim what we *are* (or *were*) (Baločkaitė, 2012, p. 41). However, the interviews have revealed that it is difficult to deal with the unwanted socialist past. Moreover, in the city there are only very few symbols idolizing socialism, while the official sites do not provide much information about them. Nevertheless, the past is hidden in social activities of everyday life that are hardly noticed by the outsider. The most visible sign is the language. Therefore, as claimed by Heller (2010), multicultural and multilingual places are a great locality for heritage tourism to explore shifts in the role of language bound to changes in local industries and effects of the commodification of authenticity. Monika Heller and her colleagues define these changes as follows:

> "Within a conventional neoliberal frame, peripheral language groups must learn to market themselves, identify the resources that can be commodified, and turn their rhetoric of political mobilization to one of marketable entertainment in complex and ambivalent ways. Thus, the sociolinguistics of tourism provides a window into understanding the emergence of new linguistically-invested forms of power which follow a logic of circulations and mobilities, and are in stark contrast with the cultural expressions of industrial capitalism, with its emphasis on territoriality and ethnonational belonging. The commodification of language and identity is then something fully consistent with the economic and cultural processes triggered by the globalized new economy". (Heller et al., 2014, p. 561–562).

In this new social reality, the city is in transformation and search of its new identity. There are many examples globally when mono-industrial places go through dramatic changes of social transformation caused by the effect of the new economy. The global new economy is usually linked to changes in language and identities (Bauman, 1997; Castells, 2000) as well as tensions between local, national, supra-national identities and language practices, and between hybridity and uniformity (Heller, 2003). The sector of tourism could become a major industry for Visaginas generating economic value as economic development is a major driving force for the community and its wellbeing. In fact, language, culture and identity as authentic phenomena may play a significant role in this new transformation. The main characteristic of the globalized new economy is the commodified value of any kind of authenticity and exoticism.

Languages are very important for services, especially in the tourism industry. The increasing numbers of foreign and local tourists in Lithuania require multilingual competences. After the Baltic States declared their independence almost thirty years ago, the status of the Russian language significantly changed. For ideological reasons, the interest in studying Russian dropped and everyone started to learn English. This situation lasted approximately from 1990 to 2000. When Lithuania joined the EU, the walls opened boosting different kinds of mobility. The job market today often looks for employees who, in addition to English, know Russian (Dabašinskienė, 2011). The education sector sees the demand for the Russian language as well. We have been watching the dynamics of the choice of foreign languages in secondary schools for several years in a row. Over 95 per cent of pupils choose English as their first foreign language, and the vast majority of schoolchildren choose Russian as their second foreign language. This motive is of economic character.

A number of different political and economic realities in the last decades have brought the Russian language on the world stage again. It became popular in the international service, especially tourism. Additionally, Russian today provides a valuable economic resource from the perspective of teaching and learning, as part of language commodification, revealing how promotion policies of learning the language can help individuals to advance language knowledge in order to meet demands of Russian-speaking tourists (Muth, 2017). Thus, we see that political discourses on national ideologies in the era of globalization, economic and demographic turbulences have changed the attitudes toward Russian-speaking tourists. Moreover, tourists from Russian-speaking countries (Russia, Belarus and Ukraine) are dominating Lithuanian incoming tourism. In this respect, we observe many strategies of commodification of the Russian language, which serves to attract and satisfy Russian tourists both locally (in Lithuania and the Baltics) and globally. Anette Pavlenko uses a special term "preferential accommodation" to explain situations where special arrangements are made for Russians, including simplification of visa regimes, acceptance of rubles, addressing in Russian and providing various media information services in Russian (Pavlenko, 2015). As business reacts fast, it ensures linguistic accommodation of Russian in the areas preferred by Russian tourists. Since Russian speakers can rarely communicate in other languages, services need to employ Russian in order to attract customers. The tourism industry reports the economic potential of Russian customers: in 2004, Russians were among the world's top ten biggest spenders, while in 2013, $53.5 billion spent abroad placed them fourth after tourists from China, the USA, and Germany (UNWTO, 2014). Tourism agencies and other tourism-oriented businesses (hotels, spa, restaurants, museums, opera houses, etc.) in countries popular with Russian tourists invest in Russian language resources and websites (e.g. http://www.visitlithuania.net/russia, https://www.visitfinland.com/ru/). They also hire a Russian-speaking staff; broadcast Russian TV channels; provide menus, newspapers, magazines, maps, travel guides and information about hotels in Russian; and in general demonstrate Russian-friendly attitudes. The developing tourism industry demonstrates that English does not always serve global needs. The arguments provided above suggest that for tourists who are ready to spend generously, local businesses have to invest into language learning (Pavlenko, 2015).

Thus, Visaginas could exploit the Russian language and soviet culture as a symbol of its authenticity. Tourist routes in Visaginas should focus not on such sites as museums, original architecture, symbolic objects, monuments, etc., but on the locality itself as an open-air museum displaying daily linguistic practices, routines, habits and attitudes, hidden and visible signs of the nostalgic soviet past, and atomic glory. Moreover, minority languages, cultures and identities could also be exploited, as they could become "means of production and [as] a product itself" (Heller, 2003). However, commodification of place, language and identity might be a challenge, as part of the community does not want their past to be "on sale". The tensions among different stakeholders regarding their vision for the city's future and its identity, from the *"city of green nature, happiness, youth to atomic tourism"*, are discussed in recent publications (Mažeikienė and Gerulaitienė, 2018). Nevertheless, the locality with its mono-industrial atomic post-soviet heritage narrative could become an object of authentic experience, especially for heritage tourism. Russian as a symbol of that time also represents the reality of today. Its authentic practice in a linguistically homogeneous country offers a valuable possibility for a unique experience attractive for tourists. In the heritage tourism market, symbolic elements marking the authenticity of the place, including the language, may contribute to the city's higher potential. The elderly people of Visaginas envisage a great potential for their own memories, as they are "walking tourist guides":

> *"It's such a pity, I feel so sorry...There are so many interesting things here, and there still are people here who remember all the corners of this city. They know why this street is here, or where the first tomb is. There are so many places and stories to tell. About the first house, the first school. It would be so interesting"* (75, female).
>
> *"Visaginas is so rich, as there are so many diverse cultures, religions, folklore. People are so friendly. We celebrate great festivals that many people come to see (….). It is the best place to live, it is so tranquil, the nature is beautiful, there are many lakes where you can go swimming and enjoy boating..."* (68, male).

It is obvious that in the globalized new economy of services language skills become a crucial factor for the development of tourist infrastructures. Thus, even though Russian might play a unique role of practice-in-place and be attractive to Russian-speakers (or older population of the region), it is not enough; languages of other ethnic minorities, including Lithuanian and global English, could offer broader linguistic resources for communication with clients (see Fig. 4). Since multilingualism is becoming an increasingly important skill in the tourism sector, the multilingual capital of the community could help construct local authenticity and manage relations with clients.

Fig. 4: The poster inside the Culture center. Most of the informants declared their positive feelings to Visaginas, their hometown.

On this view, the dominance of Russian might be conceived not as an obstacle but as an asset helping to boost the economic potential of the tourism sector and other services. Therefore, the saying "Russian is everywhere" might be interpreted differently and attract visitors to explore the new-old reality of the locality.

Conclusions

How individuals and societies cope with language and place identity in the contemporary world is an important sociolinguistic question which requires a critical analysis of linguistic practices, attitudes and understanding of the uniqueness of the locality brought about by specific historical and social circumstances.

Visaginas represents a rather peculiar place of hybrid identities, attitudes, feelings of anxiety, disappointment and hope. This is a unique place in Lithuania where diverse ethnic communities are still united by the Russian language and the feeling of "brotherhood". However, global tendencies and pragmatic attitudes toward one's future start making a difference, especially for young people. Prevailing discourses of a mono-ethnic nation state with a dominant state language create obstacles in giving rise to authentic multi-ethnic local identities in Visaginas. Language ideologies from a top-down perspective conflict with bottom-up practices and therefore require a more coherent, "softer" approach to Visaginas population. Most of the informants emphasize the richness of the city's culture, the people's creative potential, especially that of the young generation, which produces interesting projects and shows impressive skills of entrepreneurship. The multicultural, multilingual and multiethnic profile of Visaginas is perceived by the informants as an important resource for representing the uniqueness of the town.

Due to its soviet past and a strong identification with the atomic power plant, Visaginas was unable to develop consistently after 1990. The strong pro-soviet attitudes and nostalgia for the heroic past (including the construction of the most progressive nuclear plant), the unchanged ethnic composition dominated by Russian speakers, and the intense use of Russian in everyday life make this site very remote both ideologically and geographically from the rest of Lithuania. For years, it was a success story and the place to celebrate socialism with a highly progressive mono-industry, desirable living standards and a notable quality of education. After the declaration of Lithuanian Independence in 1990, the town became the site of tensions and uncertainties. Visaginas did not become what it was supposed to become, i.e., a successful project of prosperity and flourishing. The feelings of nostalgia, emptiness, longing for intimate networks, lack of multilingual appreciation and a bitter failure to accomplish the soviet utopia were strongly reflected in the sincere stories of respondents.

The future requires transformation, but the process of changes is full of ambivalent ideas for the city's search of a new identity. Different stakeholders suggest different ideas for building the de-ideologized image of *young, happy and green city* with comfortable conditions to live and raise families. Additionally, the multicultural, multilingual and multiethnic profile of Visaginas is perceived as an important resource to represent the uniqueness of the town and deliver "commercialized hospitality" through cultural and recreational tourism. The opposite idea is to position the town as a socialist city within the framework of *socialist heritage tourism* with strong emphasis on nuclear identity. The latter idea looks quite promising for incoming tourism, especially from foreign countries, in the context of the rising movement of safe energy or desire for exotic experiences while visiting "dark" tourism sites, like Chernobyl. However, the official and institutional discourse of public authorities tries to minimize the socialist past in constructing the identity of the locality. How to proceed with the two ideas for commodification of the authenticity "soviet past, nuclear identity" vs. the de-ideologized identity of being "young/green" is a challenging question for the future of the city (Mažeikienė and Dabašinskienė, 2018).

This place is exceptional due to its ethnic composition represented by forty different ethnic groups. However, the situation does not indicate the diversity of languages. The opposite is true – the prevalence of Russian, as a soviet heritage, is dominating all the generations. The state has developed and offered many language programs in promoting and teaching Lithuanian; however, the population of Visaginas, especially the elder, have not learned the language. It is obvious that the language policy introduced here was not effective and therefore failed. Unfortunately, during the thirty years of independent Lithuania no greater revisions of social, economic and linguistic policies regarding the issue of Visaginas were performed. A great number of researchers have highlighted the negative impact of top-down narratives and discourses produced by the country's politicians and journalists. These discourses, it is argued, hamper the integration process by creating incompatible identity positions between Russian-speakers and the majority (Cheskin, 2013). The attempt to remove the influence of Russian culture and various aspects of Russian identity will surely bring back discourses about marginalization and discrimination. Instead, policy makers should focus on making Lithuanian culture more accessible to Russian speakers. This research indicates that there is great potential for integrated, yet culturally distinct, Lithuanian-Russian or other hybrid identities. Indeed, a number of studies have shown that Russian speakers in the Baltic States consider themselves to be very different from Russians in Russia (Vihalemm and Masso, 2003; Zepa, 2006). Moreover, in the context of globalization and Europeanization the ethnic identity for Russian speakers does not seem to be a very valuable and important category, as young people mainly choose a pragmatic approach to their identification (Labanauskas, 2014). As pointed out by the informants, the most important factors for a successful life in Lithuania (if one considers this option instead of emigration), especially for youth and in the job market, were the knowledge of Lithuanian and social networks. The senior generation feels quite comfortable using only Russian, as there are no linguistic obstacles in communication due to the fact that "everyone here speaks Russian". However, they expressed fear for the future of the city, social insecurity for themselves and their children because the future development of the city is unclear.

The linguistic landscape and soundscape of Visaginas demonstrate varied linguistic resources of the city: from the dominance of Russian, as post-Soviet heritage; efforts to use Lithuanian, especially in formal sphere; and English in written signs. The attempts to manifest linguistic diversity as a social capital of the city are obvious and have good potential. Whatever changes in individual repertoires and group preferences will take place in the future, bilingualism at the level of society seems to be the most desirable outcome. In terms of the city's potential, it is obvious that the community possesses diverse linguistic and cultural resources to manifest multiculturalism. Any discussions related to language in place, according to Pennycook, reflect our understanding of language as action: "What we do with language in a particular place is a result of our interpretation of that place" (Pennycook, 2010, p. 2).

Thus, the current situation of Visaginas can be assessed as the process of searching for the new identity and simultaneously adhering to the established linguistic routines, local habits and attitudes. However, the unclear present obligates the community to act and cater not only for local cultural and linguistic needs but also, following Heller and Martin-Jones's (2001) claim, to get ready to encounter the "new global forms of cultural, economic and social domination". The possibilities to explore soviet nuclear heritage are attractive for the tourism industry, so taking this route might create an added economic value and enable opening a new page in the city's life.

References

Auer, P. (2013). *Code-Switching in Conversation – Language, Interaction and Identity*. London: Routledge.

Backhaus, P. (2007). *Linguistic Landscapes. A Comparative Study of Urban Multilingualism in Tokyo*. Clevedon: Multilingual Matters.

Backhaus, P. (2009). Rules and Regulations in Linguistic Landscaping: A Comparative Perspective. In E.Shohamy and D. Gorter (Eds.) *Linguistic landscapes: Expanding the Scenery*, (p. 157–172). London: Routledge.

Baločkaitė, R. (2010). Post-Soviet Transitions of the Planned Socialist Towns: Visaginas, Lithuania. *Studies of transition states and societies (STSS)*, 2 (2), 63–81. Tallinn: Tallinn University.

Baločkaitė, R. (2012). Coping with Unwanted Past in Planned Socialist Towns: Visaginas, Tychy and Nowa Huta. *Slovo*, 24, 41–57.

Bauman, Z. (1997). *Postmodernity and Its Doscontents*. Cambridge: Polity Press.

Ben-Rafeal, E., Shohamy, E., Hasan Amara, M., & Trumper-Hecht, N. (2006). Linguistic Landscape as Symbolic Construction of the Public Space: The Case of Israel. *International Journal of Multilingualism*, 3(1), 7–30.

Birka, I. (2016). Expressed Attachment to Russia and Social Integration: The Case of Young Russian Speakers in Latvia, 2004–2010. *Journal of Baltic States*, 47(2), 219–238.

Blommaert, J., & Verschueren, J. (1998). *Debating Diversity: Analyzing the Discourse of Tolerance*. London, NewYork: Routledge.

Blommaert, J. (2010). *The sociolinguistics of globalization*. Cambridge (XVI, 213). New York: Cambridge University Press.

Castells, M. (2000). *The Rise of the Network*. Oxford: Blackwell..

Cenoz, J., & Gorter, D. (2006). Linguistic Landscape and Minority Languages. *The International Journal of Multilingualism*, 3, 67–80.

Cheskin, A. (2013). Exploring Russian-Speaking Identity from Below: The Case of Latvia, *Journal of Baltic Studies*, 44(3), 296.

Cinis, A., Drėmaitė, M., & Kalm, M. (2008). Perfect Representations of Soviet Planned Space: Mono-industrial Towns in the Soviet Baltic Republics in the 1950s–1980s, *Scandinavian Journal of History*, 33(3), 226–246.

Cornips, L., & de Rooij, V. (2018). *The Sociolinguistics of Place and Belonging: Perspectives from the Margins*. Amsterdam: John Benjamins.

Dabašinskienė, I. (2011). We ar Looking for a Sales Manager in Lithuania. In *Towards a Language Rich Europe. Multilinguali Essays on Language Policies and Practices* (p. 73–87). Berlin: British Council.

Du Plessis, T. (2010). Bloemfontein/Mangaung, " 'City on the Cove'. Language Management and Transformation of a Non-Representative Linguistic Landscape". In E. Shohamy, E. Ben-Rafael & M. Barni (Eds.), *Linguistic Landscape in the City* (p. 74–95). Bristol: Multilingual Matters.

Heller, M., & Martin-Jones, M. (Eds.) (2001). *Voices of Authority: Education and Linguistic Difference*. Norwood, NJ: Ablex.

Heller, M. (2003). Globalization, the New Economy, and the Commodification of Language and Identity. *Journal of Sociolinguistics*, 7(4), 473–492.

Heller, M. (2010). *The Commodification of Language* . Annual Review of Anthropology, 39, 101–114.

Heller, M., Pujolar, J., & Duchêne, A. (2014). Linguistic Commodification in Tourism. *Journal of Sociolinguistics*, 18(4), 539–566.

Hogan-Brun, G., & Ramonienė, M. (2005). The Language Situation in Lithuania. *Journal of Baltic Studies*, 36(3), 345–370.

Hogan-Brun, G., Ozolins, U., Ramonienė, M., & Rannut, M. (2009). *Language Policies and Practices in the Baltic States*. Talinn: Talinn Univerisity Press.

Kasatkina, N. (2007). Etniškumo tyrimai: tendencijos ir esminės sąvokos. Filosofija. Sociologija 18(4), 1–11.

Kavaliauskas, A. (1999). *Visaginas 1975–1999*. Vilnius: Jandrija.

Labanauskas, L. (2014). Miesto socialinis ekonominis kontekstas ir tautinio tapatumo raiška: Visagino miesto atvejis. *Lietuvos socialinių tyrimų centras, Etniškumo studijos*, 2, 125–143.

Landry, R., & Bourhis, Y., R. (1997). Linguistic Landscape and Ethnolinguistic Vitality: An Empirical Study. *Journal of Language and Social Psychology*, 16(1), 23–49.

Lazdiņa, S., & Marten, H. F. (Eds.) (2019). *Multilingualism in the Baltis States. Societal Discources and Contact Phenomena.* London: Palgrave Macmillan.

Lichačiova, A. (2014). Dovanota vertybė ar įsiūlytas paveldas? Lietuvos rusakalbių nuostatos gimtosios kalbos atžvilgiu. *Taikomoji kalbotyra* (4), 1–26.

Marten, H. F., Lazdiņa, S., Pošeiko, S., & Murinska, S. (2012). Between Old and New Liller Languages? Linguistics Transformation, Lingua Francas and Languages of Tourism in the Baltic states. In C. Hélot, M. Barni, R. Janssens & C. Bagna (Eds.), *Linguistic Lanscapes, Multilingualism and Social Change* (p. 289–308.) Frankfurt am Main: Peter Lang.

Mažeikienė, N., & Dabašinskienė, I. (2018). Performance of Authentic Cultural Identities at Stake: Between Dominant Discourses and Theatre Performances. Post-Socialist Identities: New Approaches, In *conference 50th ASEEES Annual Convention December 6–8, 2018.* Boston, USA.

Mažeikienė, N., & Gerulaitienė, E. (2018). Educational Aspects of Nuclear Tourism: Sites, Objects and Museums. In *10th international conference on education and new learning technologies*, Jul 2–4, 2018: (p. 5692–5702). Palma, Spain: Conference proceedings.

Milroy, L. (1987). *Language and Social Networks.* Oxford: Blackwell.

Muižnieks, N. (Ed.) (2010). *How Integrated is Latvian Society? An Audit of Achievements, Failures and Challenges.* Riga: University of Latvia Press.

Muth, S. (2012). The Linguistic Landscapes of Chișinău and Vilnius: Linguistic Landscape and the Representation of Minority Languages in Two Post-Soviet Capitals. In D. Gorter, H. F. Marten, L. Van Mensel (Eds.), *Minority Languages in the Linguistic Landscape* (p. 204–224). London: Palgrave.

Muth, S. (2017). Russian Language Abroad: Viewing Language Through the Lens of Commofidication. *Russian Journal of Linguistics,* 21(3), 463–492.

Pavlenko, A. (2015). Russian-friendly: How Russian Became a Commodity in Europe and Beyond. *International Journal of Bilingual Education and Bilingualism,* 20(4), 385–403.

Pennycook, A. (2010). *Language as a Local Practice.* New York: Routledge.

Pisarenko, O. (2006). The Acculturation Modes of Russian Speaking Adolescents in Latvia: Perceived Discrimination and Knowledge of the Latvian Language. *Europe-Asia Studies,* 58(5), 751–773.

Poria, Y., Butler, R., & Airey, D. (2003). The Core of Heritage Tourism. *Annals of Tourism Research,* 30(1), 238–254.

Potashenko, G. (2010). Russians of Lithuania (1990–2010): Integration in Civil Society. *Ethnicity,* 2(3), 98–109.

Rannut, M. (2008) Estonization Efforts Postindependence. *International Journal of Bilingual Education and Bilingualism*, 11(3-4), 423-439.

Ruzaitė, J. (2017). The linguistic landscape of tourism: Multilingual signs in Lithuanian and Polish resorts. *Journal of Estonian and Finno-Ugric linguistics (ESUKA - JEFUL)*, 8(1), 197-220.

Scollon, R., & Scollon, S. W. (2003). *Discourses in Place: Language in the Material World*. New York: Routledge.

Shibliyev, J. (2014). Linguistic Landscape Approach to Language Visibility in Post-Soviet Baku. *BILIG*, 71, 205-232.

Siiner, M. (2006). Planning Language Practice: A Sociolinguistic Analysis of Language Policy in Post-Communist. Estonia. *Language Policy*, 5(2), 161-186.

Šliavaitė, K. (2012). Etninės mažumos darbo rinkoje: kalbos, pilietybės ir socialinių tinklų reikšmė (Visagino atvejis). *Etniškumo studijos/Ethnicity studies*, 1(2), 103-125.

Spolsky, B. (2004). *Language Policy*. Cambridge: Cambridge University Press.

Vihalemm, T., & Masso, A. (2003). Identity Dynamics of Russian-Speakers of Estonia in the Transition Period. *Journal of Baltic Studies*, 34, 92-116.

Vihalemm, T., & Siiner, M. (2011). Individual Multilingualism in the Baltic States within the European Context. In T. Vihalemm (Ed.), *Estonian Human Development Report 2010/2011. Baltic Ways of Human Development: Twenty Years on* (p. 135-137). Tallinn: Eesti Koostöö Kogu.

Vihalemm, T., Siiner, M. and Masso, A. (2011). Introduction: Language Skills as a Factor in Human Development. In T. Vihalemm (Ed.), *Estonian Human Development Report. Baltic way(s) of Human Development: Twenty Years on*(p. 116-118). Tallinn: Eesti Koostöö Kogu.

Schieffelin, B., Woolard, K., & Kroskrity, P. (1998). *Language Ideologies: Practice and Theory*. Oxford University Press.

Yale, P. (1997). *From Tourist Attractions to Heritage Tourism (2nd edn)*. Huntingdon: ELM.

Zepa, B., & Šūpule, I. (2006). Ethnopolitical Tensions in Latvia: Factors Facilitating and Impeding Ethnic Accord, In N. Muižnieks (ed.) *Latvian-Russian Relations: Domestic and International Dimensions* (pp. 33-40). Riga: LU Akadēmiskais apgāds.

Eglė Gerulaitienė and Natalija Mažeikienė

Energy Tourism at Nuclear Power Plants: Between Educational Mission and Retention of "Safety Myth"

Abstract: This chapter has two principal objectives: first, to conceptualize the recent development in energy tourism by giving special attention to nuclear tourism as a distinct form of energy tourism and illuminating the educational potential of nuclear tourism sites and tours. Second, the chapter provides an explorative field research study of nuclear energy tourism in Scotland and Lithuania. The case study is based on field trips and observations conducted while visiting two energy tourism attractions – the Torness Nuclear Power Station and Ignalina Nuclear Power Plant (INPP) with Visitor Centres at these nuclear industry facilities. Features of nuclear tourism at atomic reactors and their premises are described in the chapter by pointing out the expansion of activities and target groups, going beyond the concept of expert-oriented tourism and moving to experience-oriented tourism, energy literacy, STEM education and orientation to wider groups of citizens and tourists (schoolchildren, families, etc.).

The authors of the chapter shed light on energy and nuclear tourism as an open and covert corporate branding and seek to shape positive attitudes of consumers and citizens towards the nuclear energy industry. The nuclear companies deploy tourism and educational activities (STEM education) to build citizens' perception of safety and security in the industry and strengthen pro-nuclear attitudes. Taking into consideration these aims of nuclear energy industry, a critical approach towards tourist attractions and educational sites at nuclear power plants has been employed by the authors.

Keywords: energy tourism, nuclear tourism, atomic tourism, nuclear power plants, Ignalina Nuclear Power Plant (INPP), Torness Nuclear Power Station

Introduction

Extensive development of energy tourism has expanded the interest of visitors to travel to different sites of tourist destinations (wind turbines, coal mills, nuclear power plants, etc.). Tourists want "to gaze at different landscapes and townscapes that are unusual for them" (Urry and Larsen, 2011). Therefore, such objects as giant cooling towers at nuclear power plants could be experienced as tempting and fascinating by tourists. Energy tourism overlaps with other types of tourism, including cultural and heritage tourism, as well as adventure tourism and agricultural tourism. Nuclear tourism is continuing to expand and develop as a growing subset of the heritage tourism phenomenon, which has dramatically increased over the past several decades (Berger, 2006). Heritage tourism creates interest among visitors and provides enriching experiences. Extensive development of tourism is closely associated with a rise in the education demand for tourism activities. Energy tourism has the potential to improve people's energy literacy and change their behaviour in using energy towards more sustainable "energy citizenship" (Devine-Wright, 2007).

As energy tourism contributes to science and STEM education; promotes environment education; fosters the development of responsible citizenship, knowledge of heritage and history; the authors analyse the educational mission (potential) at energy sites and atomic reactors. Recently, there has been a growing debate that in addition to the educational potential and satisfying professional or learning interests, energy tourism must be classified as experience-based tourism, where tourists are invited to experience special feelings. The authors also discuss the influence of new public relations and corporate branding strategies, which have been introduced by energy companies and interest groups in order to influence public opinion about certain types of energy and energy companies. Energy companies seek to represent themselves as socially responsible and environmentally aware producers (Frantál & Urbánková, 2017). The topic of "safety myth" as a pro-nuclear narrative, created by the nuclear industry establishments together with governments, is criticized in this chapter.

The purpose of the chapter is to reveal the educational potential, which is envisaged in the existing nuclear tourism sites and tours. This chapter has two principal objectives: first, to conceptualize the wider interrelationships between energy and tourism, provide a definition of energy tourism as a new niche of special interest tourism; second, to provide an explorative field research study of nuclear energy tourism destinations in Scotland and Lithuania.

Methodology

The case studies presented in this chapter are based on field trips and qualitative observation conducted by visiting two energy and nuclear tourism attractions – excursions to the Torness nuclear power plant (Scotland, JK) and Ignalina Nuclear Power Plant (INPP) (Lithuania). Additionally, researchers conducted visits to Information/Visitor Centres at these nuclear sites (INPP and Torness) and analysed the content of these exhibitions. The researchers participated in two-hour tours inside the nuclear power plants and at Visitor/

Information centres. Additionally, these tours included conversations with guides working at INPP and Torness information centres about the narrative content of the excursions, which they provide for different tourist groups, and the topics and questions that interest visitors during the excursions. These two field research trips were arranged in 2019, the sites of nuclear tourism were explored from the perspective of a potential visitor. At the same time, experiences gained by the researchers and the content of the exhibits were interpreted by referring to theoretical considerations on energy and atomic/nuclear tourism, the role and functions of Visitor Centres at the atomic energy facilities and in relation to the concepts of publicity and openness, security and safety, and entertainment and education at the nuclear and atomic sites.

Both nuclear power plants offer two separate excursions for the visitors: one tour inside the nuclear power plants include a visit to three zones – the Reactor, the Control Panel and the Turbine Zone (from 1.5 to 2.5 hrs.); the other excursion – in the Visitor/Information Centres (from 1 to 2 hrs.), providing information on nuclear energy situation, safety, political/economic situation of the energy production and educational material.

Conceptualizing Nuclear Tourism as a Specific Form of Energy and Industrial Tourism

Nuclear tourism is viewed as a specific segment of energy and industrial tourism, which is characterized by the attractiveness of industrial sites in the country, new technologies and power plants (Otgaar, 2012; Frantál and R. Urbánková, 2017). Industrial tourism includes visits both to former, retired or regenerated sites (industrial heritage tourism) and to still operational industrial sites where some facilities have been provided specifically for tourists' use, even though the core activity of the site is not oriented towards tourism (Frew, 2008; Otgaar, 2012).

Energy tourism and its specific form – nuclear tourism – takes its shape at the junction of several types of tourism – industrial, special interest, expert-oriented, experience-oriented, cultural and heritage, adventure, farm and agricultural tourism (see Fig. 1).

What attracts visitors to operating energy production sites? On the whole, energy and nuclear tourism can be assigned to *special interest tourism* and expert-oriented tourism when certain groups (experts in energy, engineers, regional managers, students of technical universities, schoolchildren, etc.) have specific professional and educational interest in the subject related to their activities at work, in education and studies. Thus, in terms of the educational

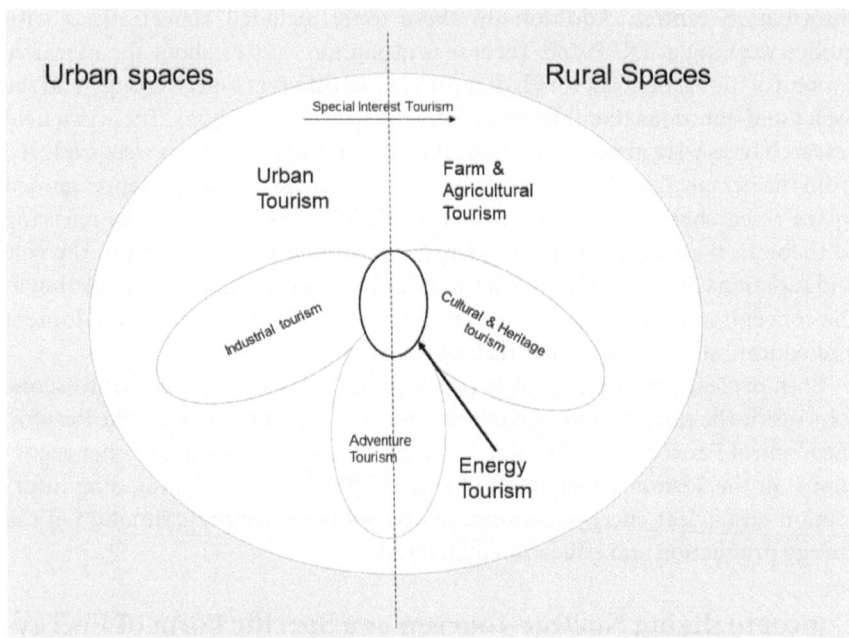

Fig. 1: Interrelationships of energy tourism and other types of special interest tourism (adapted from Frew & Shaw, 1995) by Frantál & Urbánková (2017)

impact, it should be emphasized that energy tourism (including nuclear tourism) has the potential to improve people's "energy literacy", improve the understanding of how we consume energy, and raise awareness of nuclear waste repository or potential impacts of nuclear activities on local economies (Frantál & Urbánková, 2017).

Promoting STEM education at energy sites and atomic reactors. Energy tourism contributes to science and STEM education, promotes environment education, fosters the development of responsible citizenship, knowledge of heritage and history. Regarding nuclear power plants, a certain content of knowledge and experience is exposed here: the history and construction of power plants; the principles and technology of atomic energy production and related matters, such as nuclear safety and security; topics related to radioactive waste disposal; and environmental and landscape impacts. So, the educational impact of energy and nuclear tourism is obvious.

STEM education through informal learning and "out of classroom experiences" is encouraged at most levels of the educational system as a way of strengthening students' interest and motivation. Museum visits represent one type of such experience, and museums are investigated as a means for improving the scientific literacy of both students and adults (Henriksen and Jorde, 2001).

New forms of nuclear tourism – combining environmental education, the presentation of novel technologies, interactive popular science experiments and various outdoor activities – attract schools to organize educational visits to museums to provide understandable scientific information about different topics: atoms, radiation, ionizing radiation and health, reactors, robots, physics and much more. Nuclear Power Plants (NPP) Visitor/Information Centres organize school visits, which are supplementary to the physics and chemistry curriculum. As stated in the research by Henriksen and Jorde (2001) regarding school visits to museums, studies have shown that a museum visit, in combination with pre- or post-visit experiences in the classroom, may provide considerable improvement in learning outcomes for students.

Energy sites as experience-based tourist destinations. Recently, there has been a growing debate that in addition to educational potential and satisfying professional or learning interests, energy tourism must be classified as experience-based tourism, where tourists are invited to experience special feelings. There is a new tendency in the energy tourism sector to focus more on various target groups that do not have a specific professional and study interest in energy. Energy tourism seeks to attract experience-oriented tourists of all ages (including families), for whom not only education, experience and emotions, but also entertainment and leisure are important.

Industrial infrastructure and industrial landscapes in general, and energy industry sites in particular, present tourists with an extraordinary visual experience. Tourists get fascinated and intrigued by the spectacular sizes and unusual shapes of industrial sites. Energy sites become a tourist attraction as something very specific and different from the places where the tourists themselves live, and visitors want to gaze at different landscapes and townscapes that are unusual for them (Pasqualetti, 2012; Frantál & Urbánková, 2017; Urry & Larsen 2011).

Some traditional energy landscapes are perceived as visually or environmentally polluted; that is why they potentially discourage tourists from visiting them. On the other hand, because of the difference from the usual environment, industrial sites attract tourists.

Recently, new energy tourism destinations have also attracted tourists' interest by exposing power plants with renewable energy sources (wind farms, photovoltaic, geothermal, hydropower plants, etc.). These facilities attract tourists due to unique technical nature, design and movement (Beer et al., 2017). Tourists visiting renewable energy source power plants are fascinated by the unique visual aspects of such places. As the authors (Beer et al., 2017) state, the visitors' interest is mainly caused by curiosity and in some countries by the rarity of similar types of technologies. Tourists visiting renewable energy sources sites are described as environmentally concerned visitors.

Nuclear power plants are also becoming an experience-based tourism destination. These sites evoke special aesthetic unique feelings of impressiveness and grandeur. Giant cooling towers of nuclear power plants are associated with modern cathedrals (Hecht, 1997). People even organize wedding ceremonies with a nuclear plant as a backdrop, as in the case of Temelın power plant in the Czech Republic[1].

Other authors describing thrill and excitement felt by tourists at nuclear power plants distinguish a heightened experience of being in an authentic place loaded with potential dangers (Storm et al., 2019). However, it is important to emphasize that a sense of risk arises not only from the immediate experience at the atomic facility, but also from the appeal to the broader nuclear imaginary. These are conceptions and imagination of nuclear power, which is constructed by political discourse, literature, media, cinematography, art, education, etc. According to Storm et al. (2019), visitors' imagination is fuelled by a mix of ambivalent understandings when appealed, on the one hand, to scientific utopian visions of the 1940s and 1950s, and, on the other, to dystopian fears of radioactive catastrophes and contamination.

Visitors' fear is amplified by the non-sensory qualities of radiation: it cannot be seen, heard or smelled by human senses unaided (Ibid.). So, nuclear power plants construct a peculiar perception of risk and security in the excursions and expositions at the Visitor Centres – this is where a kind of "security theatre" is arranged to calibrate the concordance between, on the one hand, calculable risk and security and, on the other hand, perceived risk, and security (Schneier, 2003 cit. Storm et al., 2019). The security theatre evoking a perception of risk

---

[1] V. Lukasova, *Cooling towers for a witness: Weddings by a nuclear power plant Temelin, Czech Republic*, 2011. Retrieved from https://veronikalukasova.wordpress.com/2011/10/04/cooling-towers-for-a-witness-weddings-by-a-nuclear-power-plant-temelin-czech-republic/

is articulated through the partially limited and also ritualized access, through security procedures and rituals. The enhanced perception of risk increases the attraction of nuclear places.

Another very specific energy tourism destination that evokes unusual feelings and experiences is dark and disaster tourism destinations (e.g. tours to the Chernobyl power plant accident area). Such objects attract more and more visitors every year since they are treated as mysterious places. Visiting such places evoke a particular existential feeling – sublime, which can be described as a mixture of ungraspable and unimaginable horror with pleasure and excitement. The sublime is "a mode of subjective experience, a broadly pleasurable sense stimulated by landscape, but combining terror and awe, which anticipates dark and thana-tourism as a leisure activity. The sublime combines fear in the face of the infinite or incomprehensible, with a transcendence of that fear" (Goatcher & Brunsden, 2011, p. 127–128)

A Shift from Special Interest Groups' Tourism to Attracting Families and Children to Nuclear Power Plants

As mentioned above, energy tourism is moving from special interest and expert-oriented tourism to experience-based tourism. At the same time, there is a shift from orientation towards energy experts to attracting families and children. Therefore, a new aim is raised to combine educational activities with recreation, entertainment and adventure. New forms of nuclear tourism – combining environmental education, the presentation of novel technologies, interactive popular science experiments and various outdoor activities – have been designed to attract not only those who are interested in energy technologies (i.e. expert energy tourism), but also family-oriented or adventure-seeking tourists (Frantál & Urbánková, 2017).

A similar trend emerging in a broader field – industrial and energy tourism – is observed. The characteristics of tourism are changing, moving from one main target group (experts, students, pupils) to wider target groups (families, individual tourists and groups looking for entertainment and adventure) (Jiricka et al., 2010).

When focusing on special interest groups (experts, students, school pupils), it is expected that this group of tourists visits energy facilities in larger groups (e.g. student classes); their visits take place during workdays; and it is a short stay lasting 1–2 days, independent from the weather. Their main interest is related to work, education, know-how transfer; and they are focused on the cognitive dimension, with aims to obtain factual and practical knowledge, technological and scientific information on energy production facility (Ibid.). The special interest groups mostly attend energy sites and indoor activities (visits to reactors, Visitor Centres). The main format of tourist activities embraces guided tours, expert lectures, discussions and seminars. Cultural activities here are just an additional incentive.

Meanwhile, the new paradigm in energy tourism to focus on families and children and the wider audience implies that the main motive for visiting is an orientation towards experience and entertainment, leisure, recreation and informal learning (Jiricka et al., 2010). Just recently, energy companies have begun offering new event and experience-oriented tourism products to attract – separately from experts, enthusiasts, or businesses – additional segments of tourists, such as young people, families with children or seniors. These tourists visit energy sites and their surroundings during weekends and holidays, in small groups, and they combine outdoor and indoor activities. The duration of their stay could be shorter (weekend trips) or longer (combination with other tourism offers, camps). It is rather seasonal and depends on the weather. To meet the needs of this group of visitors, the infrastructure of all regions is used (hotels, restaurants, camps, adventure parks, culinary, sport and entertainment facilities). According to Jiricka et al. (2010), unlike the special interest groups, this type of tourist is focused on affective (feelings) and kinaesthetic (motor skills) dimension, "learning by doing and feeling". They prefer guided offers, self-discovery offers (e.g. interpretive energy trail, energy fun parks, exhibitions and holiday camps).

Energy Tourism as a Part of Corporate Branding. Creating a "Safety Myth" at the Visitor Centres of Nuclear Power Plants

All these declared and anticipated educational benefits of energy tourism (including nuclear tourism) are questioned, taking into consideration the strong intention of energy companies supporting and organizing this tourism to influence the public opinion about the use of certain types of energy and energy providers. The companies tend to show a certain energy sector and energy sources as safe and economically useful. They seek to represent themselves as socially responsible and environmentally aware producers (Frantál & Urbánková, 2017). So, the implementation of educational benefits is contested and challenged by the "hidden curriculum" (concealed content and message as a part of public relations and corporate branding strategies of energy companies).

The nuclear industry establishment, together with governments, has been criticized for creating pro-nuclear narratives and constructing the "safety myth".

In a broader cultural and political discourse, communication and public relations of the industry, nuclear energy is represented as safe and reliable (Onishi 2011; Simon, 2019). Visitor Centres at nuclear power plants became a powerful public relations tool and tourist attraction in persuading the public about the safety and necessity of nuclear power. Nuclear tourism at nuclear power plants (NPP) is aimed to influence public opinion and shape public policy in favour of the pro-nuclear agenda. Nuclear tourism has already become a usual public relations' strategy of the nuclear industry (Sumihara, 2003). Through broader public communication in media, tourist and educational activities at the sites of nuclear plants, the nuclear industry seeks to reduce doubts, fears and anxiety in the public consciousness about the nuclear industry (Ibid.). A similar evaluation of narratives produced by the nuclear industry is given by Storm et al. (2019) in the observation that danger and unsafety related to the nuclear energy are not mentioned and represented. Anxieties of citizens are minimized during their tourist visits at nuclear reactors, and their worries shrink down; the inherent risks are rationalized through entertaining exhibits, nuclear fear is assuaged, and nuclear anxieties are transformed into nuclear acceptance via *a visit* (Simon, 2019). Visitor Centres become "paradoxically both a kind of multimedia warning sign, a marker for the danger that indicates you are entering a monitored and demarcated zone of limited access, as well as a place of tourism that promotes nuclear power, a safe space to process your nuclear fears" (Ibid., p. 64).

Scholars exploring the development of the Visitor Centers in Japan before the Fukushima accident in 2011 (Sumihara, 2003) describe how nuclear industry companies promote communication with local residents and tourists seeking to shape a positive image of the industry, impose a perception of nuclear facilities as safe and secure. Usually, the narratives about nuclear energy stress that it is an important part of the general energy landscape of the country, and nuclear energy is "clean" energy, which can create a proper response to global warming. Economic aspects are highlighted by pointing out that the supply of energy is at a relatively low cost. At the same time, harm to human beings and human costs of possible accidents are not sufficiently underlined.

Visitor Centres at nuclear power plants in Japan have gone through dramatic changes since the last decades. Starting from the mid-80s, Visitors Centres have started to reach a diverse population (not only male experts and politicians), especially women and children. Visitor Centres became an important asset to bring a sense of safety to women in general, especially those with small children. It is considered important by the nuclear industry to familiarize people from childhood as future decision-makers with nuclear power plants (Sumihara,

2003, 2019). The Visitor Centres founded in the 1970s were generally small, low cost and humble-looking buildings. They were situated at the edge of the power plant premises, which were located in remote and inaccessible areas. The visitor centres developed since the mid-1980s were very high-budget projects with a distinguished aesthetic concept – these centres were established in large-scale, fancy-looking, flamboyant and luxurious, futuristic appearance buildings of modern architecture.

These new Visitor Centres seek to attract children and young people. Interiors of the Centres are usually colourful, expositions are equipped with high-tech media, participative and interactive machines and displays, games, 3-D installations, science theatre and show-like presentations, cartoons, animation films, etc. All these audio-visual representations, music and imagery not only provide information on nuclear reactors, but also create strong emotional feelings. As Sumihara (2003, 2019) points out, Visitor Centres at Japanese nuclear plants after the late 1980s began to operate as multifunctional recreational and entertainment facilities. Visitors get access to seaside and natural parks, camping sites, outdoor open grass spaces, field athletic facilities (baseball stadium, swimming pools, tennis courts, etc.), other cultural attractions (i.e. archaeological museums). Visitor Centres started to organize sport and entertainment events (classes, competitions, picnics, etc.) for the local community, to attract people of all generations, families and children. New Visitor Centres included stores selling local commercial products, crafts, food. So, Visitor Centres performed activities such as recreation and tourism, which had nothing to do with nuclear power plants.

Sumihara (2019), comparing the concept of Visitor Centres in Japan and the UK, observed differences in narratives about a nuclear power plant. During a visit to the Visitor Centre at Sellafied in 2002, the researcher found that the UK nuclear plant did not attempt to persuade visitors to adopt a pro-nuclear stance. The exhibition avoids a one-sided depiction of nuclear power.

A debate on opinions supporting and opposing nuclear power explains the advantages and disadvantages of nuclear energy, and offers to "weigh the risks" are presented. Visitors are also introduced to the anti-nuclear movement in the country. It can be assumed that the exposition itself and a more general public discourse on nuclear energy in the UK and other countries were influenced by the antinuclear movement. This narrative is very different from the uncritical construction of nuclear power in Japan before the Fukushima accident in March 2011. From the mid-1970s onwards, public support for nuclear power fell in the UK and the US, exacerbated by the Three Mile Island accident in the US in 1979. Since the 1980s, opposition to nuclear power has been greater than support, both in the US and the UK. Similarly, in Sweden, the issue of risk related to nuclear power had begun to enter the public agenda (early 80s) (Storm et al., 2019). Public support for nuclear power was seriously damaged by the Chernobyl accident in 1986. The Chernobyl accident led to anti-nuclear energy movements and protests all over Europe and influenced governance to open the atomic industry to the broad public.

The history of the anti-nuclear movement in Europe defines the content of current communication in current visitor centres. N. Sumihara (2019) states that the myth of safety about nuclear power plants was destroyed by the Fukushima accident. Japan's Visitor Centres after 3/11 stress that their nuclear power plants have significantly improved safety measures, especially against earthquakes and tsunamis, and introduced new procedures of safety.

Nuclear Power Plants as Objects of Cultural and Historical Heritage

One more development in energy and nuclear tourism is the process of heritagization. Energy tourism presents power plants as objects of industrial heritage; that is why, the features of cultural and heritage tourism stand out here.

On the one hand, the process of heritagization process implies the process when nuclear power is displayed as an object of heritage value; on the other hand, it is presented as a promising and potent technology of the future (Storm et al., 2019). During the process of heritagization, many nuclear sites were featured as museum-like information centres which were constructed and managed by the nuclear industry. In Great Britain and France, nuclear energy facilities gained a specific cultural value since some nuclear power plants were designed by famous architects.

The narrative at these facilities is generally focused on technical descriptions. At the same time, these newly created atomic heritage objects appeal to a utopian narrative of future visions and promises of atomic power, scientific and technological achievements, and progress in the nuclear field. This heritagization process by the nuclear companies creates an image of a safe and secure industry and does not reveal the dangers and insecurities of nuclear power. As Storm et al. (2019) point out, the heritagization process promoted by the nuclear industry is rather optimistic and does not represent messages and narratives on danger and unsafety of the nuclear industry, disastrous consequences of nuclear accidents. Anti-nuclear narratives are represented in other areas that are independent of the nuclear industry – in museums, art exhibitions, and broader cultural and artistic discourse.

Heritagization process of atomic sites and the nuclear past started earlier in the military nuclear industry. Atomic museums, established in the early decades of the Cold War at America's nuclear weapons complexes have become popular tourist destinations and educational institutions in the USA (Molella, 2003). The museums underlined "technical aspects of atomic bomb development but virtually nothing about their actual uses and unimagined destructiveness" (Ibid., p. 211). Their mission has been to preserve and exhibit the memory of the atomic bomb development under the Manhattan Project, celebrate and preserve the atomic heritage of the American nation. These museums contextualized the Bomb within Cold War culture by presenting the life of ordinary nuclear workers, exhibiting popular media and music, clothing styles, toys, furniture, etc.

The process of heritagization through "heritage preservation" at museums, monuments and visitor centres could be considered as an expression of public relations arranged by nuclear industry in promoting the "safety myth" (Simon, 2019).

The opening up as a response to the pressure of anti-nuclear movement. The establishment of Visitor Centres in the USA and Europe is a broad movement throughout the nuclear industry to make the transition from the secrecy of the atomic industry to openness to the society, allowing the general public better understand nuclear energy generation. Opening up of nuclear plants for the public could be considered as a search for forms of transparency and openness to the public. It has been prompted as a result of pressure by the anti-nuclear movement and the public's demand to know more about what is going on in nuclear power plants. Excessive secrecy has been an integral part of the nuclear industry– starting from the American military atomic energy programme and becoming a part of the military industry during nuclear proliferation in the context of the Cold War (Palfrey, 1953, cited Schuck and Crowley-Buck, 2015). A unique feature of the Soviet defence-industry complex was its secret "closed cities" (Barber and Harrison, 2000). The civil use of nuclear energy was surrounded by secrecy too, especially in the early stages of nuclear development. The Chernobyl nuclear accident was named a monument to the secrecy and failings of the Cold War (Hetherington, 1997). The establishment of Information/Visitor Centres has become a powerful communication tool to realize the openness of the atomic industry; it has become an opportunity to introduce the public to how nuclear energy works, leaving as little secrecy as possible.

Visitors at nuclear power plants are allowed to have a look at what is happening in the nuclear power station; how operations of the plant are performed, monitored and checked; and how equipment and structures are managed. Information/Visitor Centres provide information for their visitors concerning the history and development of the nuclear industry; achievements in the physical science; energy development view; nuclear industry; and the relevant infrastructure construction, radiation and the measures to be taken for the protection from radiation. Visitor Centres are usually equipped with exhibits, panels and interactive displays to provide visitors with scientific and technical information about atomic energy, explaining how nuclear energy is produced and how safety is ensured.

Education in STEM for school children is among the key activities of Information/Visitor Centres. As an example of STEM education, it could be referred to Dukovany and Temelin NPP, where tourists can find a Reactor core model and Reactor model which allows them to learn what happens inside the reactor in the course of operation. In the Information Centre of Dukovany, informal evaluation of pupils' knowledge in physics is performed through "Play", where the players (visitors/learners) are invited to connect and relate images (faces) of discoverers (scientists), their names and their discoveries. Education activities are provided through simulations on how electric energy is produced in NPPs. Cloud chamber is one of a few devices enabling the observation of radiation particle trajectories.

Most Information/Visitor Centres are drawing upon various initiatives to build up better communication with visitors, including local inhabitants. One pre-condition for the educational infrastructure to work within the leisure–vacation context is that it ought to have a fun-oriented conception. First-hand experiences and learning by doing are the most important communication tools. Family-oriented offers and holiday camps are important parts of this experience-oriented tourism.

Summarizing the features of nuclear tourism at atomic reactors and their premises, the expansion of activities and target groups is distinguished, going beyond the concept of expert-oriented tourism and moving to experience-oriented tourism, energy literacy, STEM education and orientation to wider groups of citizens and tourists (schoolchildren, families, etc.).

Like many other energy companies, nuclear power plants implement open and covert corporate branding policy and seek to shape positive attitudes of consumers and citizens towards the nuclear energy industry. The nuclear companies deploy tourism and educational activities (STEM education) to build citizens' and customers' perception of safety and reliability of the industry and strengthen pro-nuclear attitudes. Taking into consideration these aims of nuclear energy industry, a critical approach should be applied while assessing tourist attractions and educational sites at nuclear power plants.

Torness Nuclear Power Station as an Educational Site and Energy Tourism Attraction

Torness is a nuclear power station – one of the eight in the UK owned and operated by EDF Energy. It is situated some 50 km east of Edinburgh, at Torness Point, near Dunbar, East Lothian, Scotland. Torness has two reactors (AGR Advanced Gas-cooled reactors), able to generate 1364 MWe electricity. The construction of Torness atomic power plant started in 1980, and it was commissioned in 1988. Torness construction faced a wide public opposition. In Scotland, direct action against nuclear power was held between 1978 and 1980 (Rudig, 1994). Antinuclear movement in Torness started in May 1978, when 4000 people from Dunbar occupied the Torness construction site. 400 groups and individuals signed the "Torness Declaration" which committed signatories to take "all non-violent steps necessary to prevent the construction of a nuclear power station at Torness" (Torness Alliance 1979, p. 4) Held in the summer of 1979, the poll found 90 % opposed to any nuclear development in the area, 8 % demanding further safety analysis and just 2 % in favour of proceeding. The Torness Alliance continued to attempt annual occupations of the site, but were neutralized by the police, so the campaign to 'Stop Torness' and the wider aim of bringing an end to nuclear power did fail (Welsh, 2001).

Torness is the last of the second generation nuclear power plants in the UK to continue working. According to the plans, Torness power plant will operate until 2030. EDF is one of the UKs largest energy providers. They generate low carbon power via nuclear, gas and coal power stations that are spread throughout the UK. EDF energy company is engaged in decommissioning and closing the old generation reactors; it also uses fossil fuel, and is oriented towards the renewable sources of energy (wind, solar, hydropower). EDF invests in building wind farms across the UK and is running a large fleet of low carbon nuclear power stations and a smaller fleet of coal and gas stations. EDF is investing in Nuclear New Build projects by building a new nuclear power station at Hinkley Point and are planning a second one at Sizewell.

The energy landscape is undergoing restructuring in the UK. The UK's electricity demands are currently met by a diverse energy mix – power generated in a number of ways: nuclear; fossil fuels like coal, gas and oil; and renewables like wind, solar and hydro. The UK needs new investment in energy infrastructure to replace old and polluting electricity generation sources and address climate change challenges. Since 2010, 26 power stations have closed, which equates to 20 % of the UK's generation capacity. By 2030, a further 35 % of the existing generation capacity will close down. EDF committed to providing a clean, secure and reliable solution to the climate change problem; at the same time, nuclear power has still an important role to play. The planned decommissioning of all "Magnox" type reactors and many Advanced Gas Cooled Reactors (AGR) means that, without a new generation of power plants, the contribution of nuclear energy will be reduced to around 5 % or 6 % of the total electricity production (till 2030).

The fact that EDF company is involved in generating different kinds of energy – nuclear, solar, wind, fossil fuel – and the whole country's power economy is in the company's hands is highly stressed in the power plant's Visitor Centre exhibition, which describes the importance of various types of power: information about fossil fuels, solar, wind and nuclear power. The exhibition introduces the structure of the national energy economy, defining the different roles of energy types and stressing the idea of energy mix. Nuclear energy is presented as part of the energy economy, and its specificity is discussed in comparison with other types of energy. The visual material of the exhibition displays information that nuclear energy is the biggest low carbon power source in the UK, producing 21 % of the nation's electricity supply in 2017. It is pointed out that solar power works best in sunny countries, but solar power still contributed 4 % of the UK's renewable electricity in 2017, and its generation was up to 87 % in the previous year. Wind power generated just under half of the UK's renewable electricity, while burning "fossil fuels" – coal, gas and oil – produced half of the electricity. It is noted that all fossil fuels are rich in carbon, so burning them releases carbon dioxide into the atmosphere, which means they are major contributors to climate change.

The enterprise takes care of the citizens' education and communication. The Visitor Centre, established in 2013, informs visitors about nuclear technology, basic properties of radioactivity and protection against ionising radiation. The guides of the Visitor Centre offer free tours in the power plant (see Fig. 2). The Centre was recently awarded five stars by "Visit Scotland" for the service quality and the range of services. The Visitor Centre is involved in educational, energy literacy and career STEM activities; introduces and discusses the use and combination of different types of energy (wind, solar, hydro, non-carbon); and attention is also given to general energy literacy. The Visitor Centre at Torness power station, opened six years ago, welcomed its 30,000th visitor at the end of 2019. In addition to offering educational tours of the power station, the Visitor

Fig. 2: Torness Nuclear Power Plant and Visitor Center (Photographs by E.Gerulaitienė)

Centre also holds themed family events, fairy and safari trails, school Christmas Cracker week and Santa's grotto, attracting people from around the area.

The review of the activities at the Visitor Centre leads to a conclusion that it aims at educating citizens and encouraging them to be active in making decisions about economic issues both in personal households and the country's economy, preparing citizens to take part in discussions on energy consumption and energy-related decisions. The Visitor Centre cooperates with educational institutions, organizes Career Days. The Torness Visitor Centre employs eight persons, who guide tours in the power plant, participate in promotional and educational activities. The EDF company, owning Torness NPP and the Visitor Centre, has developed an interactive training platform for teachers and children. This educational platform is available for visitors and clients of other EDF enterprises (NPP) and Visitor Centres. The platform offers free school lesson plans and practical tasks, organizes meetings, demonstrates films, and presents games and other information. The Torness tour guide mentions that not only excursions are organized for tourists and schoolchildren in the power plant, but outreach trainings can also be arranged at schools. Thus, well-organized educational activities are part of the EDF energy enterprise policy. A policy of openness of NPP was adopted with the agreement of the authorities. Visits to nuclear power plants have been encouraged. Such openness to the public might be considered as the common situation that has developed in the country and in the world in order to overcome isolation and to unveil the secrecy of nuclear energy, which was characteristic since the beginning of the atomic industry, to reduce and alleviate resistance to nuclear energy, and to improve the general energy literacy. This is extremely important in the countries and regions where anti-nuclear movement has been strong and anti-nuclear sentiment is quite obvious. The anti-nuclear movement resulted in changes of the information and communication policy of all NPPs. These events and processes have evidently influenced the public communication policy of EDF as well.

Visitors' participation in "security theatre" performance. A special exclusive component of tourism at the NPP is the visitor's feelings and experience. Nuclear tourism is considered experience-based tourism, where, on the one hand, a sense of security and the perception of visiting a place loaded with potential dangers are constructed (Storm et al. 2019), and on the other hand, one of the communication goals of nuclear energy companies is to show that this place is protected and at the same time safe since strict security procedures are applied. According to Schneier (2007), security is both a reality and a feeling. The reality of security is mathematical, based on the probability of different risks and the effectiveness of different countermeasures. But security is also a feeling, based on individual psychological reactions to both the risks and the countermeasures. And the two things are different: you can be secure even though you don't feel secure, and you can feel secure even though you are not really secure.

During an excursion to NPP, the visitor becomes a participant of "security theatre performance", consisting of theatrical screening procedures and security rituals, which are arranged to calibrate the concordance between, on the one hand, calculable risk and security and, on the other hand, perceived risk and security (Schneier, 2003 cit. Storm et al., 2019). It is these constructed feelings that become the key element in experience-based tourism at NPP.

Similarly, a tourist visiting the Torness nuclear power plant becomes not only a spectator of the "security theatre", but also its co-creator and co-participant. From the very first moment of communication with the Visitor Centre, the visitor understands that the excursion is taking place in an operating enterprise, where security requirements are of prime importance. Extreme security measures were introduced in all atomic power plants after 11 September 2001. Since then, security in organizing and conducting tours of nuclear power plants is strengthened.

To make the tourist understand that he/she is coming to an object which is a danger zone with an existing "security mode", before the visit (not later than a month before) he/she receives a list of questions and recommendations related to safety (a memorandum on personal protective equipment (PPE), a warning that personal belongings and equipment are not allowed in the NPP; that tourists need to inform about special needs (e.g., medicines) in advance; information that a security declaration form will need to be signed); and the requirement to have a proof of identity. When registering one month before the tour by e-mail, the tourist sends the passport data, and a presentation on the power plant is sent to him/her him/her. The procedure for organizing the excursions is very clear and carefully planned: after the confirmation of admission to the tour, a concrete time is set for the tour and its duration is defined – 90 minutes.

The second stage of security assurance takes place upon the visitor's arrival at the Torness nuclear power plant at the appointed time, where the security check procedure is performed: the visitor's passport is checked, and oral instructions are given. All personal belongings, handbags, computers and other electronic gadgets are left in the locker of the Visitor Centre. Only a pen and note paper are allowed during the tour.

The visitor's participation in this procedure creates an experience for him/her – it evokes feelings, and there are various actions and procedures, giving rise to many thoughts and impressions. The visitors can only guess why they are not allowed to take the mentioned things with them. Upon the arrival at the Visitor Centre, prior to the tour, the visitor has to sign documents: pre-tour questions, asking to assess the tourist's preparation for the tour and to self-evaluate whether he/she will be able to spend 90 minutes walking through large premises, climbing stairs, moving up and down in lifts; asking if the visitor has medical problems. The next document is signing a confidentiality agreement. The instruction part is very important, as the handrail must be observed throughout the nuclear tour (this is reminded throughout the tour); visitors are informed that when they hear the alarm, they have to follow the guide; they cannot keep their hands in their pockets and wear high heels (so as not to fall). In the Visitor Centre building, an employee (guide) and security staff check the identity documents, and upon entering the NPP itself, a personal search is performed using a hand-held metal detector and/or an explosive gas detector. Similar procedures, according to Storm et al. (2019), could be described as "security theatre" that refers to security measures taken to counteract terrorism, such as scanning bodies and luggage at airports – measures that create and construct the experience of security. In that way, it is not real, but rather a "theatre" of improved safety. Thus, security theatre performance is widely practised in atomic power plants and has a "side effect" – a special experience is created for the tourist, and the sense of danger and risk is strengthened. All of these actions, including strict staff supervision and surveillance – stripping of items, walking on a separated pavement strip, along a high-barbed wire fence, a thorough physical inspection at the nuclear site itself with detectors, and passing through inspection and detention gates – give the visitor a feeling that there are potential dangers in this place: on the one hand, a physical danger of falling, injury in a specific space; on the other hand – the physical tourist's inspection shows that someone may try to harm or cause danger to the operation and safety of the atomic power plant, to spy using electronic equipment (cameras, computers). These security measures and all the procedures indirectly create a sense of the hidden dangers of nuclear energy – radiation and its harm, which can be exploited by criminals (spies and terrorists), who need cameras, computers and explosives for carrying out their evil intentions. On the other hand, as stated by the authors analysing security theatre (Schneier, 2007, Storm et al., 2019), such exceptional attention to security is needed not only for the real security when visiting an atomic power plant, but for creating perceived security, for calibrating psychologically the real and perceived danger. However, real and perceived risks do not always coincide. Schneier argues that in some cases we tend to exaggerate risks, such that the security theatre may then help to adjust feelings and reactions to match the calculable level of threat. In other cases, the risks are greater and the security lower than we recognize; the reason for this might be traced to performances of a security theatre that is an expression of power abuse – that is, to actually disguise risks (Schneier, 2007). It is difficult to decide how in this concrete case real danger and risk are calibrated with the security theatre performance played for the visitors, and the authors of this chapter could not claim that the performed risk procedures at the Tornes NPP exaggerate real risks; however, it can be said with great certainty and conviction that all these actions and rituals really create an exciting experience for the visitor and it is an important part of nuclear tourism. The tourist begins to realize the exceptional importance of the object itself, the dangers that lie here (the danger of radiation and the potential crimes (terrorist attacks) that are prevented here). These safety rituals at nuclear power plants are exceptional, unknown to many people, unusual and in this sense exotic, thus creating completely new long-lasting experiences and feelings for the tourist.

The excursion in the nuclear power plant premises, after passing through the inspection post, proceeds in the controlled power plant area: reactor hall, engine/turbine hall, and unit control panel. The tour takes place in an operating enterprise, so there are certain rules that are observed by the guide. The guide takes visitors along the intended path-route. While moving along the corridors and looking round the premises and halls, the tour participants meet and see the company employees, who move among units and perform work tasks. The only staff member who the tourists can contact directly is the guide. This is foreseen and regulated in the memos and is followed during the tour itself as a physical-communicative pattern so that the visitors cannot move wherever they want and initiate contacts with the employees. A tour in this enterprise is a tightly controlled, supervised action, when the trajectory of the visitor's movement around various planned areas, contacts (communication) are quite strictly guided, supervised and surveilled. The movement of the visitors' group is supervised – the guide opens and closes the door, visitors are invited to follow her/him along the corridors, and they take the lift with the guide, who constantly explains and shows by hand where to go and how to go (repeatedly reminding to hold on to the handrail). This physical and communicative guidance and supervision are also part of the security theatre performance, creating a distinctive unusual experience that is undoubtedly an integral part of this tourist destination attractiveness. The whole course of the tour, when the visitor feels controlled, directed and guided, at the same time monitored and followed, enhances the tourist's sense and perception that he/she is in a place where strict security requirements are taken while protecting something, and the tourist himself/herself is, on the one hand, protected against injuries, on the other hand, is considered a potential source of danger, the one who can cause harm (because of which all the personal belongings were taken away, personal and passport data were checked, documents against collecting and disseminating information about the company were signed, talks with the staff were forbidden). All this constitutes a perhaps completely incomprehensible experience for the tourist himself.

When walking along the corridors of the power plant, the guide tells the visitors by demonstrating the visual educational materials on the walls, about the history of the atomic power plant, the construction, posters to commemorate the start and finish of the construction, dates, educational material about the gas cooled reactor type, etc. A separate topic is storing of the used fuel, transfer to Sellafield for reprocessing and recycling into new fuel.

In the reactor hall, in order to better acquaint the visitors with the principles of the reactor's operation, there is a glass room from which a view of the huge spacious reactor hall opens from above. This room also features training material, reactor parts and an explanatory text on the reactor design, commented by the guide. Next to the control panel room, there is also a large glass window, through which visitors are allowed to observe the technological processes taking place there, i.e., computers, buttons and the complex monitoring of the nuclear power plant. The tour lasts approx. 1.5 hours. However, visitors have an opportunity to ask additional questions, which are not part of the standard planned narration. Therefore, depending on the visitors' education, interests and activity, the story can be easily expanded and guides willingly provide

Fig. 3: Interactive exhibition in the Visitor Centre (Photographs by E.Gerulaitienė)

additional commentaries, direct towards other topics. For example, the authors of this chapter initiated a discussion with the guide about the technical factors that caused the Fukushima disaster, how technological processes are arranged in the Tornes reactor and how disaster prevention mechanisms work. The guide explained the differences between the technological processes in Tornes and Fukushima in detail. It is important to mention that talk about accidents in nuclear power plants is not a usual part of a tour narrative; however, when it is initiated by tourists, guides are ready for a competent discussion.

Educational Mission at the Torness Visitor Centre. The Visitor Centre is not large, and it covers about 80 sq. m. It is divided into educational spaces (desks, chairs, information about atomic energy and the atom, interactive game screens, etc.). The Visitor Centre contains a lot of stand information about the reactor, how it functions, how energy is generated and how much of it is produced at the Torness atomic power station. Great attention is given to the description of how the power plant safety as well as the removal and storage of used fuel are ensured. There is also a full stand with handouts, information material on used reactors, renewable energy, used waste management and storage, etc. (see Fig. 3).

Lectures and demonstrations about nuclear technology and radiation protection are intended primarily for school students. Occasionally the visitors are pupils from lower grades of primary school and even kindergarten children. For these children, lectures about power plant operation and nuclear technology

Fig. 4: Interactive displays and exhibits to familiarize visitors with nuclear energy

are appropriately adjusted according to the level of their knowledge. Therefore, various interactive displays and demonstrations are held in the Visitor Centre (see Fig. 4).

The Visitor Centre can boast of a very interesting educational exhibit/experiment: measuring the radiation level by selecting which material is most likely to inhibit the spread of radiation. The photo shows two overlapping wheels, the blue one of which contains various radioactive materials (dinosaur bones, uranium glass, granite, welding rods, etc.), while the orange wheel contains materials that block the action of radiation as a shield (wood, clothing, glass, aluminum, lead). The radiation level is shown at the top (see Fig. 5).

Another interesting exhibit demonstrates the operation of the advanced gas-cooled reactor (AGR) and how much electricity the electric reactor can generate at what capacity. The process is demonstrated quite simply: on the left side there is a hand which regulates the reactor capacity, then the process of generating electricity starts. To the right are houses with light bulbs that turn on when electricity is started; if the reactor power is pressed low, then one or two houses turn on the lights, while if the reactor is started at full capacity, then electricity appears in all the houses, thus allowing the visitor to understand how much

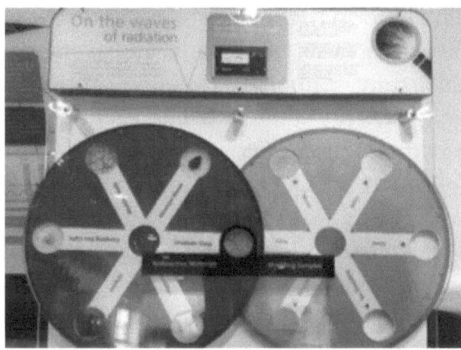

Fig. 5: Interactive educational exhibit – measurement of radiation level

energy is needed to have electricity in the house. This exhibit is meant for younger children who enjoy physical interactive activities. At the same time, the operation of energy is demonstrated here, when the energy power generated by the reactor transmits electricity to households.

One of the topics at the Visitor Centre presented for target groups is management and storage of used fuel. The sequence of the whole process of managing used fuel is introduced.

Another significant educational environment is the **virtual online space**. The Internet website https://www.edfenergy.com/energy/education contains educational films about how energy is generated, safety issues are discussed and the future of energy industry is described. Altogether, five films are offered. By using virtual reality and CGI animation, the visitor is suggested to trace the journey of an atom to the core of a nuclear reactor. The films show how nuclear fission happens inside power stations and how the energy is then used as steam to turn turbines – producing electricity for people's homes.

The Pod is an award-winning school programme, launched back in September 2008, and has since become the largest programme that lays a particular focus on energy, waste and biodiversity, and also covers a range of other environmental topics. The Pod presents topics on energy, waste transport, biodiversity and climate change. The emphasis on learning through science, technology, engineering and mathematics is provided in www.jointhepod.org – an interactive website for teachers, community group leaders and children. All of the resources are curriculum linked, free and are aimed at 4–16 year olds. The Pod provides lesson plans, practical activities, assemblies, films, games and information packs, as well as posters, stickers and badges. The analysis of the materials on the Pod platform highlights the main topics: energy information pack, which describes energy use, energy mix, global energy mix, UK's current energy mix, energy demand; minimising the effects of climate change; reducing our demand for energy; energy sources; advantages and disadvantages of different types of energy production; curriculum links; plastic waste information pack; and history of food packaging timeline. Each topic encloses a separate information pack with theoretical and practical tasks and games.

Another interesting communication and education trend at the EDF is attracting and retaining female talent to the energy industry. The programme aims to spark the imagination of 11–16-year-old girls, encouraging them to pursue STEM subjects at school and in their careers. This programme is called "Pretty Curious" and aims to inspire teenage girls to imagine a future where they use STEM – science, technology, engineering and maths – to help make a difference (https://www.edfenergy.com/prettycurious). The platform also includes the STEM career quiz, which can be tried by any child. The quiz game shows whether the child is interested in science, technology, engineering or maths.

In summary, the Torness Visitor Centre performs an educational mission to educate the public, especially school-age youth, not only by providing them with knowledge about nuclear energy and its production, safety, technological solutions, future renewable energy resources in connection with formal education, but also by allowing them to experiment, to touch, to experience very closely and practically how energy is produced, to simulate processes, to take a close look at the grandiosity and complexity of a nuclear power plant that was hidden from society for many years. The Visitor Centre carries out educational, energy literacy and career STEM activities; presents and discusses the use and combination of different types of energy (wind, solar, hydro, without coal), with a focus on and care of general energy literacy. The Visitor Centre also promotes and develops citizens' involvement, political participation and energy literacy.

Ignalina Nuclear Power Plant: Exploring Education and Tourist Facilities at the Enterprise Under Decommissioning

In 1974, preparatory work for the construction of INPP began. At that time, the most powerful NPP in the world, INPP, was built not only to meet the needs of Lithuania, but also those of the integrated North-West energy system of the former Soviet Union. In 1974, the INNP atom city-power plant satellite monoindustrial town Sniečkus (present Visaginas) was founded. The first reactor unit was put into operation in 1983. The second unit started operating in 1987. After the Chernobyl accident, the construction of the third unit of the

INPP was preserved, and in 1989 stopped completely. After the restoration of Lithuania's independence in 1991, the INPP became an important part of the country's energy system. In 1993, the INPP produced about 88 % of the electric power needed by the state. Following the Chernobyl accident, a number of detailed investigations and safety assessments were carried out at INPP, and additional safety measures were introduced. However, the operational risk of RBMK-type reactors could not be reduced to the level that would be safe enough to operate for a long time. The first unit of INPP was shut down in 2004, the second in 2009. An important political aspect of the closure of the INNP was that it was the condition of Lithuania's accession to the EU.

After Lithuania closed its nuclear power plant, gas and oil, which are imported energy sources, accounted for the largest share of energy consumption. Lithuania has inherited technologically inefficient and resource-guzzling centralized energy sector, long time relying on Russia's natural resources. After the closure of the INPP in 2009 and finally terminating the operation of the nuclear energy, Lithuania focuses more on expansion of renewable resources of energy, green energy. Production of bio-fuel, bio-mass, bio-gas and wind energy is being developed. In 2020, Lithuania will produce 35 % of the needed electrical energy from renewable resources.

According to Lithuania's National energy independence strategy (2018), the strategic directions of the Lithuanian energy sector will be followed by implementing the outlined objectives: the breakthrough – more electricity produced in Lithuania than imported – should take place in 2030, when electricity import will decrease twice and Lithuania will produce 70 % of the necessary electricity. By 2050, complete energy independence from imported energy resources will be ensured – all consumed electricity (100 %) should be generated in Lithuania (see Tab. 1).

In INPP, two water-graphite nuclear reactors RBMK-1500 operated. These were the most powerful energy reactors in the world. The thermal capacity of one unit was 4800 MW, the energetic capacity – 1500 MW. After the accident in Chernobyl NPP, the capacity of the RBMK-1500 reactors was reduced and they were allowed to operate only with a maximum thermal capacity of 4,200 MW. In INPP, as in all power plants with RBMK reactors, a single-circuit thermal scheme was used: saturated 6.5 MPa pressure steam supplied to the turbines was generated simply by boiling light water circulating through the reactor in a closed circuit. The power plant had only two power units. Each power unit had premises for nuclear fuel transportation systems and control panels. Common to both power units were the engine hall, rooms for gas cleaning and water treatment systems.

Tab. 1: The strategic directions of the Lithuanian energy sector (Lithuania National energy independence strategy, 2018)

2020 Energy Security	2030 Competetive Energy Market	2050 Energy Sustainability and Independence
Objectives 1. Integration of the energy system in the EU energy system 2. Improvement of efficiency of energy consumption 3. Balanced and sustainable RES development 4. Optimisation and modernisation of energy infrastructure	**Objectives** 1. Energy price in the industry sector will be the lowest in the region (compared to other (Baltic, Scandinavian and Central and Eastern European countries); for citizens – a decreasing share of energy expenditure compared to average income 2. Smooth transition from fossil-based energy sources to RES	**Objectives** 1. 80 % of the country's energy needs is generated from non-polluting (zero emissions of GHG and air pollutants) sources 2. 100 % of local electricity production in the country's gross electricity consumption

According to the data of 1 January 2019, Ignalina Nuclear Power Plant had 1901 employees, of which 985 employees were Russians, 314 – Lithuanians, 195 – Belarusians, a similar number of Poles and Ukranians (https://www.iae.lt/apie-imone/statistika/62). This specific demographic composition of the company's employees reflects the circumstances of how nuclear energy was developed in Lithuania in the 1970s. Workers, engineers and nuclear energy specialists from other nuclear power plants in the Soviet Union were sent to build and work at the enterprise. The current Lithuanian city of Visaginas (INPP satellite) also has a very specific demographic profile, with about 18 % of the population (according to 2011 data) belonging to the titular nation (Lithuanians), the main part of the population being Russians, Belarusians, Poles, Ukrainians and others. The Russian language in this city is the lingua franca, next to Lithuanian as the state language. In 2018, this city had a population of about 18 thousand, over 30 nationalities.

Interest in excursions at the INPP has grown significantly since May 2019, after the premiere of the HBO series Chernobyl. INPP received huge attention from tourists, as the series, which became popular, was filmed here. In total,

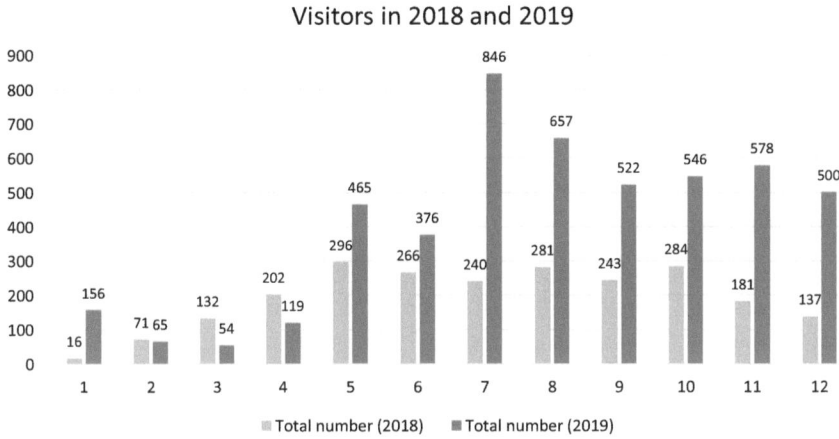

Fig. 6: Statistics of INPP visitors in 2018–2019 (based on data by INPP Communication Unit)

in 2019, almost 5 thousand people visited INPP from Lithuania and foreign countries, and about 500 excursions were organized (See Fig. 6). For comparison, in 2018, INPP organized 240 excursions (which is twice less), comprising over 2 thousand people. During the excursions in the controlled INPP territory, visitors have a unique opportunity to visit the reactor hall, the turbine hall currently in the process of dismantling, and the unit control panel room. Excursions are organized for both individuals and groups of visitors interested in the history of the INPP the principle of operation, decommissioning work, radioactive waste management, radiation safety requirements, etc. Since the opening of the INPP Information Centre in 1995, about 100 thousand visitors from around the world have taken part in the excursions.

The organization of excursions in INPP reflects the goals of public communication carried out by nuclear power plants around the world – to inform the general public about the safe work of the enterprise. Prior to the closure of the nuclear power plant as a part of the energy industry, the excursions served the purpose of realizing the idea of openness to the public, demonstrating operational safety, and carrying out STEM education. In the context of the INPP decommissioning mission, the main content of the communication includes a message on the safe and efficient implementation of decommissioning of the nuclear power plant with RBMK-type reactors, safe and reliable management of radioactive waste. The INPP communication on decommissioning declares that the project under implementation has no analogues, and Lithuania is the first country in the world to dismantle an RBMK-type reactor and related systems after the unloading of spent nuclear fuel. The society's support and information about the work being done and planned is a significant part of this process.

Visit to Ignalina NPP Closed Territory – "Security Theatre" and the *Masquerade*

The excursion after the checkpoint takes place in the controlled (closed) territory of the INPP: reactor hall, engine room, unit control panel. The tour lasts approximately 2.5 hours. The excursion, unlike in Torness NPP, is paid – the cost of organizing the excursion per person is about 60 Eur. Actually, the fee does not apply to companies controlling INPP activities, members of the Seimas and Government of the Republic of Lithuania, official delegations, journalists, representatives of educational institutions, students, public and non-governmental organizations and other persons visiting INPP on official visits. Excursions are organized for groups of up to 15 people on weekdays. Due to the increased interest in the object after the release of the series Chernobyl in 2019, INPP organizes two excursions per day. Participation is prohibited for persons under 18 years of age and pregnant women. Before arriving at INPP, a memorandum is sent to INPP visitors about prohibited items and general internal rules, which state that the INPP nuclear power facilities are equipped with a physical security system. Safety is ensured by armed security officers. The territory is monitored by video cameras, and videos are made. Access to the territory of INPP facilities is possible with permits issued by INPP. Persons, vehicles, transported objects and cargo are inspected at the control point.

When entering the main administrative building, tourists have to wait for a guide; they are not allowed to take photos, videos or use the Internet in the building itself. It should be noted that most of the guides are people working in the INPP Communication Department, but when there is a high flow of visitors, INPP employees from different departments lead the tours. Upon arrival of the guide, a permit to enter the territory is issued at the box, but before that, visitors must present passports or personal ID. Before passing the inspection, the issued permit must be attached to the outer garments at the chest level so that it is clearly visible. Visitors may enter the premises and be in the territory of different objects only if they are accompanied by a designated person. The visitor is informed that if he/she loses the permit, he/she must immediately inform the accompanying person. Before passing the control room, it is mandatory to remove metal items from the pockets and place all the possessions on the conveyor of the baggage control X-ray machine. Personal belongings and items for which a permit has not been given to take them to the INPP site may be left in the lockers.

Visitors must comply with all legal requirements of the accompanying person, employees of the INPP Physical Security Organization Service and officers of the Security Unit. When visiting INPP, passing through security posts takes quite a long time because each visitor scans his/her permit, enters the code, then passes through the metal rotating turnstile and waits for others to perform the same action. The equipment (camera) owned by the guides, who are INPP employees, is also checked. There are signs on the doors of the building prohibiting photography, filming and even using the Internet. After meeting at the Permit Office, the participants and the guide – (INPP employee) are issued permits. Before that, the visitors could put their belongings in lockers for storing things, because computers, equipment and cameras cannot be carried inside.

After passing through security detectors (similar to the airport), the guide informs that all employees have their own movement routes within the INPP territory, i.e., not everyone is allowed to move around the whole area; usually the staff, depending on the specifics of his/her work, get a designed route they can move along. The person is not allowed to enter the zones outside the movement route because no permit has been granted. The tour guide must also withdraw a permission to lead the participants along the established trajectory in the enterprise.

One of the greatest impressions of the tour for the visitors is the application of physical safety measures – to avoid taking radioactivity and contamination outside the controlled area. The participants of the tour change their clothes, putting on special clothing, the same as that worn by all the employees of this company inside the company. This procedure is specific at RBMK reactors: since there is no container over the reactor, or protection layer, radioactivity is more dispersed in the controlled area and the worker's or visitor's own clothes could be contaminated during a longer contact with the body. That is why there is a requirement to change the clothes before entering and leaving the INPP. So, the visit to the controlled area of the RBMK reactor can be considered more risky than in reactors of other types. White professional clothing at RMBK reactors and in the Soviet atomic industry can illustrate the symbolic connection with science culture – nuclear physicists were using white robes for protection purposes. Additionally, white robes and uniforms are a widespread symbol depicting scientists working in laboratories and medical doctors working in hospitals. This association of white robes and uniforms of the Soviet nuclear industry staff with medical doctors has been noted while analysing the iconography of the Chernobyl accident. "On Soviet TV, the workers in the front lines of the clean-up at the Chernobyl nuclear plant all wear white cloth caps, white mouth guards and white uniforms. They look eerily like surgeons operating with bulldozers and precast concrete blocks instead of scalpels and surgical thread" (Barringer,1986).

Thus, visitors at the INPP continue to participate in the "safety theatre" performance by adding a new attraction – dressing, which becomes a kind of carnival or masquerade. Radiation measurement and control procedures on potential radioactive contamination give tourists a sense of thrill experiences, health and life risks, which is a valuable experience for this type of tourism. On the other hand, the fact that it takes place as part of tourism creates a sense of safety..

After the security checkpoint, the visitors are taken to the "Sanitary Sweeper", i.e., the changing rooms for the staff. Visitors take off their clothes and put on white underwear pants and a white shirt. Then they pick up beige socks and tuck the pants into the socks. Then, wearing blue rubber slippers, they go to another room where they put on top white clothes – pants and a jacket, pick up white shoes from the closet (similar to medical staff hospital clogs, but with closed heels) of each size, put on a white hat, a helmet and tie hair with rubber bands; the visitors are explained that this is done so that their hair does not absorb radioactive substances (see Fig. 6). At the end, they put on white gloves. Each visitor receives a dosimeter that they put in their pocket. The dosimeter shows how the radiation level, measured by the dosimeter, has changed during the tour.

The tourist route in the NPP includes a visit to three zones: Reactor zone, Control panel and Turbine zone. The move towards the starting point – the Reactor area – makes a big impression, as it goes primarily through the glass transit corridors. The guide also stops in the corridors and tells about the specificity of INPP: it consists of 6 generators for 1 unit. During the tour, in which the author of this chapter Eglė Gerulaitienė participated (see Fig. 7), the story of the INPP was linked to the well-known nuclear accidents in Fukushima and Chernobyl. The guide commented on the low probability of an accident at INPP compared to the Fukushima accident, pointing to certain technical details (INPP generators are 25 m above ground level, while in Fukushima it was underground). A lot of attention was given to the Chernobyl theme, mostly about filming the HBO series at this enterprise. The guide commented on where certain scenes in the series were filmed. It was noted that during the filming, the filming team faced the challenge of participating in all security procedures,

Fig. 7: Excursion in Ignalina Nuclear Power plant: EDUATOM project researchers Ilona Tandezegolskienė and Eglė Gerulaitienė

when both the filming team members and all filming equipment and props had to be thoroughly inspected according to the usual security protocols here.

During the tour, visitors walked along long, dimly lit stuffy corridors, climbed stairs to the sixth floor. Upon entering the Reactor area, visitors had to put disposable bags on their shoes. Before entering this zone, there was an apparatus for measuring how much the hands and feet were contaminated. The green colour would indicate that they were clean, uncontaminated, yellow colour would show that the hand or foot were more contaminated, and the red colour would indicate high contamination. When standing in the reactor area, the visitors are greatly impressed by the size of the room and the height of the ceiling. Visitors are explained about the removal and replacement of fuel cartridges, about special safety measures (during the operation, removal of the cartridges used to take place at night, when it was unlikely that many people would be injured in case of an accident).

In summary, a visit to the nuclear power plant makes a really strong impression, enhanced by the ritual of changing clothes, safety assurance rituals, and long dark corridors along which the visitors walk for two hours.

Exploring Narratives of INPP at Information Centre

The INPP Information Centre introduces tourists to information about the former most powerful INPP in the world. The centre is located in a separate building, which is not subject to any inspection procedures. Visitors register for a tour of this centre in advance. The exposition includes several rooms (one larger and two smaller rooms). Guides provide information about INPP, its history, construction, operating principles, safety and decommissioning, waste storage and atom functions; but very often visitors also ask additional questions, are interested in economic issues, e.g., what is the cost of decommissioning, what is most expensive and what is the share of the EU. The Information Centre is meant for various visiting groups, from schoolchildren (physics, biology teachers with pupils), students from various universities (especially engineering and technology profile) and families, to representatives of public authorities, politicians and energy experts, who are interested in decommissioning and financing.

At the beginning, a 10-minute educational film is demonstrated in Lithuanian or English. The guide tells about how the nuclear power plant works, what is uranium, and the technical capabilities of the RBMK 1500 (high power reactor, channel reactor) work. A total 17 such reactors have been built around the world.

The INPP Information Centre has a working model of the INPP unit, a model of fuel assembly, a model of a spent nuclear fuel storage container CASTOR and models of short-lived radioactive waste storage or disposal of low and medium radioactivity to be built. During the tour, a live view of the INPP interior – the reactor room, the turbine hall and the spent fuel storage hall – can be observed on the TV screen. Video films about INPP and other educational films about radioactive radiation (in Lithuanian or Russian) can be shown upon request.

Throughout the tour, visitors are shown and explained in detail how the dismantling takes place, how the technical management of radioactive waste procedures takes place.

The Information Centre also mentions the Chernobyl accident, as this power plant's reactor is the same as it was in Chernobyl (RBMK 1500), explaining to visitors the reasons associated with the power experiment at the Chernobyl power plant at that time.

As the Information Centre focuses on narration and communication about the decommissioning of the power plant, it is natural that most of the narrative is devoted to telling about various dismantling work, discussing radioactive waste management mechanisms and financial aspects (see Fig. 8).

It should be noted that the educational material in the INPP Information Centre is oriented towards senior gymnasium classes, technology students or specialists who work and are interested in the field of nuclear energy – the history of the nuclear power plant, the principle of operation, safety and

Fig. 8: INPP Information Centre

decommissioning. It can be assumed that the presented visual material, posters with educational information and layouts are sufficiently complex and incomprehensible for younger school-age children. Besides, there are no interactive educational solutions here that visiting students could try, feel and understand.

To sum up, four topics can be identified, which are reflected in the Torness and Ignalina Information Centres and nuclear power plants (see Fig. 9):

Energy Literacy. This topic is reflected through presenting general knowledge about energy as an economic branch at the country's macro level, i.e., tourists gain knowledge about the energy sector, how energy is produced in the country, how nuclear energy contributes to the country's economy, what is the role of nuclear energy, and how environmental issues, CO_2, energetic and technological parameters arise. Visitors learn the story about the nuclear power plant in the context of the nationwide energy, e.g., how many NPPs there are, how many are obsolete, what happens to the decommissioned NPPs, and what are the related economic processes, environmental, waste disposal and technological aspects.

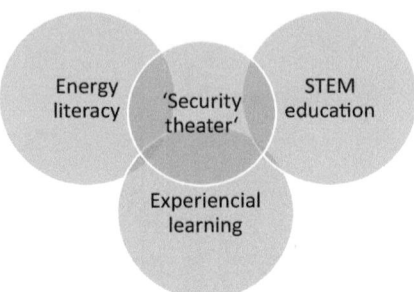

Fig. 9: Topics reflected in Torness and Ignalina Information Centres and Nuclear Power Plants

STEM Education. The topic is reflected in the analysis of the technological aspects of a nuclear reactor. Various information is provided about the reactor, a nuclear power plant as an aggregate in the territory of which different technological processes take place, e.g., in the reactor hall, engine room, unit control panel (which combines information about the atom, gas, turbines, rods, graphite, the whole Mendeleev table). All this includes various sciences – physics, chemistry, technologies, engineering and mathematics.

Experiencial learning is the third very clearly observed aspect in both power plants and their Information Centres. It is related not only to the ability to focus on kinaesthetic (motor skills) dimensions, where "learning by doing and feeling" or first-hand experience is involved in the learning process, but most importantly – affective (feelings) or common phenomenological sensation, because visiting nuclear power plants disclose their greatness, security, organization and control. The very complexity of the nuclear power generation process opens up; tourists see lots of buttons, wires, turbines, mechanisms large and small; and guides try to explain it all in simple words that will help at least bring the tourist closer through experiential learning to greater knowledge.

"Security Theatre" performance – an important element in enhancing the impression for nuclear power plant visitors. As mentioned before, security is both a reality and a feeling. A member of an excursion to NPP becomes a participant of "security theatre performance", consisting of a theatrical checking procedure and safety ensurance rituals, which are arranged to calibrate the concordance between, on the one hand, calculable risk and security and, on the other hand, perceived risk and security (Schneier, 2003 cit. Storm et al., 2019). When visiting the Torness NPP, the tourist becomes not only a spectator of the "security theatre", but also its co-creator and co-participant. Safety requirements are extremely important both in the active Torness Nuclear Power Station and in INPP under closure.

Conclusions

Turning a nuclear power plant into a tourist attraction goes in line with the main aims of public communication of nuclear energy companies – to open nuclear energy generation facilities to citizens; to "spread the veil of mystery" and remove the secrecy; to acquaint with the principles of operation of a nuclear power plant, technical aspects of ensuring safety; and to conduct STEM education and develop energy literacy. At the same time, through tourism, nuclear industry companies also accomplish corporate branding to shape positive pro-nuclear public opinion about nuclear energy. The aim is to create an image of the nuclear industry as safe and reliable.

Both Torness Nuclear Power Station and INPP are undergoing the process of heritagization of the atomic industry. The expositions of Visitor and Information Centres present the history of the nuclear industry and the companies and tell how these power plants were built. The story about the construction of the satellite city Visaginas occupies an important part in the INPP exposition. It is discussed and presented in the expositions how nuclear energy played an important role in the general industrial history of the country and in the period of 1980–2000.

Today, the Visitor/Information Centres carry out an educational mission to educate the public, especially school-age youth, by providing them not only with knowledge about nuclear energy and its production, safety and technological solutions, but also about future renewable energy sources. At the same time, nuclear companies seek to provide an opportunity to visitors and learners to see from a very close range the grandeur and technical complexity of a nuclear power plant.

The Torness NPP exhibition aims to develop schoolchildren's energy literacy by presenting information and knowledge about nuclear energy together with the role and importance of renewable energy sources in the further development of the country's economy and the important impact on sustainability.

After analysing the expositions of INPP and Torness NPP, it can be seen that they aim at presenting technological processes to the public, showing how safely and reliably nuclear power generation processes are carried out here, and safety is ensured. Torness NPP is an operating company producing energy, so the aim is to show through nuclear tourism that the processes are smoothly organized and controlled. The content of the exposition of the INPP Communication Information Centre and the excursion inside the company create a narrative about the safety of decommissioning processes, professionalism and high reliability of the work performed.

In the excursions of both analysed NPPs, facts about failures (Chernobyl, Fukushima accidents) are only partially included. Stories about these accidents are presented to reassure visitors that the technological processes and safety are significantly different from those at the Chernobyl and Fukushima power plants, and therefore, similar accidents cannot occur here. This topic about nuclear accidents is of great interest to visitors, who ask questions and initiate discussions. It should be noted that in both power plants the topics concerning non-safety of nuclear energy are not initiated by the companies themselves. The analysis of these two tourist destinations reveals the features of the atomic industry narrative, identified by Storm et al. (2019) that messages on danger and unsafety of the nuclear industry are usually not presented in the expositions.

The main communication message of the INPP is to tell about the safety of the decommissioning and waste management procedures. However, the challenges and problems of waste management are not extensively presented to the visitors. In the exposition of the Information Centre, one of the last exhibits mentions the general problem of nuclear waste, noting that the decay of nuclear elements has been going on for hundreds and even thousands of years. Yet, the real and topical technological and environmental waste management problems faced by the company and the nuclear industry as a whole are not raised or discussed. The story of Chernobyl in the current INPP tours appeared due to the interest generated by the HBO series Chernobyl. The staff of the nuclear power plant included this topic in the story of their tour after the premiere of the series. However, the tour guides focus more on filming the series on the INPP site than on INPP's links with the Chernobyl nuclear power plant, both of which operated the RMBK reactor with an inherent design unsafety.

INPP communication in the exposition and excursion of the Information Centre follows the line that despite the unsafety of the RMBK reactor, all necessary improvements were made at the INPP power plant after the Chernobyl accident, a new incident was prevented, and INPP was closed due to political decision, when the EU politicians raised the issue of closure as a condition for Lithuania's accession to the EU. This narrative is widely developed by the company itself, as well as by the public media, in political discourse, and formal education content (textbooks).

After analysing the excursions of the two power plants, it becomes clear that in addition to the common goals of nuclear tourism (development of energy literacy and STEM, corporate branding of energy companies, development of a narrative on nuclear safety and reliability), this aspect of tourism becomes more evident: it becomes experience-based tourism, creating an exceptional and striking visitor experience. Being in these facilities, visitors have a unique opportunity to admire the grandeur of the objects in this industry and the complexity of the nuclear power generation process – visitors view a lot of buttons, wires, turbines, large and small mechanisms. Visitors feel the grandeur when looking around, spotting huge production halls and seeing enormous-size aggregates.

A separate memorable experience for visitors is participation in security screening and assurance procedures, what researchers (Schneier, 2008, Storm et al., 2019) called security theatre, described earlier in this chapter as security theatre performance. Visitors experience special feelings that allow them to understand what the nuclear power is, what the invisible risks and dangers associated with radiation are, and thus indirectly understand and feel how safety procedures seek to prevent harm to human health and the environment. This part of tourism becomes the most important part for the person who visits the nuclear area itself, causing affective (feelings) or common sensations. Participation in security screening procedures and rituals creates a sense and understanding of security, management and control. The function of the Visitor Centre is usually ancillary and secondary to the main excursion inside the power plant.

Torness Visitor Centre, compared to INPP Information Centre, lays greater focus on affective (feelings) and kinaesthetic (motor skills) dimensions "learning by doing and feeling", first-hand experience, touching interactive displays and playing simulated energy games.

Tornes Visitor Centre exposition demonstrates transition from expert-based tourism, where the main target groups are experts and students, to attracting other groups of tourists and visitors. Torness Nuclear power station offers educational tours and holds themed family events, fairy and safari trails, school Christmas Cracker week and Santa's grotto, attracting people from around the area. Integration of additional attractions and entertainment takes place, thus expanding target groups, involving families.

It must be acknowledged that the communication and education of these power plants with old-generation reactors goes in line with the general contraction and declining role of this industry. Lithuania has closed its nuclear power industry, Scotland is undergoing a major restructuring of its energy industry to reduce the share of nuclear power by closing old reactors and investing in renewables.

However, the development of STEM skills and energy literacy at the Visitor/Information Centres demonstrate the educational potential of power and nuclear tourism. At the Torness Nuclear Power Station Visitor Centre, the staff are actively working with the local community, with local and regional schools organizing events, and the exhibition itself demonstrates an orientation towards younger children. Despite the decline in total nuclear energy in the UK, Tornes NPP is part of the EDF company, which supplies all types of electricity and operates the energy sector, has an interest in building a positive reputation and brand, in forming positive attitudes of young people towards the enterprise generating electricity, in developing general energy literacy and ensuring the training of specialists by contributing to the joint STEM training in the region and the country.

References

Barringer, F. (1986) White-Clad Workers Cleaning Up Chernobyl, *The New York Times,* https://www.nytimes.com/1986/08/22/world/white-clad-workers-cleaning-up-chernobyl.html, accessed 16 June 2020.

Barber, J., Harrison, M., Simonov, N., & Starkov, B. (2000) The Structure and Development of the Soviet Defence-Industry Complex. In J. Barber & M. Harrison (Eds.), *The Soviet Defence-Industry Complex from Stalin to Khrushchev* (pp.3–32). London and Basingstoke: Macmillan.

Beer, M., Rybár, R., & Kaľavský, M. (2018) Renewable energy sources as an attractive element of industrial tourism. *Current Issues in Tourism,* 21:18, 2139–2151, DOI: 10.1080/13683500.2017.1316971

Berger, J. (2006). Nuclear Tourism and the Manhattan Project.*Columbia Journal of American Studies,* 7, 196–214.

Devine-Wright, P. (2007). Energy citizenship: Psychological aspects of evolution in sustainable energy technologies. In J. Murphy (Ed.), *Governing technology for sustainability* (pp. 63–89). London: Earthscan.

Frantal, B., & Urbankova, R. (2017). Energy tourism: An emerging field of study. *Current Issues in Tourism,* 20(13), 1395–1412.

Frew, E. A. (2008). Industrial tourism theory and implemented strategies. In A. Woodside (Ed.), *Advances in culture, tourism and hospitality research, issue 2: Industrial tourism theory and implemented strategies* (pp. 27–42). Bingley: Emerald.

Goatcher, J., & Brunsden, V. (2011). Chernobyl and the Sublime Tourist. *Tourist Studies,* 11(2), 115–137.

Hecht, G. (1997). Peasants, Engineers, and Atomic Cathedrals: Narrating Modernization in Postwar Provincial France. *French Historical Studies 20:3* (Summer 1997), 381–418.

Henriksen, E. K., & Jorde, D. (2001) High School Students' Understanding of Radiation and the Environment: Can Museums Play a Role? *Science Education,* 85(2), 189–206.

Hetherington, K. (1997). *Badlands of Modernity: Heterotopia and Social Ordering.* London: Routledge.

Jiricka, A., Salak, B., Eder, R., Arnberger, A., & Pröbstl, U. (2010). Energetic tourism: exploring the experience quality of renewable energies as a new sustainable tourism market. *WIT Transactions on Ecology and the Environment,* Vol. 139, WIT Press, www.witpress.com, ISSN 1743-3541 (online), doi:10.2495/ST100061

Palfrey, J. (1953). The Problem of Secrecy. *The Annals of the American Academy of Political and Social Science, 290,* 90–99.

Lithuania National energy independence strategy (2018), Resolution No XIII-1288 of the Seimas of the Republic of Lithuania of 21 June 2018. https://e-seimas.lrs.lt/portal/legalAct/lt/TAD/e7fcc2608f1f11e8aa33fe8f0fea665f?jfwid=-2y4hgvql9

Molella, A. (2003). Exhibiting atomic culture: the view from Oak Ridge. *History and Technology,* 19:3, 211–226, DOI: 10.1080/0734151032000123954.

Onishi, N. (2011) '"Safety myth" left Japan ripe for nuclear crisis', *The New York Times,* www.nytimes.com/2011/06/25/world/asia/25myth.html 25 June, accessed 16 June 2020.

Otgaar, A. (2012). Towards a common agenda for the development of industrial tourism. *Tourism Management Perspectives, 4,* 86–91.

Pasqualetti, M. (2012). Reading the changing energy landscape. In S. Stremke & A. Van Den Dobbelsteen (Eds.), *Sustainable energy landscapes: Designing, planning, and development* (pp. 11–44). Boca Raton, FL: CRC Press.

Schneier, B. (2003). *Beyond fear: Thinking sensibly about security in an uncertain world.* Berlin: Springer.

Schneier, B. (2007). In praise of security theater. *Wired.* Retrieved from www.wired.com/2007/01/in-praise-of-security-theater/, accessed 20 November 2020.

Schneier, B. (2008). *The Psychology of Security* (Part1). Retrieved July 24, 2020 from https://www.schneier.com/essays/archives/2008/01/the_psychology_of_se.html.

Schuck, M. J., & Crowley-Buck, J. (2015). *Democracy, Culture, Catholicism: Voices from Four Continents.* New York: Fordham University Press.

Simon, S. (2019). A Visitor Centre for the Next Nuclear Disaster. *Performance Research, 24*:5, 61–64, DOI: 10.1080/13528165.2019.1671718

Storm, A., Krohn Andersson, F., & Rindzevičiūtė, E. (2019). Urban nuclear reactors and the security theatre The making of atomic heritage in Chicago, Moscow, and Stockholm. In: H. Oevermann & E. Gantner (Eds.), *Securing Urban Heritage: Agents, Access, and Securitization* (pp. 111–129). New York: Routledge.

Sumihara, N. (2003). Flamboyant Representation of Nuclear Powerstation Visitor Centers in Japan: Revealing or Concealing, or Concealing by Revealing? *Agora: Journal of International Center for Regional Studies,* No. 1, 11–29.

Sumihara, N. (2019). How the "Anomaly" of Nuclear Power Plants Has Been Explained Before and After the 3.11 Disaster in Japan: An Observation Through Power Company Visitor Centers in Japan and England. In: H. Nakamaki, K. Hioki, N. Sumihara, I. Mitsui (Eds.), *Enterprise as a Carrier of Culture: An Anthropological Approach to Business Administration* (pp. 49–67). Translational Systems Sciences, Volume 16. Singapore: Springer.

Torness Alliance (1979). *Torness Alliance Occupiers' Handbook.* Edinburgh: Torness Alliance.

Urry, J., & Larsen, J. (2011). *The Tourist Gaze 3.0.* London: Sage Publications.

Welsh, I. (2001). Anti – Nuclear Movements: Failed Projects or Heralds of a Direct Action Milieu? *Sociological Research Online 6*(3), 36–50. DOI: 10.5153/sro.642

Natalija Mažeikienė and Eglė Gerulaitienė

Chernobyl Museum as an Educational Site: Transforming "Dark Tourists" Into Responsible Citizens and Knowledgeable Learners

Abstract: The chapter focuses on the concepts of dark tourism and disaster tourism, which could be viewed as a cultural representation of disaster. These tourist destinations provide visitors as potential learners with an understanding of the social, political and cultural causes of technogenic disasters. The chapter analyses what meanings of the disaster are created, how the disaster is constructed in the collective consciousness, and how social and cultural construction of the disaster takes place in media, popular culture, museums, and other cultural and social spaces. The cultural turn and cultural perspective on disasters allow understanding cultural framing of disaster, how disasters are framed and interpreted, collectively imagined, remembered and memorialised, represented and portrayed through folklore, songs, movies and other media (Webb, 2018). The exposition of the National Museum "Chernobyl" and the Chernobyl Exclusion Zone tours are viewed as specific versions of cultural and social representations of the Chernobyl disaster and can be related to and compared with cultural representations of Chernobyl in the media, films, literary and fine art works. The authors of this chapter conducted an ethnographic study of the exhibition of the National Museum "Chernobyl". The exhibition's overall space, the choice and positioning of objects, texts and labels were analysed. The exhibition was treated as a specific means of education and communication. The research aimed to carry out a study of the representation of the Chernobyl accident as a disaster by analysing verbal content, visual materials and symbols. The question was raised how the Chernobyl disaster is constructed in the Museum and what educational potential this disaster tourism destination has for visitors as potential learners.

Keywords: The Chernobyl Museum, educational nuclear tourism, dark tourism, disaster tourism, disaster studies, chronos, kairos, heroisation

Introduction

Nowadays, Chernobyl is becoming an attractive place for tourists seeking exclusive existential experiences. Disaster attracts tourists because it conveys a specific sensory and aesthetic perception of the environment and landscape. The disaster place disrupts usual and enjoyable images of the landscape (Miller,

2007). It has a "high emotional impact" and is associated with death and other atrocities. Disaster tours are organised in areas affected by natural and technogenic disasters. Tourists' interest in this topic and trips to the Chernobyl area and visiting the National Museum "Chernobyl" could be seen as disaster tourism. Chernobyl is one of the largest human-made technological disasters.

The better understanding of the disaster tourism in the Chernobyl Exclusion Zone and the cultural construction of disaster at the National Museum "Chernobyl" was made possible by referring to ideas and approaches presented in disaster studies. This cross-disciplinary field views disasters as a social and cultural phenomenon, when disasters are presented through the cultural imagination, cultural representations and meanings of disaster (Holm, 2012). Disasters are not just a physical reality phenomenon where society and landscape undergo physical harm. Disasters are social disruption. Therefore, the disaster should be viewed as a social and cultural phenomenon when it is analysed in terms of what meanings of the disaster are created, how the disaster is constructed in the collective consciousness, and how social and cultural construction of the disaster takes place in media, popular culture, museums, and other cultural and social spaces. According to Webb (2018), the cultural turn and cultural perspective on disasters allow understanding cultural framing of disaster, how disasters are framed and interpreted, collectively imagined, remembered and memorialised, represented and portrayed through folklore, songs, movies and other media.

Referring to this concept of cultural perspective on disasters, the exposition of the National Museum "Chernobyl" and the Chernobyl Exclusion Zone tours are viewed as specific versions of cultural and social representations of the Chernobyl disaster and can be related to and compared with cultural representations of Chernobyl in the media, films, literary and fine art works. In this sense, disaster tourism is associated by visitors with other cultural experiences of the Chernobyl disaster (reading books about this disaster, watching films, following the discourse in the media).

The Cultural Construction of Disaster in Tourism Destinations

The cultural turn in analysing a disaster allows delineating forms of collective representations and "myths" of disasters. Holm (2012) distinguishes several historically stable symbolic forms of how disasters are imagined and represented. One of the cognitive schemes of how a disaster is perceived is *sublime* as a specific aesthetic sense experience about the terrible, awe-invoking beauty of disasters; it is 'the violent sense experience overwhelming the observer who, stricken with terrified dumbness and bodily stupor, experiences a masochistic blend of pain and pleasure' (Holm, 2012, p. 24). The cultural symbolic form of *trauma* represents the disaster through images of people's psychic health threatened by a "shock" or "post-traumatic stress syndrome". One more form of the collective representation of a disaster is a *"state of emergency"* which focuses on the breakdown of legal and normative structures caused by a disaster with possible reference to asocial behaviour and social chaos. The collective representation of *apocalypse* represents a disaster as the end of the world. According to Holm (2012), when we perceive a disaster through the cognitive scheme of *imbalance* and sustainability which was developed mostly by the ecological movement, we focus on the imbalance between human and biophysical systems causing disaster. Other cultural representations of disasters are associated with the idea of God and with theological and mythological images of the disaster as caused and influenced by god, evil and fate. The cognitive scheme *the blessing in disguise* constructs the disaster as "world fire" and "purification", the ground for new growth. *Theodicy* represents disaster as an expression of the gods' justice and will.

Disaster tourism and other cultural representations of disaster allow tourists understand the social aspects of the disaster, how the disaster affected people's lives and how social structures are related to the disaster. This aspect of social structures is analysed in disaster studies, where, in addition to cultural representation of accidents, disasters are viewed from a structural perspective when analysing their impacts on social structures (how they are disrupted and their functions are distorted) and how these social structures respond to large-scale systemic disruptions; also, how these social structures respond to large-scale systemic disruptions (Fritz, 1961; Kreps, 1989, as cited in Webb, 2018). Visitors to disaster sites and people who get to know disasters more closely through cultural activities and tourism become familiar with the structural approach which reveals how organisations (police, army, fire departments and special emergency divisions) and communities mobilise response efforts, adapt their structures, alter their tasks and create new response-related tasks to meet the demands of disasters (Webb, 2018).

Authors analysing cultural representations of disasters emphasise the role of culture as a source of resilience that protects communities from the impacts of disasters since it helps to make sense of the world (Webb, 2018). Disaster tourism can help understand how communities have tried to cope with disaster by creating the meaning. Disasters are viewed not only as physical damage, but it is also a disruption of social meanings when a vacuum of meaning occurs, when meaninglessness or absurd opens. Therefore, it is very important to find cultural tools to create meanings that provide the basis for resilience during the disaster and in the post-disaster period. A disaster is viewed by disaster studies as a cultural phenomenon when disaster-stricken communities go through processes of social reconstruction, regeneration and recuperation; when new values and norms, new disaster communities and subcultures emerge in the disaster and post-disaster environment (Ibid). The authors writing on Hurricane Katrine's Disaster Tourism in New Orleans emphasise that this tourism aims to expose tourists to revival, recovery, the signs of hope and rebirth rather than a decline (Miller, 2007).

In this context, it is important to recognise how disaster memorials take place and how they play a significant role in the process of recovery from a disaster (Eyre, 2006). Cultural representations (memorials, museums, tourist routes, films, books) are dedicated to memorialise a disaster, pay tribute to victims and recount stories of loss and heroism, revealing the response of social structures to disasters and giving us a better understanding of how social and political structures worked during and after the disaster.

Moreover, one of the more important aspects analysed in disaster studies and what can be learned and discovered by tourists and visitors in disaster tourism is an understanding of how the technogenic disaster emerged, how society and culture became the cause of the disaster. Webb (2018), discussing cultural perspective on disasters, mentions the role of culture as a source of vulnerability. "The origins of disaster lie not in nature, and not in technology, but rather in the ordinary everyday workings of society itself" (Webb, 2018). Referring to the topic of causes of catastrophe, questions are raised who is to blame; what caused the disaster; what social, political institutions; and what social and cultural values determined it. In that regard, it is explored what behavioural patterns, values and attitudes of individual social and political structures, organisational cultures led to a disaster. It is analysed what social structures, organisational cultures have been dysfunctional, dangerous. Such disaster-inducing behaviour may have been of structural secrecy, whereby certain institutions and socio-political structures concealed information or provided misinformation; knowledge and materials may not be intentionally concealed but ignored or neutralised because different units or departments failed to communicate (Vaughan, 1999, as cited in Webb, 2018). Disasters may have occurred even though socio-political systems, organisations had their lack of imagination and their "failures of foresight", which lead them to underestimate the potential (Turner, 1976, as cited in Webb, 2018).

Thus, based on disaster studies, disaster tourism could be viewed as a cultural representation of disaster, which can provide visitors as potential learners with an understanding of the social, political and cultural causes of technogenic disasters. Disaster tourism has the potential to evolve into environmental, social and political sciences education, i.e., to teach people how to think about society, culture and structures that create risk, lead to disasters.

The authors analysing dark tourism also highlight its educational potential. The main difference between heritage tourism and dark tourism is that the latter is often associated with some atrocity (Sharpley & Stone, 2009). By their nature, dark tourism sites, museums can elicit strong emotional reactions (Seaton 2009; Weaver et al., 2017; Podoshen, 2013). Broadly, the term refers to tourism focused around sites of death and suffering. The "darkest" tourism sites are generally solemn and highly sanctified places of actual suffering where ideologically mediated narratives serve instrumentally to attract empathy, contemplation and transformation (Weaver et al. 2017). The idea of the "aura", or emotion or mood conveyed, is an important theoretical construct in dark tourism (Seaton, 2009). Even though travel to places associated with death, disaster and dark tourism is not a pleasant experience, tourists are now more knowledgeable, critical, highly selective, highly segmented and more discerning than they were (Wu & Cheng, 2018). It is difficult to understand the essence of why one undertakes what may be unsettling, awkward, contestable or unpleasant experiences, but we still find in the existing literature that learning and education, as per Biran, Poria, and Oren (2011), Preece and Price (2005) and Kang et al. (2012), are situated as primary motivators. As Grebenar (2018) states, dark tourism sites must offer tourists the chance to consume death within the accepted boundaries of modern tourism and taste (Young & Light, 2016; Stone, 2012) whilst simultaneously facing the elements of taboo relating to death in our society (Stone, 2012; Stone, 2009). Dark tourism sites afford an opportunity to "write or rewrite the history of people's lives and deaths, or to provide particular (political) interpretations of past events" (Sharpley & Stone, 2009, p. 8).

The study of Wu and Cheng (2018) proposes eight constructs from the perspective of dark tourism research: participation, innovation, experiential risk, experiential memorability, experiential satisfaction, experiential trust, experiential involvement and supportive intentions towards dark tourism sites. There is a base of literature which explores dark tourism as a distinct phenomenon, including concepts such as site presentation (Yoshida, Bui & Lee, 2016; Friedrich & Johnson, 2013), heterotopia (Stone, 2013) and commodification within tourism (e.g. MacCannell, 2011; Sather-Wagstaff, 2011) and, specifically, dark tourism (Seaton, 2009).

The study "A Dark Tourism Spectrum" by Stone (2016) provides observations and research on dark tourism products which are multifaceted, complex in design and purpose and diverse; it is perhaps clear that the universal term "dark", as applied to tourism, is too broad. Therefore, it is perhaps prudent to argue for an analysis that accounts for multiple shades of dark tourism, concerning identifiable product traits, characteristics and perceptions. Stone (2006) distinguished seven classifications of dark suppliers along the dark tourism spectrum (darkest to lightest) (see Fig. 1): from the "darkest" – sites of death or suffering with an educative and historic approach alongside a strong authenticity of product and location – to the "lightest" – sites associated with death or suffering with an emphasis on entertainment.

Miles (2002) proposes there is a crucial difference between sites associated with death and suffering, and sites that are of death and suffering. So the experience and education at the Chernobyl zone is conceivably darker than the one at the National Museum of Chernobyl in Kyiv. As a result, questions have been raised about the distinction between authentic and inauthentic history. Indeed, one of the main contentions is how "dark sites", such as the Chernobyl zone, with a dominant commemorative ethic are portrayed as real, whilst it is more and more linked with a commercial orientation and a tendency to seemingly romanticise and thus distort past dark deeds. Since a diverse and fragmented set of dark tourism suppliers exists, so equally diverse are the motives of tourists who visit and consume these products. A recognised and structured framework of dark tourism supply is required to aid the identification and subsequent research of potential visitors and their experiences to these dark tourism products.

According to Miles (2002), a "darker–lighter tourism paradigm" does indeed exist; he suggests that dark touristic sites must engender a degree of empathy between the sightseer and the past victim (or product). He has it that recent death and tragic events that may be transported in live memory through survivors or witnesses are perhaps "darker" than other events that have descended into the distant past. Thus, those dark events which possess a shorter time frame to the present, and therefore can be validated by the living and which evoke a greater sense of empathy, are perhaps products which may be described as "darker" (Miles, 2000).

Methodology of the Research

The authors of this chapter conducted an ethnographic study of the exhibition of the National Museum "Chernobyl" in January 2019. The exhibition's overall

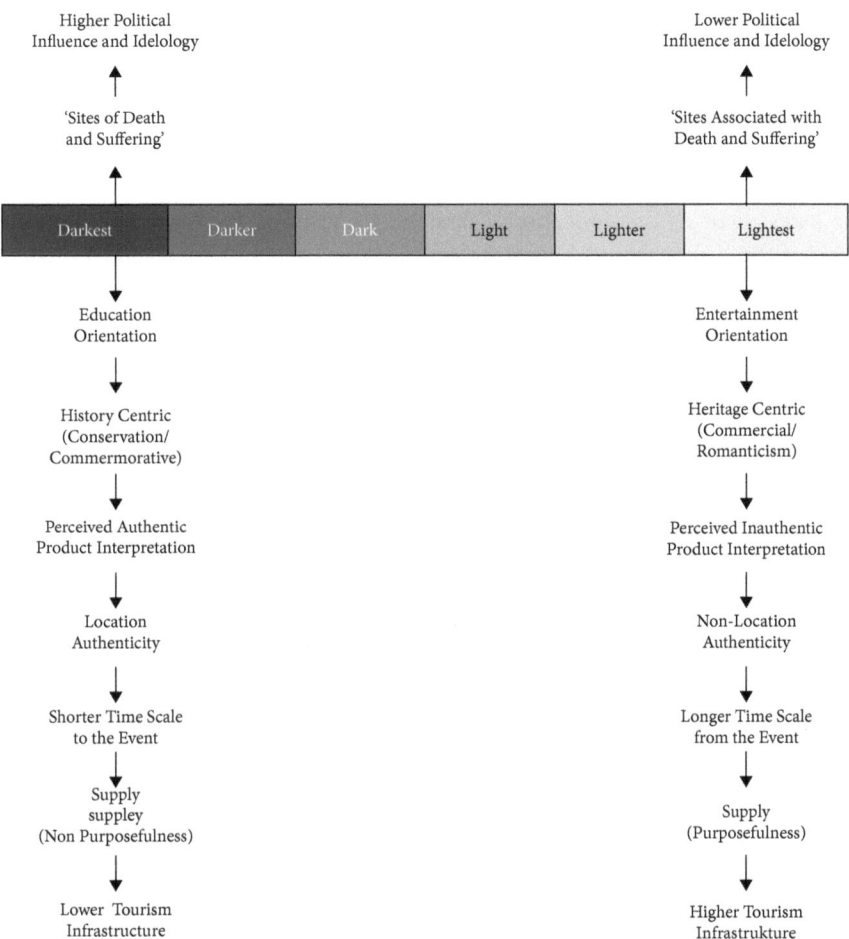

Fig. 1: A Dark Tourism Spectrum: Perceived product features of dark tourism within a 'Darkest–Lightest' Framework of Supply by Stone (2006)

space, the choice and positioning of objects, texts and labels were analysed. The exhibition was treated as a specific means of education and communication. The research aimed to carry out a study of the representation of the Chernobyl accident as a disaster by analysing verbal content, visual materials, symbols. We have been raising a question of how the Chernobyl disaster is constructed in the Museum and what educational potential this disaster tourism destination has for visitors as potential learners. We acknowledge the limitations of this research object and field: we conducted the content analysis of the exposition and grounded on our insights regarding a potential educational effect. This analysis does not refer to interviews and conversations with curators, educators and other staff of the Museum. Moreover, this investigation did not encompass our analysis of educational programmes delivered by the Museum as well as assessment of the educational demand by visitors. We leave the description of these elements of Museum's activities and research on them to further investigations conducted by us and other authors in the future.

Chernobyl Museum as a Disaster and Dark Tourism Destination

The disaster. On the 26th of April 1986, at night, a disastrous accident happened in reactor 4 of the Chernobyl nuclear power station in Ukraine, the Soviet Union. The radioactive fallout that leaked out from Chernobyl spread over large parts of Europe. Following the winds, it reached Belarus, Sweden on the 27th and 28th of April. The increased level of radioactivity was first recognised at the nuclear power plant in Forsmark, Uppsala county, Sweden. At first, the source was suspected to be within Forsmark, but the Soviet news agency TASS confirmed the Chernobyl accident in the evening of the 27th (Hultkrantz & Olsson, 1997). On the 18th of May, the World Health Organisation, the WHO, declared that the accident in Chernobyl would have no medical consequences to people outside the Soviet Union. Already on the 7th of May, the WHO declared that radiation in Europe, except for the area in the immediate neighbourhood of Chernobyl, was of no danger to people's health. However, the papers reported that, in spite of that, tourists outside Europe cancelled trips to Scandinavia and other parts of Europe. The explosion destroyed the reactor number 4 at Chernobyl, killing two plant workers that night of the accident, and, further, 28 operators and fire-fighters died within 5 weeks as a result of acute radiation sickness (WNA, 2012).

In the aftermath, up to 600,000 people, including soldiers, miners, plant workers and fire-fighters from all across the former Soviet Union – referred to as "liquidators" – were drafted in to decontaminate the site. P. R. Stone (2013) named Chernobyl post-apocalyptic place and analysed the number of deaths attributed to the disaster, which is still growing, partly due to lack of accurate records and politically contested criteria to determine Chernobyl-related mortality; a Greenpeace report suggests approximately 270,000 cancer cases within the affected region have been caused by the accident (Greenpeace, 2006).

Greenpeace also concludes that since the disaster, 60,000 people in Russia and 140,000 people in Belarus and Ukraine have died as a direct result of the incident. Stone (2013) examined reports on ongoing health impacts of Chernobyl and argued that radiation from the disaster had had a devastating effect on survivors, including the clean-up workers ("liquidators"); damaging the immune and endocrine systems, leading to accelerated ageing, cardiovascular and blood illnesses, psychological disorders, chromosomal aberrations and an increase in foetal deformations (Greenpeace, 2006). In 2011, the 25th anniversary of the disaster, the Ukrainian government sanctioned official tours to the site as well as to the abandoned "ghost town" nearby, Pripyat. Arguably, therefore, Chernobyl has become a destination associated with dark tourism and the "darker side of travel" (Sharpley and Stone, 2009; Stone, 2011).

The Museum. The National Museum "Chernobyl" is a multifunctional institution combining scientific, cultural and educational activities with a modern museum and archive, documenting, preserving and conveying the history of the Chernobyl nuclear disaster as the most severe radioecological disaster of the 20th century. The after-effects of it have no analogy and differ from other natural or human-made catastrophes. The museum opened to the public on the 26th of April 1992 in Kyiv, the capital of Ukraine, 150 kilometres away from the epicentre of the disaster. The current exhibition of the Museum has three exhibition halls with a total area of 1,100 square metres and more than 7,000 exhibits; there were only 200 exhibits in 1992. Large number of exhibition items, classified documents, maps, photos, special equipment, printings of that time, historic relics from the Exclusion Zone and other authentic materials, that are calling for reflections about the most burning problems of today's life and about ecological, social and spiritual consequences provoked by the Chornobyl disaster. Numerous records testify liquidators who sacrificed their health and even life, demonstrated courage and heroism to save the Earth from a global catastrophe. The Museum strives to become the centre of education of ecological culture, the culture of a safe life. Traditional tours, ecological lessons-excursions, lectures that aspire for new knowledge, conferences for those who are testing their own erudition, seeking answers to a controversial problematic question are organised at the Museum; moreover, ecological holidays are arranged as well.

The Museum exhibits modern audio and video records, information technology, allows expanding the chronological and thematic boundaries of the Museum and ensuring the authenticity of existing exhibitions. It presents a three-phase diorama "Chernobyl Nuclear Power Plant before and after the Accident", work model 4 of the power unit, an electronic book of Chernobyl liquidators, unique documentaries, computer programs on disaster and its consequences. It is claimed that emotional and philosophical art devices are very important in the exhibition. They bring their message and help them understand the 20th-century global tragedy. The idea of creating such devices is the result of collaboration between Museum scientists and artists, inspired by expeditions to the exclusion zone.

The function of the Museum is not only to document history, but also to mobilise sentiments and evoke emotions which range from anger to pride. In other words, strong emotional reactions result from the visit to the Chernobyl Museum because it has expressed feelings of sublime and fear (for what happened and can happen in the future), sorrow, sympathy, depression and appreciation (for the peaceful present), and doubts about the future (energy and environmental concerns). This potential of museums to evoke complex and contradictory feelings is illustrated by the Chernobyl Museum's motto "Est dolendi modus, non est timendi", which welcomes the visitor at the entrance of the exhibition. Translated from Latin it means "suffering has its limit, but fears are endless".

The education at the museum. Museums and other historical sites could be presented as institutions that preserve, interpret and memorialise the past and suggest pedagogical strategies for different groups of learners. Museums are sites to learn about history, geography, biology, citizenship education, ethics, literature, philosophy and other subjects of the formal curriculum. Museums offer opportunities to promote learning of different subjects taught in the school curriculum. The artefacts they display, narratives they tell and re-creations of the past they exhibit potentially engage students with content in ways unavailable through classroom activities or textbook reading (Marcus, 2007). The site such as Chernobyl Museum may develop students' historical empathy by allowing them experience the history and make personal connections to people. Museums also create opportunities for students to think critically about the past and history as a discipline by analysing how museums interpret and present the past; and this construction is subjective, evolving and influenced by many factors. Visiting museums as part of educational experiences is critical because, long after students finish their formal education, they will continue visiting and interacting with the past at museums and memorials. Thus, facilitating students' skills in interpreting and understanding museums and memorials is essential (Marcus, 2007).

The official website of the Chernobyl Museum (http://chornobylmuseum.kiev.ua/en/about-us/) states that the museum aims to become an educational centre to teach about environmental issues and promote the culture of an environmentally safe life. Efforts of specialists (teachers, scientists, scholars) of Ukrainian and public organisations from all over the world are put together developing various educational and socio-cultural programmes for all types of visitors.

Construction of the Nuclear Nation and Nuclear Belonging at the Chernobyl Museum

The Chernobyl Museum creates a relation to the past; it is a site of memory, an imaginative reconstruction of the past. Memories are reconstructed through practices of storytelling. Museums contribute to the construction of the national identity by exhibiting and presenting national and cultural heritage by educating the public (McLean, 1998). This museum performs a memory work and, at the same time, negotiates and constructs meanings of the collective and national identity in presenting the Chernobyl event. It is a national museum, and, thus, in one way or another, reflects the state's memory policy in interpreting this historical event.

One way to construct a national identity is to instil a sense of national pride in their citizens. We discover this aspect at the Chernobyl Museum as the presentation and highlighting of the heroism of the liquidators, when the actions of the people involved in the emergency work are presented as heroic feats and are intended to exhibit commemoration of their heroic deeds, to pay homage and remembrance of the "Impossible Mission".

Based on the literature review, a tendency is recognised that nuclear issues are incorporated into the process of construction of the national identity. In this context, we would like to mention the Manhattan Project in the USA, which was devoted to atomic energy research into the creation of an atomic bomb, which later developed nuclear energy in the USA. Researchers (Masco 2006, Gerster 2013) point out that the Manhattan project has been an American nation-building project that began after World War II and was developed during the post-Cold War period and contributed to the "nuclear nation" and the formation of "nuclear nationalism". The Manhattan Project, an intensive tourist destination at present, introduces the atomic bomb-making process, providing knowledge of industrial nuclear heritage, yet at the same time giving shape to the perception that the country was a leader in entering the Atomic Age (Gerster, 2013).

The collective experience of Chernobyl becomes a significant element in the formation of the national identity in the post-Soviet period. The construction of the national identity during and after the Chernobyl event, which takes place in a wider public discourse, also is represented in the Chernobyl Museum. The Chernobyl event itself became a marker and trigger for the collapse of the Soviet Union. The representation and narrative of Chernobyl were created already during the collapse of the Soviet Union and incorporated into the nation-building process of Ukraine, Belarus and other countries in the post-Soviet area. The identity structure of these countries can be recognised as "nuclear belonging" (Briukhovetska, 2016) when, on the one hand, several post-Soviet countries (Belarus, Ukraine, Russia) "inherited" nuclear weapons and nuclear testing sites (Semipalatinsk in Kazakhstan) before they belonged to the Soviet Union. Chernobyl as the worst nuclear reactor accident site (contaminated nuclear landscape) "belongs" to Ukraine, Belarus and Russia. These countries have become nuclear nations in this sense. In addition to the physical, biological and environmental consequences for these countries, Chernobyl is becoming part of the process of constructing a national identity. For Ukraine, it even has become a country brand, because of Chernobyl; this country is known to others and "appears" on the world map as a place of Chernobyl (Briukhovetska, 2016).

One important aspect of the Chernobyl experience as nuclear belonging is that it is a collective traumatic experience and memory. The "nuclear trauma", which as Briukhovetska (2016) has it, formed the core of the group identity. In the psychological sense (as individual and collective trauma) as well as in the cultural, symbolical sense (as "master signifier", "key symbol" and "myth"), the Chernobyl experience had to pass through a process of "nationalisation", to be integrated into the narrative of construction of the national identity. This collective trauma through memory work finds its expression in museums, fiction, cinematography and other areas of symbolical creation. The psychological and existential side of Chernobyl is presented in films, books and art projects.

The exposition of the Chernobyl Museum performs a function of citizenship education, since it presents a specific version of one of the largest and most significant events of the modern history of the nation. The exposition tells how divisions and participants (clean-up workers) of appropriate structures stopped and curbed the catastrophe, how they saved the world after the most terrible technogenic disaster in the world has happened. The idea of heroisation is also related to the idea of nuclear nationalism, when nuclear energy is treated as a national project, part of national history and shared collective identity of the nation. No doubt, the liquidators are introduced as "our heroes", not only as citizens of the Soviet Union, but also as citizens of our country (Ukraine) who sacrificed themselves for the wellbeing of the country, nation and entire humankind. Heroisation of the narrative on the Chernobyl disaster complies with the logic of creation of the national identity, when the history of the nation is being created through the actions of heroes during the events which are significant to that nation. The heroisation prevailing in the narrative on the said disaster is a means to give meaning to the suffering and deaths of the liquidators, to assure that these deaths were not in vain.

In the process of the nation building, "biological citizenship" formed, the social-political, civic process took place, when citizens (especially Chernobyl sufferers) in the post-soviet Ukraine joined together for the process to fight for their rights (social protections and compensation as a form of payment for past damage), to employ bio-scientific knowledge (social statistics, radiobiology, health physics, etc.) to the aftermath's data-producingin the establishment of new policies guaranteeing safe living, social equity and human rights (Petryna, 2002). It was sought to create financial and moral obligations which would strengthen a national bond between sufferers and non-sufferers. Therefore, nuclear belonging and nuclear identity in post-soviet Ukraine formed not only as a group trauma, but also as a group civic process oriented towards an active political and civic action, implementation of own social rights.

Besides heroism, the Chernobyl Museum creates the collective and national identity through victimhood, when appealing to collective experience of the trauma, and the traumatic historical event becomes the basis for bringing the group, nation together. For instance, the Holocaust is treated as an experience and historical event bringing the nation together, as a historical trauma that brings together and "creates" modern Jewish nation, and museums dedicated to the Holocaust perform this role throughout the world. The National Museum of the Holodomor-Genocide in Kyiv, Ukraine, is also attributed to a similar construction of the group identity through group victimhood.

This creation of the collective and national identity through the victimhood complies with another example of the construction of the nuclear nation, i.e., building post-war national identity in Japan. The so-called atomic-bomb nationalism in Japan is promoted by creating the "victimization narrative" and unifying national feelings of suffering and being victims after defeat in the war (Schäfer, 2016). The educational tourism site at the Hiroshima Peace Memorial Museum as the atom bombing place plays an important educational role in citizenship and history education; it is considered as an essential element of the national project in constructing the collective memory and national identity in Japan. Additionally, the anti-nuclear stance and striving for peace is one more constituent element of the post-war identity in Japan, besides collective mourning and grief. These messages of suffering, the traumatic experiences of victims during the bombing and aftermath, anti-nuclear discourse and promotion of peace are strongly reflected in the part of the Hiroshima Peace

Memorial Museum which was built first and expresses a traditional approach in representations of history and identity.

Epic Heroic Narrative in Commemorating Heroes – Clean-up Workers (Liquidators)

Considering the Chernobyl Museum's exposition, we recognise these two discussed conceptions (heroism and victimhood). The first idea deals with heroism, sacrificial actions of specialists and citizens who participated in liquidation of the disaster (clean-up workers-liquidators), helped ceasing the fire, stopped the spread of radiation and cleaned up the territory. On the one hand, this heroic construction of the narrative on Chernobyl meets the representations of heroism which are deep-rooted in the Soviet tradition of heroisation. On the other hand, representatives of the disaster studies mention heroisation as a typical strategy applied in cultural disaster construction depicting how social structures responded to large-scale systemic disruptions during and after disasters, how organisations (police, army, fire departments, special divisions) and communities mobilise response efforts, adapt their structures and alter their tasks, and create new response-related tasks to meet demands of disasters (Webb, 2018).

The major narrative line in the exposition halls 1 and 2 deals with occurrence of the teams of liquidators-heroes throughout the disaster event. First, engineers of the nuclear power plant, who were the first attempting to identify what was happening, to find out the actual situation of the disaster, appear. The first heroes were fire-fighters who were extinguishing the fire and underwent lethal doses of radiation. Later, other groups of rescuers joined in: these were miners, helicopter pilots, soldiers attending obligatory military service, who cleaned the reactor's roof from graphite, other officers, and civilians called in from military reserve to clean up the territory, construct the roof (sarcophagus). Scientists, physicians and other medical staff are also presented as the heroes. The presented heroism of the liquidators involves a dominating style of presentation of authentic details and documents. In this part of the exposition, a visual style of documents prevails; texts including facts, detailed descriptions, black-and-white photographs of the participants, images of the disaster and acting liquidators are presented; in single cases, documentary video recordings (for instance, rendering how soldiers using shovels are removing graphite from the reactor's roof), authentic details of liquidators' clothing and separate units of equipment are displayed. Besides photographs and descriptions of single participants (heroes) and groups of them, documents evidencing state awards (letters of acknowledgement, medals), in single cases, information and related documents on undergone dose of radiation, degree of injury, and history of disease or death are presented.

Heroism of the liquidators is demonstrated and acknowledged by presenting many state awards for heroism (letters of acknowledgement, medals, honorific names of heroes) given to people-liquidators. These awards not only formally acknowledge and enhance the narrative of heroism, but also construct the relation between heroes and state which awarded them. At the same time, grounding on the critical perspective to construction of Soviet heroism, a question is raised whether the medals and state awards could be considered as sufficient expression of respect of the state in this case (if no other kind of support, i.e., treatment, social guarantees, and financial allowances, is given together with the medals). It is worth noting that, besides the awards, information on undergone doses of radiation and the fate of the heroes, many of whom died from the effect of radiation or are presently severely ill, became people with disabilities, is presented.

Comparing Heroic Narratives of the Chernobyl Museum with Non-heroic Representations in Other Texts: An Intertextual Reading

Emphasis on the heroic discourse in halls 1 and 2 of the museum can be "read" and interpreted in connection to other texts, through intertextuality, when texts of the museum's exposition are read by collating them with widely publicly known and read (by creators and visitors) texts on the topic. Svetlana Alexievich's famous book *"Voices from Chernobyl: Chronicle of the Future"* (1999) is one of the significant texts making an impression and conception comparable to the exposition of the Chernobyl Museum. In this book, we find another, "non-heroic", conception of Chernobyl as a version of an experienced, undergone event, which became "a monument to suffering and courage", reporting traumatic experiences and exploring how these events affected the lives of people, representing the new human condition of the trauma (Marchesini, 2017). Working as a journalist and using collective testimonies about traumatic events for 10 years, S. Alexievich collected more than 500 interviews of witnesses, including fire-fighters, liquidators, physicists, physicians, politicians and ordinary citizens. Depicting Chernobyl, she deeply explores not the disaster as a sequence of historical facts and events (even though, while reading the book, we, readers, reconstruct the proceeding of the disaster and what happened aftermath); she is more interested in an existential and psychological measure of this event – how people survive and "measure" this event by their existence, feelings, how they undergo suffering and losses, what psychological and philosophical meanings they construct decades later after this accident. This narration is not heroic; participants of the accident are depicted as people who suffer from physical, psychological, existential pain, raise philosophical questions and tell their doubts about their self-sacrifice. In this book, participation of liquidators is testified by their wives narrating about liquidators' mortal suffering and death. The "non-heroic" narrative of S. Alexievich's book manifests in a way that the liquidators, participants of the event, see themselves not as heroes but rather as victims of the Soviet regime and radiation, tell how absolute majority of them were called in and transported to the site of the disaster not by their will – specialists, members of paramilitary statutory divisions, army soldiers and reserve soldiers; many of them note that back then they did not understand, did not know and were not informed about the extent of radiation on the site (about obtained doses) and the effect on their health; how presently they suffer from diseases, were witnesses of many of their comrades facing "terrible, characteristic to the liquidators" death. The liquidators, their wives and relatives tell about the psychological and spiritual pain and suffering when watching physical pain (oncological diseases) of their children.

Besides S. Alexievich's book "Chernobyl Prayer: A Chronicle of the Future" (2016) emphasising not heroism but rather existential suffering, other fiction texts analysed in scientific literature as contrasting to the heroism-based narrative of Chernobyl and presenting the existential perspective to Chernobyl, how the Chernobyl disaster is experienced by the participants, through the existential perspective of their daily lives can be pointed out. A film "Innocent Saturday" (*V Subbotu*) released by a Russian director Aleksandr Mindadze in 2011 is analysed as a case of construction of the non-heroic narrative; in it, A. Mindadze depicts Chernobyl in terms of an "existential zone", through existential dilemmas, revealing the existential impact on ordinary peoples' lives (Lindbladh, 2012). This "existential action movie" depicts ordinary people struggling with their ambivalent and complex thoughts and feelings, wishes and fears. In the course of the entire film, the protagonist tries to escape from Chernobyl on the first day after the explosion (on Saturday) after he finds out about the disaster that happened earlier that night. The hero fails to leave Chernobyl; one can anticipate that he will suffer and finally die from radiation. The film avoids telling about the Chernobyl disaster from the perspective of the theme of Soviet heroism where heroic self-sacrificing deed and heroic death, the brave actions of Soviet people fighting against the catastrophe (Lindbladh, 2012), are depicted. The protagonist of the film is an anti-hero, who is trying to escape from the place of the accident and fight against inner feelings, suffer the inner crisis. It should be noted that when the film was created it was not shown in Belarus because it was considered as an offence to the memory of the heroic deed of liquidators.

The above-discussed Chernobyl narratives present in the museum and other texts (S. Alexievich' book, A. Mindadze's film) reveal different strategies of construction of the collective identity while constructing the Chernobyl disaster. The museum more emphasises the tradition of heroisation; whereas other two mentioned fiction texts are oriented to existential representation of physical and spiritual suffering.

Structural Approach to the Chernobyl Disaster: Learning How the "Soviet System" Worked

On the one hand, the strategy of heroisation is a way to commemorate participants, to express gratitude and respect to them. On the other hand, heroisation is a means of construction of the collective identity, when a disaster as a significant event becomes common experience of a group or a nation; it brings together residents and citizens to make one nation. Heroes and their heroic deeds create the present, the post-disaster world and the nation. Besides heroisation, the highlighting of the role of clean-up workers allows museum visitors understand how a disaster is presented from a structural approach, when attempts to understand and depict how social structures mobilised and adapted their tasks to meet demands of large-scale disasters (Webb, 2018) are made. In the case of the Chernobyl disaster, a visitor has an opportunity to perceive how the system of the Soviet Union whose resources were massively employed in response to this large-scale disaster operated. By seeing in the exposition that vast numbers of people who took part in the liquidation operation (over 350,000 male liquidators were involved between 1986 and 1987 and till 1992 totally 650,000 liquidators took part in the liquidation process), how many clean-up workers (fire-fighters, soldiers, helicopter pilots, miners, called in military reserve troops, scientists, physicians, etc.) in teams operated compatibly were organised and performed according to the system regulation, how much of materials, technical mechanisms and means were raised and utilised, visitors can have an impression and understanding of the general extent of the crisis and emergency management in the Soviet Union, when resources of impressive scale were allocated and retrieved from throughout the entire Soviet Union.

Together with the heroisation narrative, this moment may raise surprise in visitors due to the extent and scale of the entire action and process. Interesting to note that this narrative about high efficacy of emergency management in the Soviet Union is similar to representation in the HBO series "Chernobyl", where also a strong emphasis is put on depiction of heroic actions of teams of liquidators and demonstration of large-scale managerial, scientific and technical capacities of the Soviet Union allocated for liquidation of the disaster, paying less attention to the repressive and forced character of the emergency management model.

However, it is important to compare another cultural disaster representation, the text of S. Alexievich's book *"Chernobyl Prayer: Voices from Chernobyl: A Chronicle of the Future"* (2016), with the narrative of the Museum's exposition. The non-heroic narrative in the discussed Alexievich's book is developed by demonstrating the repressive operation of the Soviet system during the liquidation of the disaster. In this documentary book, former liquidators tell how they were called in (e.g. from military reserve) and threatened by court martial or other punishment, were sent under obligation to the site of the disaster. Others note that even when going by their own choice they, nevertheless, were impacted by the ideological system and cultural values as well as developed attitudes "to perform heroic deeds", "carry out significant deeds", "show/prove their masculinity", "to do civic duty" or simply could not act or think in other way than in line with that system and people of that (Soviet regime) time.

In the discussed book by S. Alexievich and the HBO series "Chernobyl" based on it, we find quite strong criticism of the Communist Party and the Soviet system, when the system is criticised for not providing information on the scale of the disaster and impact of radiation, protection measures (e.g. there was no information on the taking of iodine tablets) and compulsory mobilisation to the site of the disaster. Narratives and disaster interpretations in both Alexievich's book and series "Chernobyl" slightly differ from the narrative available in the Chernobyl Museum, where heroism of the liquidators is underlined, their self-sacrifice is presented as necessary, unavoidable and meaningful, there is much less of criticism expressed towards the local regime and party actions.

Comparing with these two texts (S. Alexievich's book and HBO series "Chernobyl" based on the book), the Museum's exposition does not provide any deeper and stronger criticism of the Soviet regime explaining why the disaster happened and how liquidation of it proceeded by employing a repressive mode of the system.

A museum visitor gets acquainted with the criticism of the ideological system of the Soviet Union at a lesser degree (comparing to the narratives of the book and series on Chernobyl). Museum visitors have an opportunity to have a short insight of common cognition of the political and ideological context of the Soviet Union, when the scale of a powerful emergency management campaign is introduced; however, there is no introduction to the general ideological and repressive character of the system.

The regime of the Soviet Union is criticised at the beginning of the Museum's exposition by short information telling that the fault for the explosion of the nuclear plant was attributed by the Soviet government exclusively to the managing bodies and the dispatcher team of the nuclear plant as well as "the human factor in the management of the nuclear plant". The Museum's narration briefly introduces that the human factor as a problem of the actions of the nuclear plant's personnel and management decisions was identified both during the demonstrative trial of the causers of the disaster and in public explanation of the reasons of the disaster; whereas constructional drawbacks of the nuclear plant's RMBK reactor as well as general mistakes in development of nuclear energy science and industry were not publicly identified and recognised. A short text presented in the exposition highlights that the main fault for the accident was attributed to the team of the nuclear power plant without reasoning; whereas the drawbacks of the constructional reactor were not investigated and explored in public. To compare, we could mention the narrative of the HBO series "Chernobyl" where, in line with heroic depiction of liquidators' performance, much attention and major emphasis of the film is paid to showing how faultily the nuclear energy and nuclear energy science serving it operated. In this sector, scientific knowledge was being created under the conditions of strict control and secrecy. A belief in Soviet nuclear science allowing no errors ("Soviet nuclear reactors do not explode") was being ideologically constructed by scientists, party activists and citizens themselves. The systems of the Communist Party caused a specific interaction between party decisions (party nomenclature personnel), scientific knowledge (scientists), etc.

Nevertheless, the narrative of the Chernobyl Museum presents criticism of the Soviet regime when secrecy that covered the events of the disaster is revealed. The exposition of the hall 1 tells about the reticence of the fact of the accident and the true extent of the disaster by the Communist Party as well as hiding this information at both national and international levels. This narration is developed through exhibits, introducing the demonstration held on the 1st of May 1986 in Kyiv. The annual festive event dedicated to celebrate the 1st of May, the Labour Day, having high ideological significance as one of the largest celebrations of the year, was not cancelled to prevent from panic among residents and to demonstrate that the accident was secure to the people.

In such a way thousands of Kyiv residents (adults and children) were exposed to a very high level of radiation. Only nineteen days after the Chernobyl accident, Mikhail Gorbachev publicly announced the catastrophe. An impressive exhibit of the Museum contains two spread pages of a major daily issued on the first days after the accident, including a small part encompassing short, laconic information on the Chernobyl accident as a text telling that the situation is being managed and under control.

The Museum narration suggests that the news on the explosion was not publicised for international community, and local residents were not informed about the degree and extent of the danger and possible damage; there were no information and actions which would reduce the damage to the residents (e.g. information on affected territories, contaminated and forbidden for consumption agricultural produce). The Museum exposition dedicates a separate narration to the hiding of information from international community and other states that underwent radiation (Sweden, Finland). The Museum visitors find out the circumstances of how Swedish scientists at the nuclear-power plant in Forsmark, in Uppsala county, Sweden, discovered the increased level of radioactivity. The Soviet news agency TASS confirmed the Chernobyl accident in the evening of the 27th (Hultkrantz & Olsson, 1997). On the 6th of May, the Minister of Health of Ukraine informed the population that they took all necessary precautions for increased radiation. This announcement occurred only ten days after the explosion.

The Chernobyl disaster which contaminated Ukraine, large swaths of Eastern Europe, Scandinavia and neighbouring states was inseparable from the slow "social and political unravelling" of the Soviet Union (Petryna, 2002, p. 21). These facts which are exhibited in the museum have to be interpreted by visitors themselves, and this amplifies the importance of teaching interpretation and analysis skills to students. The 1986 nuclear disaster has come to embody the demise of the Soviet era not only in the way the accident itself contributed to the sudden implosion of the internally vulnerable Soviet system (Van der Veen, 2013), but also in the way that the Exclusion Zone today has become a frozen microcosm of late-Soviet everyday life (Davies, 2013).

Thus, expositions of the Museum reveal some important features of emergency management during disaster within the Soviet system – reticence, hiding and secrecy of vital information. Also, one can recognise a common feature of nuclear energy that was briefly pointed out, though not broadly developed in the Museum's exposition – secrecy, when information on the very objects of nuclear energy and smaller incidents as well as large-scale accidents taking place there is hidden, classified as secret. Scientists analysing history of the nuclear energy (Brown, 2013) underline that knowledge of risk was a closely guarded in nuclear industry. Information was hidden in all earlier emergency actions, which had all played out in Ukraine before in 1951, 1953, 1955, 1957 and 1967 in the Urals. "The compartmentalization of information, the secrecy, the failure to inform the public of radiation dangers, the evacuations that occurred with critical delays, the deployment of expendable prisoners and soldiers on the most dangerous jobs, the failure to inform these "jumpers" and other employees of ways to protect themselves, the unpredictability of radioactive fallout in concentrated hot spots outside the neat zones of concentric circles – all were eerie repetitions of the plutonium disasters of the previous four decades. The only new feature in 1986 was that the catastrophe occurred while the cameras were running." (Brown, 2013, p. 285).

To summarise, the Chernobyl Museum provides knowledge on history, political sciences and sociology, and civic education. Analysing social and political context presented by the museum exposition, the students could get knowledge about the Soviet Union as a political, ideological system which praises itself for the controlled and managed most severe radio-ecological disaster of the 20th century. Visitors to the Museum learn about the Soviet regime, Soviet emergency-and-crisis-management system, technological, political, ideological aspects dealing with the nuclear energy industry. The Museum provides a rather weak critique of the Soviet regime in comparison with other above-mentioned literary and cinematographic texts of Chernobyl. It amplifies the importance of teaching interpretation and analysis skills to students. Exploring the political and social contexts of how the Chernobyl Museum was created and maintained enhances students' ability to deepen their critical understanding of memory work at museums which is a part of broader processes of politics of memory. The Chernobyl Museum is a historical source that needs to be critically analysed and evaluated. Teachers can encourage a critical reflection of museums and memorials as interpreters of history and recognise the political, social and economic factors that influence them.

Learning about Radiation in the Contaminated Nuclear Landscape

Environmental and nuclear geography are another learning subject represented in the Chernobyl Museum exposition. Nuclear geography includes critical geographies of nuclear energy, waste mobilities, nuclear geopolitics, or more-than-human interactions with ionising radiation (Alexis-Martin and Davies, 2017). The Museum exposition presents and describes the nuclear landscape in the Chernobyl Exclusion Zone. Moreover, presented geography of nuclearity covers a narrative on how radiation has affected the territory of neighbouring states and Europe in general. The fallout from the accident covered 150,000 km2 of Europe, affecting Belarus, Ukraine and the Russian Federation in particular (UNSCEAR, 2000). Due to the enforcement of a 2,600 km2 "Zone of Alienation" (Зонавідчуження Чорнобильської АЕС) around the epicentre, about 350,000 people had to evacuate, and 2.1 million Ukrainians still inhabit the land officially designated as affected by the accident (Davies & Polese, 2015). The radioactive landscape of the Chernobyl Zone is a place of invisible danger. The nuclear landscape reverses the old adage "what you can't see, it won't hurt you" and blurs the boundary between "contaminated and safe", "seen and unseen", "formal" and "informal". As such, those living in Chernobyl-affected territories can be viewed as "bare life" (Agamben, 1998). The Chernobyl landscape is a place infused with contested meanings: for some, a rural idyll tarnished by the invisible spectre of radiation; and for others, simply "a place called home". Instead, they live on in the memories, photographs and everyday lives of those who call this nuclear landscape "home" (Davies, 2013). These efforts to treat nuclear landscapes as "used and lived in" (Cresswell, 2003, p. 280; Cram, 2016) offer reminders of the connections among bodies, homes, states and colonial networks – the uneven geographies of nuclearity. Davis and Polese (2015) point out that people's, who live in Chernobyl-affected territories, lives are stripped of the protection of the law and abandoned through insufficient welfare and compensation protection to an uncertain fate; their potentially damaged biologies are placed outside the responsibility of the state to face the hidden violence of abandonment (Davis & Polese, 2015).

The topic of environmental geography and health issues due to the nuclear disaster is presented in the second exhibition hall. The Museum presents artefacts, such as medical tools, equipment, photographs of physicians helping Chernobyl victims, medical records and personal belongings. There is a wide range of discussions on the scope of contamination of the uninhabited area, exclusion zone, cancer disease and how radiation affects human beings. The Museum exhibition provides knowledge on how high radiation in the contaminated zone caused mutations of animals: a skeleton of a newly born pig which has six legs is exposed. At the same time, there is some lack of information in the Museum on how the disaster affected nature, plants, forests and wild animals.

The exposition at the Chernobyl Museum dedicates part of the second hall to environmental issues, presenting the information about the people and territory which used to be "lived in" and the geographical history of the development of the region in different centuries attracting people from all over Europe to live in and contribute to economic and cultural development of this territory. The map of the Chernobyl-affected territory is exhibited in the second hall with presentation of silent stories of people who lived there.

Their everyday life is presented as ceased moments that will never happen again. After the Chernobyl disaster, they became a nuclear community which shared the experience of radiation and became communities medically and economically affected by the disaster. These communities are being stigmatised for their association with a polluted place (Davis & Hayes-Conroy, 2017) or a sense of pride and resilience in their communal ability to survive in such an environmentally hostile situation (Davies, 2015; Stawkowski, 2016). A nuclear community is defined as any group that is associated with ionising radiation, which offers a significant scope for exploration of different perspectives, demographics and geographies (Blowers, 2016; Butler, Parkhill, & Pidgeon, 2014). Next to the map of the Chernobyl-affected zone, visitors can find an interactive display (see Figure 6), with the detailed information about towns and settlements devastated by radiation. This information could be an important learning material for geography subject, as it provides information about location and history of all towns falling to the Chernobyl-affected zone. The information includes the region of the town, the history, population at the time the catastrophe has happened and current situation, the date of evacuation, the place of resettlement, and the radiation background on the day of evacuation, including information on how many thousands of times it exceeded the acceptable norm and radioactive contamination comparing the data of 1986 and 2006.

Another interesting exhibit is situated in the third hall: it is a computer display demonstrating the material depicting the geographic trajectory of the radioactive cloud movement across Europe and the countries that have received high levels of radiation in ten days from the start of the disaster.

It is worth remembering that disaster studies deal with how a disaster affected communities of inhabitants not only physically, but also what that disaster meant to these communities in a social sense, how that disaster altered their social and emotional world and how the meanings are being created after the disaster. Usually, cultural representation of disasters presents how communities tried to cope with a disruption of social meanings during the disaster, how it went through processes of social reconstruction, regeneration and recuperation, how new values and norms, new disaster communities and subcultures emerge in the disaster and post-disaster environment (Webb, 2018).

In the Museum, quite many exhibits are dedicated to the nuclear communities: depicting how the disaster affected local residents; here, social aspects of nuclear geographies are revealed. At the start of the exposition, at the entrance of the Museum, the villages and settlements of the Chernobyl Zone contaminated with radiation, no longer inhabited and abandoned by people, are displayed for visitors: 76 names of the towns and settlements where people lived before the explosion and disappeared from the map after the explosion; these residents were evacuated within 10 days. Several years after the disaster, 92 Ukrainian towns and 303 towns in Belarus were additionally evacuated.

The view of abandoned territories is presented in the initial part of the Museum (at the entrance) by using a metaphor of a road sign: hundreds of road signs with inscribed names of settlements hang from the ceiling at the entrance. These road signs are used for marking exit from a settlement (a name of a settlement in white on a black field crossed through with a red line). In its metaphorical form, this road sign means that these settlements are abandoned, uninhabited, there is nobody in there and there is no road leading to these settlements. In the third hall of the Museum, visitors find symbolical depiction of inhabitants and settlements affected by the Chernobyl catastrophe. A large part of the walls in this hall depict facades of empty, abandoned village cottages. Visitors are invited to feel the no-longer-inhabited area – this is the feeling which overwhelms when exploring windows of empty cottages, noticing absence of residents (hosts, women, children). The sense of abandonment and uninhabitedness is enhanced by neglected things (child bicycle and other belongings) scattered around. Here, one observes a specific resemblance (intertextuality) to the iconography of the Pripyat town how the town is depicted after evacuation of its inhabitants – toys left or scattered around in a hurry, household items showing signs of physical decay: broken, damaged, rotten, no longer needed to anyone.

Existential Conceptualisation of Time at the Chernobyl Museum: Multiple Temporalities in the Interplay Between *Chronos* and *Kairos*

The construction of tourist's and visitor's experience is a complex process, like an exhibition, a museum, a tourist site, etc.; it evokes specific feelings, experiences and, in such a way, the co-creation of experiences proceeds. The museum presents the event-related memories which deal with understandings of time and temporalities. Going deeper into the understanding of the temporal structure of museums' narrative, one can find numerous temporal modalities. Analysing the construction of time in pieces of art, literature, cinematography, museum expositions, the time *chronos* and *Kairos* is singled out. These two ancient Greek divisions of time allow to decide how temporalities intertwine in narrative, how tension evolves in them, which temporality prevails, which is expressed less.

Chronos means time which is measured as a sequence or an order; it is literally chronological, like in the passing of time indicated on a clock (Hannam & Ryan, 2019). *Chronos* means the quantitative experience of time; it is a sense and representation of time as the time of history, the narrative duration-time, the age of an object, event or person (Metcalfe, 2006). An exposition structured by a *chronic* articulation of the time is presented as a historical chronological sequence of events, as the homogeneous, orderly and seemingly objective flow of time (that is why *chronos* is measured and represented by clock and calendar). Time as *chronos* in specific modern understanding is presented and experienced in a linear manner – as a linear time with a sense of historical continuity. In comparison, in the Ancient concept, the time of history was represented as a cycle. The non-cyclical arrow-like trajectory of linear time normatively valuates future as progress (Hom, 2018).

Kairos refers to specific decisive and life-changing moments and turning points in time, it is a critical and decisive, "right" time to act, time that is taken or grasped, or a perfectly timed opportunity which has value (Hannam & Ryan, 2019, p. 2). The notion of Kairos stresses a sense of time as occasion and the opportunity to seize the moment and take timely actions. "*Kairos* meant appropriateness, timeliness, the right and judicious moment to act, the season or point in time at which something appropriate happens that cannot happen at any time, in other words, *kairos* involves a much more qualitative notion of time" (Metcalfe, 2006, p. 247).

Kairos is a kind of time of action to be accomplished, to a decision to be reached or to an initiative to be undertaken (Cipriani, 2013). These *karoitic* moments of possibility and taking an opportunity become interruptive in relation to *chronos* as continuity and flow of history. *Kairos* represents interruptive time as disruptive moments which can open the space for agential possibility, the capacity of individuals to act in a given context agency; *kairotic* interruptions have the transformative potential to subvert the *chronic* logics and continuity and open opportunity for agency of people, what is important in the time of crisis that requires a response or a decision to be made and an action to be taken (Winderman, 2017). The development of a unified global *chronotic* imaginary is frequently interrupted by *kairotic* considerations. Kairos allows us to see history as not simply "one damn thing after another", but as endowed with a trajectory and purpose. In "kairos", a unique event is seen to create, arrest or change time rather than endure it (Rao, 2019).

In some texts, the notion of *kairos* emphasises the centrality of the subjects' experiences, interpretations, understandings and narratives to the life-course or historical sequence of events. *Kairos* denotes experienced and significant time which may have an existential and sacred meaning (Cipriani, 2013). *Kairos* temporality reveals existential feelings and transformations that not only interrupt the flow of the *chronos*, but are at the same time existential transformations in a mythical space (in some cases – in a post-apocalyptic space) that is temporarily untied from historical time (*chronos*). In this sense, *Kairos* can also be a way of constructing the apocalyptic narrative, where the apocalyptic temporality of *Kairos* is connected with surviving existential transformations (revelation, rebirth, awakening) in a post-apocalyptic space which is interruption (rupture) and as the final end of a linear chain of events, temporal interruption of a chronological, historical narrative (Lindbladh, 2019). "The temporal structure of this apocalyptic narrative can be described in terms of *Kairos* rather than *chronos*, which means that the radical event is represented in relation to its impact on the characters in the present, contributing to their self-transformation, rather than as an end "set in the future, in a chronological view", thus revealing "the ultimate meaning of history" (Oppo 2013, 24 cited from Lindbladh, 2019, p. 14).

Interplay and tension between *chronos* and *kairos* are not only ontological qualities of time which were pointed out by Ancient Greeks. They are distinct and interweaving ways to structure the narration.

The Chernobyl Museum could be analysed in terms of interrelation of these different temporal modalities. Temporality construction reflects timeline as an organising structure of museum expositions to present and understand the past, when events are organised in the timeline, which is narrative by itself, a way to tell a story, which is recognised by a visitor even when it is being constructed – when separate stages and divisions of time are not marked (Lubar, 2013). A linear chronological structure of timeline, when events and exhibits are set out in the space according to a line of historical sequence, is the most traditional way to construct the timeline. The exhibits are linked to a physical space, when, physically moving, a visitor goes through a sequence of events of the past being constructed – materials of the exposition situated in the space following the logic based on the historical sequence, from earlier to subsequent times, from the past to the present; the visitor in the museum timeline walks through history (Lubar, 2013). The situation of the timeline as chronology reflects *chronos* temporality, when exhibits are set out in compliance with the logic of a historical timeline. However, contemporary museums, as Lubar (2013) has it, seek constructing even more complex flows, when, for

Fig. 2: Road signs depicting abandoned and uninhabited towns after the Chernobyl disaster

instance, part of an exposition is set out according to chronology, whereas other parts follow other principles. Specific strategies are used (for instance, hypertext) to create a more sophisticated and more open structure of the narrative, which is a "less coercive kind of chronology".

After entering the Chernobyl Museum, a visitor occurs in a lobby which, besides its functional purpose (ticket office, cloakroom), is also a specific exposition: the entire lobby and stairs leading to major exposition halls upstairs display road signs similar to those used along roads when leaving settlements (a name of a settlement is crossed over with a red line). Names depicted on these road signs are the names of actual villages and settlements in Ukraine (see Fig. 2).

The lobby of the Museum symbolically marks the consequences of Chernobyl, when thousands of hectares of land left abandoned, and many settlements remain neglected and unsuitable for living due to contamination with radiation. Additionally, there is a place in the lobby where temporal expositions are situated. At the time of our visit (January 2019), there was a mini exhibition (photographs) from Japan depicting the disaster of the Fukushima Nuclear Power Plant. When considering temporality of this part of the Museum, one may notice that the entire exposition of the Museum starts not from the moment of explosion of the nuclear power plant (April 1986), but, when entering the Museum and occurring in the lobby, we start moving in the physical space from exhibits which represent the Chernobyl disaster aftermath (abandoned villages) and neglected empty, contaminated territory of Ukraine. The exposition moves a visitor from Fukushima to the year 2011, which is 25 years past the Chernobyl accident; the connection between the largest disaster at the nuclear power plant and another large accident in Fukushima is revealed. Thus, the beginning of the exposition is the post-disaster period.

The next point in the moving around the exposition as well as another experience of time deal with an exhibit of a clock showing the time of the accident: 1:23. The clock has stopped (broken). This is one of the most prominent symbols directly signifying the conceptualisation of time in the Museum. This clock marks the time of the accident, when a mechanical device no longer works due to mechanical damage – explosion of the reactor. The exposition starts with this metaphor of a stopped time demonstrating the disjuncture, rupture of a historical flow of time (*chronos*). The accident itself along with the stopped time and the broken clock are depicted as interruption and breaking of a historical timeline, a rupture, when history of the entire humankind like a ceaseless moving – a road towards technological progress and brighter future of the humanity stops (a metaphor of a stopped/broken clock), the historical timeline being created by humanity for years and meant to lead to that likely future which had to be better than the "present" is broken.

The stopped clock symbolises an interrupted and stopped flow of ordinary time and tame reality; the flow of the life which was lived by the state, humanity before the accident, and stopped time which will no longer exist – like interruption of the lives that proceeded earlier (see Figure 9). Along with it, the limit of the stopped time indicates the end of a specific historical time (*chronos*) – the time before the world changes – enters the irretrievable period of time, the period with no way back to the time where this apocalypse has not happened yet. The stopped clock is a particular metaphor of rupture of hegemonic temporality and unified global chronotic imaginary, dealt with by Hom (Hom, 2018). The ceased hegemonic temporality, in this case, the historical time of human evolution, is being constructed in a cultural, political and social way; the flow of time is depicted as an arrow-like line, where development of nuclear energy, along with development of entire science, was related to the technological progress. As Hom (2018) puts it, talking about hegemonic temporalities, rupture as temporal disjunction is shocking and unprecedented moments of radical discontinuity; it "disrupts" hegemonic logics. A disruptive moment (i.e. an event) disrupts the "present" and "history" which were constructed as dominating/hegemonic temporality (in this case, it is constructed history of technological progress with the "bright" future of nuclear energy).

The explosion of the Chernobyl Nuclear Power Plant becomes this disruptive event in the hegemonic temporality (unified global chronotic imaginary). After entering Museum's hall 1, starting with the first exhibits, they describe what has happened in the nuclear plant and that the explosion happened. Since the *chronos* time stopped (the stopped clock as a metaphor), the meaning of the *Kairos* time becomes highlighted. The accident itself is a karoitic event which stopped the flow of *chronos*. Moreover, the stopped/broken clock symbolises the end of time as such; this is the entering of the eternity where time is no longer counted (is eternal "now"). By stopping, the clock shows that not only ordinary, "happy" time of millions of people who took part in the liquidation and directly suffered from radiation has ended; the time stops for the entire relationship between humanity and nucleus, a naïve attitude towards the peaceful nucleus ends, before entering the new post-apocalyptic reality where lives proceed alongside the Chernobyl Zone, hundreds of people who fell ill from radiation, the territory contaminated for hundreds of years and a doubt that occurred forever about the safety and suitability of the "peaceful nucleus".

The first hall of the Museum displays activities of nuclear power plant's operators and engineers, whose performance led to the explosion, by presenting a text on the proceeding of causing the accident. Being aware of the whole flow of further events, we know that this moment was highly important; we know that these actions of operators and nuclear plant's engineers became decisive and life-changing moments, turning points in time (in this sense, it is the kairotic moment), which actually changed the entire *chronos* time, symbolised by the clock, and the flow of the historical time (reflected by a calendar).

On the one hand, no doubt, actions of dispatchers and managing bodies that led to the disaster are described in the text of the exposition, in such a way demonstrating what has become the beginning of the significant event at Chernobyl. On the other hand, we can observe that much lesser attention is paid to the description and presentation of this trigger event of the disaster, as a karoitic interruptive event, in comparison to other available cultural representations of the accident. To compare, let us remember the BBC documentary film "Chernobyl Nuclear – Surviving Disaster (BBC Drama Documentary)", where five minutes of the text at the beginning of an important film are allocated to the drama in the dispatcher office (when the actors playing major characters perform not only actions but also the drama of actions and feelings that took place). The drama in the dispatcher office that led to the accident, as a karotic interruptive event, is given much attention in another important text on Chernobyl making up intertextuality of the Museum – the HBO miniseries "Chernobyl" (2018). Here, even two episodes are dedicated to this event (everything what happened "before that symbolical clock had stopped"): the first episode shows "how this all started" and the last, fifth, episode, after presenting the drama of the liquidation of the disaster in earlier episodes, returning to the decisive moment that changed the history and stopped the time and clock, back to depiction of the actions of the dispatcher team. Later in this episode, actions of the team that led to the disaster are commented while repeatedly showing the trial of the causers of the Chernobyl disaster, explaining through a scientist Legasov how the reactor operated and how the accident had happened when members of the dispatcher team specifically acted and made particular decisions. Comparing with mentioned other representations of the Chernobyl disaster (HBO miniseries "Chernobyl", BBC documentaries), the Museum pays quite little attention to the analysis of the actual event of the accident. As depicted in the exposition, the event of the accident itself that broke the historical timeline of events is presented quite in brief, by mentioning the participants, shortly identifying their actions. Slightly more attention in this exposition is paid to what has happened after the accident, by discussing how the investigation, trial proceeded (closed procedure of the court), what punishment was imposed on the managing bodies and remaining alive direct participants who were present at the dispatcher control desk; their health condition is discussed and causes of death (radiation sickness) are mentioned.

When discussing the presentation of the accident moment in the Museum and comparing it with other above-mentioned fiction and documentary texts, one can notice that the technical and human emotional dramatic moment/episode of living the very first minutes and hours of that historical and human experience in the dispatcher office is presented in brief (textual factual information), without engaging a visitor into emotional experience of the participants involved in those decisive minutes and hours of actions and feelings in the dispatcher office that changed the world. We suppose that representation and construction of temporality of this moment that happened in the dispatcher office can be differently constructed in a text. We see that even though this moment is represented in the Museum as karoitic (that disrupted the planned and anticipated flow of historical time), this karoitic moment is depicted as a disruptive moment which actually happened because the space for an agential possibility was opened (actions of the dispatcher team that led to the accident were that negative and ill-fated agency which aimed at performing extraordinary successful testing turned into a fatal error that changed the flow of history). The karotic moment as the capacity of individuals to act in a given context agency and having the transformative potential to subvert the *chronic* logics and continuity, in this particular case, opened the Pandora's Box and created a new trajectory and direction of the historical flow line. To present and reveal this decisive moment, the Museum allocates quite a small part of the exposition, and, to compare with other analysed documentaries and films, this moment is not emotionally, existentially and dramatically presented. This could be linked to the very genre of museums, which has different possibilities to present dramas, other than those provided by cinematography and literature.. This karoitic moment becomes disruptive because it opens a new historical line – it breaks explanatory/interpretive frameworks (constructed successful history of the nucleus of the state and the humankind), and in this sense it no longer allows "anticipating" the future.

According to Hom (2018), rupture links closely to the "trauma time"; this fractured moment creates incapacity in acting and interpreting, since the fall of hegemonic temporality creates absence of interpretation. The future becomes radically open and this disruptive moment as a radical break, *rupture*, the event which disrupts the "present" and "linear time", *destabilising* hegemonic temporality (as a dominating politically and culturally constructed idealistic vision about the move of history and the humankind towards progress in line with the myth of safety of nuclear energy) and future as an alternative which is unknown and *"something completely different"*. The event 9/11, when planes piloted by terrorists smashed into the World Trade Centre and the Pentagon, opens the unknown and cannot be integrated into the existing dominating hegemonic historiography, and, thus, is an event of a similar kind. Back then it could not be placed in a larger historical context and narrative structure; they were disruptive singular events (Hom, 2018).

The drama in the dispatcher office at the Chernobyl Nuclear Power Plant as a rupture can be explained in a similar way. After the accident happened (as a result of what has happened in the dispatcher office), the unknown, new likely horrible trajectories of the history line-to-come open, which were beyond the people's ability to explain and consider; in this sense, the "trauma time" opens. It can be treated as a karoitic interruptive event; only in this case what is disrupted is the line being constructed in the future drawn by *chronos* as the homogeneous, orderly and seemingly objective flow of time; the future-to-come opens after the explosion of the reactor, which has no interpretative framework of how to explain. Later, this rupture causes a new flow of events – fire, penetration and spread of radiation; contamination of Ukraine, Belarus and large part of territories across Europe; harm to people, nature and ecosystems.

To prevent the opening of these new historical apocalyptic lines (of newly constructed *chronos*), a whole liquidation of the disaster as one grand event comprising smaller events was created. After the explosion, actions of the teams of liquidators directed to liquidate the accident are presented in the exposition as separate karoitic moments. A special attention in the exposition is focused on heroic liquidation of the consequences of the accident, the actual events that allowed creating measures to stop, reduce the consequences (extermination of fire, construction of the sarcophagus, cleaning of the territory). Referring to Hom (2018) dealing with the perception of time during disruptive events, after the rupture happened (actions of the dispatcher team that caused the accident, the explosion) the agency of other people – liquidators – started operating. Another karoitic moment, which created the *rupture of rupture*, started. The campaign of liquidation of the consequences of the disaster is a heroic deed which allowed blocking, reducing the newly occurred trajectory of the flow of events, a new apocalyptic historical line. The liquidation is treated as actions of the *rupture of rupture*, as an attempt to terminate the termination, an attempt to control the times of rupture, "transforming it from a description of traumatic and unliveable conditions to the foundation of a novel ethics that insists we 'remain with uncertainty' and 'hope that something different' will emerge" (Hom, 2018, p. 327). These liquidation actions are treated as karoitic moments which become the disruptive event as a response to the new apocalyptic vision of the history with unclear and horrific future, which is created and drawn by the accident itself.

In the first part of exposition at the Chernobyl Museum (halls 1 and 2), a visitor gets acquainted with a presented proceeding of the liquidation campaign as one grand event which becomes that actual karotic interruption (*rupture of rupture*). This one grand event is divided, presented to a visitor as a sequence of smaller events. Separate teams of heroes – fire-fighters, helicopter pilots, miners, soldiers cleaning the reactor's roof from graphite, all other teams – physicians and other participants, are introduced. Each team is introduced separately; emphasising the character of their work; singling out family names of individual people, their actions; and highlighting the radiation doses they obtained. Almost in every case, the harm to health is mentioned, and further proceeding of actors' lives is noted – usually, the year of death or a degree of affected health is described. Thus, a visitor sees that a large army of liquidators is divided into smaller professional groups, in such a way rendering these groups some individuality, exceptionality. The whole group is characterised: its role and approximate number of the participants; also, the harm underwent by the entire group is underlined. Inside each group, single actors are chosen (their names, family names, further destinies are mentioned). Such division of the liquidation operation as the whole into smaller groups, singling out separate heroes-individuals in each group, is a way to show the entire historical event through a more individualised perspective, by revealing separate groups and single heroes, demonstrating heroism of these people. In this sense, the exposition attempts to point out the karoitic moments – a specific aspect of time, when a grand historical event is presented through actions and lives of individual people.

Heroic actions of the liquidators can be viewed as the karoitic moments, when the *rupture of rupture*, the interruption/suspension of the flow/line of that apocalyptic history that could have had happened/occurred after the accident proceeds. This is the *rupture of rupture* moment, as a karoitic moment, when an opportunity for agency of people was opened up, when in the time of crisis a response was required and a decision to be made, an action to be taken (Winderman, 2017). These heroic actions became the decisive and life-changing moments; this was the "right time to act" with transformative potential to subvert the *chronic* logics and continuity, which, in this particular case, is set/drawn by the already happened accident and the *chronos* logic which had to lead to the apocalypse. On the one hand, an extraordinary contribution of all teams of people to the liquidation reflects the role of human agency in creating a new line of history (rupturing the rupture). This was achieved, thanks to heroism (presenting completed technical operations, showing technical challenges and difficulties) and self-sacrifice of these people. On the other hand, when comparing the narrative of the Chernobyl Museum with other fiction-documentary texts dedicated to Chernobyl (S. Alexievich's book "Chernobyl Prayer", HBO series "Chernobyl", BBC documentaries featuring actors), we can observe that fates of the people, their existential experiences, are demonstrated in the Museum in a rather narrow manner. In her book, S. Alexievich constructs the experienced liquidation of the disaster and the post-accident period through magnifying experiences and suffering of individual participants (liquidators, wives, scientists, physicians); separate experienced moments are singled out by phenomenologically profoundly magnifying/revealing them as if looking through a magnifying glass; the life events are described in detail by mentioning actions, feelings and philosophical pondering. We find quite many such moments in the HBO series "Chernobyl" as well; the series were created grounding not only on documentary materials of archives, but also Alexievich's book – when people's experienced moments, many artistic symbols and metaphors, and tragic and dramatic suffering of the heroes resulted in high emotional response of audience, readers.

Whereas despite the above-mentioned single moments introducing individual personalities and names of people, mentioning their life facts (disease, death), the Chernobyl Museum does this without revealing description of one person's life moment in full, as it is done in other already discussed pieces. Such non-disclosing of people's existential experience indicates the domination of *chronos* temporality in this part of the exposition (halls 1 and 2). The explosion, accident and liquidation of the consequences are being constructed as an event situation in time, having its specific linear chronology and timeline, when the historical event, i.e., the explosion of the nuclear plant and liquidation/reduction of its consequences are presented as a sequence of actions and events situated in the timeline – from the explosion and liquidation of the consequences, having the sarcophagus over the reactor and cleared up Chernobyl Zone as an outcome. A visitor creates this logic of chronological order and line while physically moving from introduction of one team of liquidators-heroes to another in the exposition. While presenting the consequences of the explosion of the nuclear power plant and demonstrating actions of the nuclear power plant's personnel (engineers, shift of dispatchers, managers) (halls 1 and 2), the time represented in the narration is being slowly "counted" and introduced minute by minute (here, the karoitic element is quite strong); whereas while presenting actions of fire-fighters who are extinguishing the fire the time is counted by hours.

Later, the narration "moves faster", and the chronology is measured by days – joining in of other groups of rescuers and their actions are already being counted by days (miners, helicopter pilots, soldiers cleaning the reactor's roof from graphite); further, the time is counted by months and years (clean-up workers – soldiers, military officers). At the end of the second hall (physically moving around the hall and travelling "in time"), a visitor finds introduction of the role of physicians and scientists, which also points out not only the role of medical staff at the start of the liquidation, but also the role of physicians and scientists when treating the victims several decades later, investigating the long-term effect on people's health and biological environment. In such a way, a new historical line and the chronological ordering of the time flow/structure (*chronos* time) are created. The events attributed to the time of the several first days (first 3–4 days), i.e., actions of the governing bodies of the Soviet Union being reticent in public (in major dailies, like "Pravda") about the accident; dedicating a very brief press release; a small patch of text on a large sheet of the newspaper and the non-cancelled, arranged in Kyiv; festive demonstration dedicated to celebration of the 1st of May, "as usual"; and aiming to prevent from panic and pretend that nothing extraordinary has happened.

Later on, the timeline of the presented events is counted by weeks and several long first months aftermath: when during the first month all brigades of liquidators (helicopter pilots exterminating the fire, miners digging the tunnel) join in, the exposition of the timeline of several first months introduces soldiers on the military duty equipped with shovels throwing pieces of graphite down from the reactor's roof. From a very accurate counting of time by seconds there is transition to the representation of the time flow by days, months and years. In the first part of the Museum intended to reveal the role of the liquidators (halls 1 and 2), a visitor gets acquainted with the period, starting with the explosion, liquidation works taking place (extinguishing the fire, removing radioactive graphite from the roof (how many months) and covering the reactor/constructing the sarcophagus over the nuclear power plant (how many months/years), cleaning, "deactivating the territory" and creating what presently is known as the Chernobyl Exclusion Zone (as an isolated territory which is currently uninhabited, but the works of de-activation/liquidation proceeded there). Besides already mentioned participants, i.e., nuclear power plant's personnel, fire-fighters, helicopter pilots, miners, soldiers and military officers, at the end of the first part of the exposition, the role of medical staff, scientists of nuclear physics and medicine is presented. Even though the major intention of the first part of the exposition's narrative is to show the role of various groups of liquidators and in such a way to commemorate their role and pay tribute to them, we can also recognise the chronological narrative along with the major narration, when the entire mission of clean-up workers and liquidators is set up as a chain of separate events with logic of its flow.

Thus, the timeline as a way to structure a narration is quite complex, sophisticated in the Museum, when, besides traditional chronological representation of the proceeding of events inserting them into the historical timeline (*chronos* temporality), there are attempts to present another modality of time – *kairos*, showing interruptive moments of the flow of history, revealing people's agency potential, attracting attention to single details, and experiences of individuals' lives. In comparison to other earlier mentioned fiction texts on Chernobyl (the book and series), these elements, however, have quite limited *kairos* as existential experience of people and emphasise the *chronos* modality of time more strongly in the first part of the exposition (halls 1 and 2) dedicated to underline the role of liquidators as heroic deeds. The Chernobyl Museum becomes a memorial commemorating the historical event which happened at a particular historical point, having perhaps changed the history of the humankind forever.

The Museum's expositions narrate a story which disrupted that other imaginary and anticipated history and other chronology (that possible alternative history of the Soviet Union, Ukraine and humankind) which could have happened/formed if the accident had not happened. Authors analysing the timeline in exhibitions (Lubar, 2013), dealing with the chronological ordering of time of traditional expositions underline that timelines highlight before and after, cause and effect, and linear progression (Lubar, 2013). When exploring the chronological logic of the narrative of the "event" and phenomenon of the explosion of the nuclear plant and liquidation, this logic "before and after" (one action follows the other, there is proceeding of liquidation as a sequence, cause and consequence of events; liquidation works and heroism of liquidators as a cause of the sarcophagus covering the reactor and rescuing the humankind from doom) is revealed; and there is the Chernobyl Exclusion Zone itself as a "cleared up" (partly de-activated, washed out) territory where radiation has been localised and is not spreading out. The idea of progress and advancement that is implicitly characteristic to chronological depiction of the time flow is presented in the exposition of the Chernobyl Museum in quite a complex way. As investigators of the timeline in expositions emphasise (Lubar, 2013), the chronological timeline is characteristic of presentation of history as a line that leads to progress. On the one hand, the accident itself and its consequences (contaminated Ukrainian territory for hundreds and millions of years, damage done to health of inhabitants and liquidators) are depicted in the exposition as impossibility of the lines of history being constructed by the humankind, the lines that led to progress before the accident – this "pre-accident" line of the historical flow was disrupted by the accident. The stopped clock symbolises impossibility of the progress imagined in the past (before the accident) and planned, approaching "better future" (it is impossible to return to the pre-accident situation).

On the other hand, the extinguishing of the fire, construction of the sarcophagus and partial de-activation of the territories, and safeguarding of other territories (those of Europe and entire humankind) from even larger catastrophe are considered to be a specific progress (improvement) in the chronological presentation of the liquidation of the accident. This is the "small" progress which we recognise in a general Chernobyl-related idea of the history of the eschatological "fall" and regression.

Analysing timeline in museum expositions, Lubar (2013) discloses the strategies applied by contemporary museums allowing creation of a more sophisticated and more open structure of the narrative, which is a "less coercive kind of chronology". In the case of the Chernobyl Museum, this more sophisticated strategy is implemented through the merging of two lines depicting history, when the disruption of the historical line of the happened accident is combined with regress of the historical process, also including depiction of "afflatus" – depiction of reduction of the consequences of the accident as logic of chronology and hidden afflatus/progress. Reduction of the consequences of the accident, as a sequence of events having their own chronology and achieved result (sarcophagus, localisation of radiation within the Zone), is directly related to another important narrative of the exposition – the narrative on liquidators as heroes.

Exactly these endeavours and actions of the liquidators organised in a chronological order have created a new quality and new reality after the accident: the fire was extinguished, the sarcophagus was constructed, and the territory was cleaned up. Sophistication of the narrative in this first part of the exposition dedicated to the liquidators is implemented in compliance with a very specific strategy: information on the future of single individuals and separate groups is presented when chronologically organising the proceeding of the happened accident and liquidation of its consequences, presenting the works and introducing single people and separate groups of liquidators (nuclear power plant's personnel, fire-fighters, pilots, miners, soldiers, medical staff, etc.) who actually started acting from the very first hours, months and throughout the entire period of liquidation; the narration is on when and how these people died (separately presented are the cases of deaths immediately after the accident (firefighters), in other cases the same fate happened several months, years or decades later (after they got ill and died from the radiation)). Thus, there are two principles of classifying and organising exhibits – the liquidation works (this is reflected by presented groups of liquidators by professions – fire-fighters, helicopter pilots, miners, soldiers, etc.) and the chronological principle, when these works are organised across time.

Thus, from the point of view of construction of the proceeding of the events, this is quite a complex way to show the "historical" event that has been lasting for 30 years – while depicting it in full precision, arranging events and people in a chronological sequence by minutes-hours-weeks-months-years. On the other hand, in this clear line of chronology of events, "ruptures" occur – these are excurses to the future of the liquidators (different circumstances and dates of diseases and deaths). Even though it is possible to recognise the chronology of arranging exhibits (characteristic to traditional museums) in the Chernobyl Museum, still the road from the past to the present which is constructed by a chronological linear narration is being constructed in a sophisticated manner.

Kairos in the Symbolic, Philosophical and Religious Narrative on the Disaster

A special construction of the narrative and temporality is discovered in the third hall of the Chernobyl Museum. The exposition of this hall demonstrates the attempt to move away from the heroic narrative and chronological temporality towards depiction of existential suffering and emphasising the *kairos* temporality. The construction of time becomes more sophisticated than the chronological organisation of time (*chronos*) and a multiple hetero-temporality occurs, which allows problematising clocks, calendars and heroic state narratives.

Here, another strategy and logic of organisation of the exposition are followed: the effect of the accident on residents, children and women, and victims of Chernobyl in the Chernobyl Zone villages are depicted. Exactly this hall is an attempt to create a narrative as a more complex, sophisticated organisation than chronological. Authors, analysing the timeline in museum exhibitions (Lubar, 2013) have it that in contemporary museums there is an attempt to de-emphasise chronology and time as the organising structure of the exhibits and to use common human experiences as the thematic framework – to approach history from the standpoint of common human experiences of family, work, community and sense of place. The third hall reflects the aspect mentioned in the disaster studies – cultural representation of a disaster aims at depicting how communities after disasters live; how disaster-stricken nuclear communities have tried to cope with disaster by creating meaning, by maintaining resilience; going through processes of reconstruction, regeneration and recuperation; and creating new values and norms. Researchers dealing with disaster studies (Webb, 2018) reveal that a disaster-stricken community creates values and meanings reflecting revival, recovery and the signs of hope. In the Chernobyl Museum's hall 3, we find symbolical philosophical and religious meanings of the nuclear disaster.

If the first part of the exposition (halls 1 and 2) is dedicated to the liquidators-heroes and a clear chronological logic of narration is recognised, in the case of the third hall, a visitor finds an artistic exposition including many symbols dedicated to commemoration of citizens-victims, emphasising extraordinary suffering of village inhabitants and children. This part is not chronologically organised. On the one hand, in the aspect of time, all human experiences undergone by people throughout thirty years after the explosion (without singling out detailed and precise chronology) would belong here. On the other hand, exhibits and installations presented in all this appeal to the sense of non-temporality and eternal time – here, religious, existential lived experiences, and eternity and infinity of time are appealed to. In the hall 3, a visitor discovers a place of experiencing existential, sacral disaster. Here, one should remember symbolic forms of representations of disaster – theological and mythological motif of theodicy and divine involvement, as pointed out by Holm (2012). The halls 1 and 2 narrating about heroism of the liquidators reveal more social and political aspects of the disaster, whereas in the third hall the disaster is constructed as a religious, existential and philosophical phenomenon.

This part of the exposition deals with cultural construction and representation of the Chernobyl disaster presenting physical, psychological and spiritual suffering of Chernobyl sufferers – resettled persons and inhabitants of contaminated territories. In this case, the collective identity is being constructed not through heroism but rather through belonging, commonness in suffering and belief. Here, experiences lived by the community several decades after the Chernobyl disaster are presented. This hall has specific iconography. The first and second halls are characteristic of a documentary genre: the event of the accident is chronologically described in detail; particular people having their family names are introduced; and historical artefacts and documents are displayed. Black and white colours prevail in the first two halls, corresponding to the visual style of exhibited documents (black-and-white photographs and documentary films, blackish grey colour of the burning reactor, dark colours of uniforms worn by statutory officers and those of technical mechanisms; all materials (photographs, texts) are organised in a line on walls at the visitor's eye level, without more complex, sophisticated spatial and colour solutions); whereas the third hall presents an artistic project including many artistic-religious symbols, without displaying any historical documents and historical materials. The visual style is characteristic of colours, illumination effects and complex arrangement of exhibits in space (on walls, floor, ceiling). The purpose of this artistic installation is to create emotions, to represent suffering of people (residents and community). Artistic installations of this hall encompass many mythological and religious symbols. These mythological-religious symbols are used together with already traditional iconographic images of nuclear energy and nuclear disaster. One of such images widely used in various photographs, films is the image of a liquidator wearing a special protective green uniform and a gas-mask – a de-personified man without face and eyes, an unrecognisably veiled persona (who may even be not a man in this iconography at all, but an animate robot-function of the post-disaster techno society). In this hall, liquidators are depicted next to religious and spiritual symbols (large crosses, icons) (see Fig. 3).

Exploring this hall, we find strategies and techniques of representation which are described in museums aiming to render the suffering of victims and also seeking avoidance of direct frightening depiction of atrocity. These aesthetics techniques are applied in Jewish museums which commemorate the Holocaust; unique aesthetic and spatial strategies draw on particular aesthetic techniques of representation to evoke specific experience and sensations of the sacred by demanding a "particular form of contemplation" (Hansen-Glucklich, 2016). It is choreographed within museums and their exhibits through a number of

Fig. 3: Images of a liquidator wearing a special protective green uniform and a gas-mask next to religious and spiritual symbols

techniques, including spatial design, the use of symbolic materials and forms (such as water, rock and light), and the manipulation of the visitor's movement through space and passage.

A golden gate with two sides to enter – white from one side and black from another, situated at the entrance to the third hall – symbolise the gate between heaven and earth. The floor at the entrance to the hall is a chessboard, with white and black colours symbolising a game between life and death, balancing between black and white, good and evil, and life and death.

Another exhibit having a religious-spiritual function in pondering on the suffering of victims after the Chernobyl catastrophe is a candle flame demonstrated on the monitor's screen. A visitor has an opportunity to approach the monitor and light a candle, watch the flame wavering on the screen. The symbol of a candle has many meanings; however, in this context, a visitor, having lit a candle and immersed in the contemplation state, can see this flame of a candle as a sacred fire, the light of God, the light that illuminates the path for the dead in their journey; it can be seen as a sign of illumination and hope.

A special place of this exposition is allocated to the suffering of children. One of the most memorable exhibits of the hall is a large photographic/arts installation depicting the Fuel Assemblies of Nuclear Reactor from RBMK reactor's Central Hall: photographs of children are placed in all separate elements of the Fuel Assemblies. These children were born in 1987–1988 into families of people evacuated after the Chernobyl disaster and people who were working on the sites of the catastrophe. This exhibit has a purpose to emphasise that the nuclear disaster had effect on children's health, fates, and took away thousands of children's lives.

In the middle of the third exposition hall, there is a boat with toys and plush animals. Two angels – white and black, hang above the boat; these are symbols of life and death. The boat is a religious symbol mentioned in Christianity and other religions; on the one hand, it depicts a journey and a voyage of life, when the boat carries people through life's shifting currents; on the other hand, it is a symbol of safety, security and refuge, when God protects believers. In the context of the Chernobyl catastrophe, the boat with the angels of life and death can be treated as a religious symbol telling that souls of ill or dead Chernobyl victims' children are fostered and safeguarded by God. Separate exhibits of this hall highlight the effect of the Chernobyl disaster on the entire humankind. The catastrophe altered the fate of the whole world's nuclear energy for ever – this disaster shook the understanding of the nuclear energy as reliable and secure throughout the world.

In this hall, next to the exhibits of the boat, burning candle, gates of life and death, visitors find/see a map of the world created as an installation of lamp bulbs. Looking at the ceiling, one can see lights which represent nuclear power plants operating in Europe; having turned his/her head, one sees South and North Americas, Asia, Australia and Africa in the centre. The map is supplemented with the data from the International Atomic Energy Organization (2012): in 2012, 34 countries of the world had 435 operating nuclear reactors. Almost half of them were built in 1970–1980; two of them have had irreversible consequences for society, i.e., Chernobyl disaster in the 20th century and Fukushima explosion in the 21st century. This artistic-visual depiction of global energy as a geographically situated industry, including all religious-philosophical symbols in this hall, creates a narrative which expresses some doubt about further development of nuclear energy. The subjugated and employed to human needs nuclear energy brings together largest disasters and suffering. In such a way, here a narrative of development of science, economy and industry is criticised and questioned while juxtaposing to philosophical and religious-spiritual consideration of the suffering from the nuclear disaster.

Conclusions

Analysing the educational potential of the Chernobyl Museum, we attempted to view the exposition from the perspective of disaster studies. In its specific manner, the Chernobyl Museum constructs, frames and interprets the Chernobyl disaster; it is a manifestation of collective imagination and memory work. Aiming to fully reveal and use the educational opportunities of the exposition, it is possible to construct in the education process (school curriculum and non-formal learning) a stronger educational effect through intertextuality – by employing additional texts (using fiction, documentary and feature films). As demonstrated earlier, the Museum's exposition can be "read" and used as an educational text/teaching aid in interaction with other known texts on the topic, e.g., S. Alexievich's book, HBO series, BBC documentaries on Chernobyl and other creative artistic projects. On the one hand, juxtaposition of these texts allows better understanding of the specificity and uniqueness of cultural, political and social interpretation presented by the Museum; on the other hand, having combined the Museum's exposition and other texts, a multiple image of the catastrophe makes up.

Considering that the Chernobyl Museum is a national museum, it would be beneficial to analyse in the education process how exposition of this Museum is the manifestation of the institutional memory policy. Such explanation would be a particular attempt to deconstruct the narrative, to recognise not only the content of the narrative but also to reveal what institutions and how create it. No doubt, when creating the Museum's exposition, there was collaboration with the liquidators and various organisations; thus, the exposition itself is a result of happened "negotiations" among different groups (national memory policy, liquidators and their relatives, organisations, various supporters). It would be interesting to analyse different cultural representations of the Chernobyl disaster (Alexievich's book, HBO series "Chernobyl") and compare them as stances on explanation of the past carried out by different organisations and groups during history, social sciences and geography classes for senior form students. Here, it is worth noting that criticism towards Alexievich concerning books on Afghanistan and Chernobyl is expressed in public. These critics say that she presents a point of view which strongly differs from the official, institutional interpretation of the mentioned events emphasising heroisation, necessity and meaningfulness of self-sacrifice. The HBO series "Chernobyl" is also treated as a specific version and interpretation of the events by American creators much grounding on the said Alexievich's book, various documentary sources. Thus, an additional educational effect is created by an opportunity to compare the content of different narratives on Chernobyl, their relation to official institutions of the memory policy and various groups of interests in the education process.

Besides critical deconstructing attitude towards the content of the Chernobyl Museum's exposition, for educational purposes, it would be meaningful to analyse together with students the structural approach towards disasters which reveals how organisations and communities mobilised response efforts. Alongside with other mentioned texts, the Museum presents an image of how the Soviet Union structures operated in a *"state of emergency"* and how the Soviet emergency crisis system functioned. The version of how to cope with the disaster consequences presented by the Museum may be integrated into the school curriculum through the content of taught subjects – history, public sciences and geography. Besides the understanding of how the emergency system functioned (including heroic deeds of citizens, clean-up workers), the Museum's exposition combined with other sources (HBO film and BBC documentaries, Alexievich's book, etc.) may render knowledge and understanding of how the science of nuclear physics operated as a social institute under the conditions of constraint and secrecy in the Soviet regime, how it impacted the development of the nuclear energy, and, finally, how the Chernobyl disaster induced the collapse of the Soviet Union and made an effect on the fate and future of nuclear energy worldwide.

Environmental and nuclear geography is another important topic of the educational impact. This is the content related to the learning about radiation, the contaminated nuclear landscape and nuclear communities. The Museum reveals the impact of ionising radiation and nuclear contamination on people's bodies and lifestyles, animate nature and landscape. The Museum's exposition presents geography of nuclearity which covers topics of how radiation has affected the territory and communities. It is important to underline that such learning about contamination does not limit itself with knowledge coming solely from biology, nuclear and radiation physics, chemistry, medicine and physical geography. The Museum develops the topic of nuclear communities and presents how the disaster affected communities of inhabitants not only physically, but also what that disaster meant to these communities in a social sense, how that disaster altered their social, emotional and spiritual world. The exposition illuminates how communities tried to cope with a disruption of social meanings during the disaster, how it went through processes of social reconstruction, regeneration and recuperation.

The explanation about the interaction between Chronos and Kairos displayed earlier in the text demonstrates how through *kairos* temporality there is an attempt to show, create the existential-philosophical-religious sense of time and reality, how there is an attempt to perform the sense-making of existential suffering. This aspect of the exposition allows combining the attendance of the Museum as non-formal education with the school curriculum while integrating studies of literature, ethics, art, philosophy and religious aspects. Again, in this case, it would be meaningful to relate the attendance of the exposition as part of the education process to the analysis and experiencing of other fiction texts (literary works, films, art projects and exhibitions).

Analysing the Chernobyl Museum as a dark tourism site according to the dark tourism spectrum (darkest to lightest) presented by Stone (2006), we would attribute the Chernobyl Museum to the category "dark". In this case, the very exposition of the Museum is not directly a place of suffering or death (it is not the actual Chernobyl Exclusion Zone); it tends to create association with death and suffering. The exposition implements the feature of dark tourism – it renders a strong educational and commemorative orientation, and there is no emphasis on entertainment. One of the strongest narratives of the Museum – heroisation and commemoration of heroic deeds of liquidators, mitigates the "darkness" of the event, highlighting meaningfulness of heroes' self-sacrifice. Artistic installations in the third hall dedicated to the religious and philosophical contemplation on the disaster aim at transforming the experiences of death and suffering characteristic to dark tourism towards a moving aesthetical, existential and spiritual experience. Having used a broad arsenal of additional texts, spectrum of additional educational activities (likely being assisted by a teacher), Museum visitors have an opportunity to create a unique existential experience of learning.

References

Agamben, G. (1998). *Homo Sacer*. Stanford, CA: Stanford University Press.
Alexievich, S. (1999). *Voices from Chernobyl: Chronicle of the Future.* Translated by A. Bouis. London: Aurum Press.
Alexievich, S. (2016). Chernobyl Prayer: Voices from Chernobyl: A Chronicle of the Future. London: Penguin Modern Classics.

Alexis-Martin, B., & Davies, T. (2017). Towards nuclear geography: Zones, bodies, and communities. *Geography Compass.* https://doi.org/10.1111/gec3.12325

Biran, A., Poria, Y., & Oren, G. (2011). Sought experiences at (dark) heritage sites. *Annals of Tourism Research, 38*(3), 820–841. DOI: 10.1016/j.annals.2010.12.001

Blowers, A. (2016). *The legacy of nuclear power.* Oxford: Routledge.

Briukhovetska, O. (2016). "Nuclear Belonging": "Chernobyl" in Belarusian, Ukrainian (and Russian) films. In: S. Brouwer (Ed.), *Contested Interpretations of the Past in Polish, Russian, and Ukrainian Film. Screen as Battlefield* (p. 95–122). Leiden, Boston: Brill Rodopi

Brown, K. (2013). *Plutopia.* New York: Oxford University Press.

Butler, C., Parkhill, K. A., & Pidgeon, N. (2014). Energy Consumption and Everyday Life: Choice, values and agency through a practice theoretical lens. *Journal of Consumer Culture,* 19, doi:10.1177/1469540514553691

Cipriani, R. (2013). The many faces of social time: A sociological approach. *Time & Society, 22*(1), 5–30.

Cram, S. (2016). Living in dose: Nuclear work and the politics of exposure. *Public Culture, 29*(3), 519–539. https://doi.org/10.1215/08992363-3511526

Cresswell, T. (2003). Landscape and the obliteration of practice. In: K. Anderson, M. Domosh, S. Pile & N. Thrift (Eds.) *Handbook of Cultural Geography* (pp. 269–282). London: SAGE.

Davies, T. (2013). A visual geography of Chernobyl: Double exposure. *International Labour and Working Class History,* 84, 116–139.

Davies, T., & Polese, A. (2015). Informality and survival in Ukraine's nuclear landscape: Living with the risks of Chernobyl. *Journal of Eurasian Studies,* 6, 34–45.

Davis, S., & Hayes-Conroy, J. (2018). Invisible radiation reveals who we are as people: environmental complexity, gendered risk, and biopolitics after the Fukushima nuclear disaster. *Social & Cultural Geography, 19*:6, 720–740, DOI: 10.1080/14649365.2017.1304566

Eyre, A. (2006). *Literature and best practice review and assessment: Identifying people's needs in major emergencies and best practice in humanitarian response.* UK Government, London. Retrieved 15February 2021, from https://assets.publishing.service.gov.uk/government/uploads/system/uploads/attachment_data/file/86357/ha_literature_review.pdf

Friedrich, M., & Johnston, T. (2013). 'Beauty versus tragedy: thanatourism and the memorialisation of the 1994 Rwandan Genocide'. *Journal of Tourism and Cultural Change*, 11 (4), 302–320.

Fritz, C. E. (1961). *Disaster and community therapy*. Washington, DC: National Research Council, National Academy of Sciences.

Gerster, R. (2013). The Bomb in the Museum: Nuclear Technology and the Human Element. *Museum & Society*, 2013. 11(3), pp. 207–218, 2013.

The Chernobyl Catastrophe: Consequences on Human Health. Greenpeace,(2006). Amsterdam: Greenpeace; 2006. Retrieved May, 2019, from Greenpeace International website, http://www.greenpeace.org/international/en/publications/reports/chernobylhealthreport/

Grebenar, A. (2018). *The Commodification of Dark Tourism: Conceptualising the Visitor Experience*. Doctoral thesis, University of Central Lancashire. Official URL: https://core.ac.uk/download/pdf/159754327.pdf, Accessed 15 February 2021.

Hannam, K., & Ryan, E. (2019). Time, authenticity and photographic storytelling in The Museum of Innocence. *Journal of Heritage Tourism*, 14:5–6, 436–447, DOI: 10.1080/1743873X.2019.1622707

Hansen-Glucklich, J. (2016) Poetics of Memory: Aesthetics and Experience of Holocaust Remembrance in Museums. *Dapim: Studies on the Holocaust*, 30:3, 315–334, DOI: 10.1080/23256249.2016.1240844

Holm, I. W. (2012). The Cultural Analysis of Disaster. In: C. Meiner and K. Veel (Eds.), *The Cultural Life of Catastrophes and Crises*. (pp. 15–32). Berlin: de Gruyter.

Hom, A. R. (2018). Silent Order: The Temporal Turn in Critical International Relations. *Millennium: Journal of International Studies, Vol. 46*(3), 303–330.

Hultkrantz, L., & Olsson, C. (1997). Chernobyl effects on domestic and inbound tourism in Sweden — A time series analysis. *Environmental & Resource Economics, vol. 9*(2), 239–258.

Kang, E.J., Scott, N., Lee, T. J., & Ballantyne, R. (2012). Benefits of visiting a 'dark tourism' site: The case of the Jeju April 3rd Peace Park, Korea. *Tourism Management, 33* (2), 257–265.

Kreps, G. A. (1989). Description, taxonomy, and explanation in disaster research. *International Journal of Mass Emergencies and Disasters*, 7, 277–280.

Lindbladh, J. (2012). Coming to Terms with the Soviet Myth of Heroism Twenty-five Years After the Chernobyl' Nuclear Disaster: An Interpretation of Aleksandr Mindadze's Existential Action Movie Innocent Saturday. *Anthropology of East Europe Review 30* (1) Spring 2012, pp. 113–126.

Lindbladh, J. (2019). Representations of the Chernobyl Catastrophe in Soviet and Post-Soviet Cinema: The Narratives of Apocalypse. *Studies in Eastern European Cinema*, 10:3, 240–256.

Lubar, S. (2013). Timelines in Exhibitions. *Curator: The Museum Journal*, Volume 56, Number 2, 169–188.

MacCannell, D. (2011). *The Ethics of Sightseeing.* Berkeley, CA: University of California Press.

Marchesini, I. (2017). A new literary genre. Trauma and the individual perspective in Svetlana Aleksievich's Chernobyl'skaia molitva. *Canadian Slavonic Papers*, 59:3–4, 313–329, DOI: 10.1080/00085006.2017.1379128

Marcus, A. S. (2007). Rethinking Museums' Adult Education for K-12 Teachers. *Journal of Museum Education, 33*:1, 55–78, DOI: 10.1080/10598650.2008.11510587

Masco, J. (2006). *The Nuclear Borderlands: The Manhattan Project in Post-Cold War New Mexico.* Princeton NJ: Princeton University Press.

McLean, F. (1998). Museums and the construction of national identity: A review. *International Journal of Heritage Studies*, 3:4, 244–252, DOI: 10.1080/13527259808722211.

Metcalfe, A. (2006) 'It Was the Right Time To Do It': Moving House, the Life-Course and *Kairos. Mobilities*, 1:2, 243–260, DOI: 10.1080/17450100600726621

Miles, W. (2002). Auschwitz: Museum Interpretation and Darker Tourism. *Annals of Tourism Research, 29*, 1175–1178.

Miller, D. S. (2007). Disaster tourism and disaster landscape attractions after Hurricane Katrina. An auto-ethnographic journey. *International Journal of Culture, Tourism and Hospitality Research*, Vol. 2 No. 2, 115–131.

Oppo, A. (ed.) (2013). *Shapes of Apocalypse: Arts and Philosophy in Slavic Thought.* Boston, MA: Academic Studies Press.

Petryna, A. (2002). *Life Exposed: Biological Citizens After Chernobyl.* Princeton; Oxford: Princeton University Press

Podoshen, J. S. (2013). Dark tourism motivations: Simulation, emotional contagion and topographic comparison. *Tourism Management*, 35, 263–271.

Preece, T., & Price, G. G. (2005). Motivations of participants in dark tourism: A case study of Port Arthur, Tasmania. In C. Ryan, S. Page, & M. Aitken (Eds.), *Taking tourism to the limits: Issues, concepts and managerial perspectives* (pp. 191–197). Oxford: Elsevier.

Rao, R. (2019). One Time, Many Times. *Millennium: Journal of International Studies* 2019, Vol. 47(2), 299–308.

Sather-Wagstaff, J. (2011). Heritage that hurts: tourists in the memoryscapes of September 11. *Heritage & Society*, 5:2, 281–283, DOI: 10.1179/hso.2012.5.2.281.

Seaton, T. (2009). Purposeful Otherness: Approaches to the Management of Thanatourism. In: R. Sharpley & P.R. Stone (Eds.), *The Darker Side of Travel:: The Theory and Practice of Dark Tourism* (pp. 75–109). . Aspect of Tourism Series (41). Bristol: Channel View Publications.

Schäfer, S. (2016). From Geisha Girls to the Atomic Bomb Dome: Dark Tourism and the Formation of Hiroshima Memory. *Tourist Studies*, Vol. 16(4), 351–366.

Sharpley, R., & Stone, P. R. (Eds.) (2009). *The Darker Side of Travel: The Theory and Practice of Dark Tourism*. Aspect of Tourism Series (41). Bristol: Channel View Publications.

Stone, P. R. (2006). A dark tourism spectrum: Towards a typology of death and macabre related tourist sites, attractions and exhibitions. *Tourism: An Interdisciplinary International Journal*, 54(2), 145–160.

Stone, P. R. (2009). Making Absent Death Present: Consuming Dark Tourism in Contemporary Society. In: R. Sharpley & P.R. Stone (Eds.),*The Darker Side of Travel: The Theory and Practice of Dark Tourism* (pp. 23–38) Aspect of Tourism Series (41). Bristol: Channel View Publications.

Stone, P. R. (2011). Dark tourism experiences: Mediating between life and death. In: R. Sharpley & P.R. Stone (Eds.), *Tourist experience: Contemporary perspectives* (pp. 21–27). Abingdon, Oxon: Routledge.

Stone, P. R. (2012) Dark tourism as „mortality capital": The case of Ground Zero and the significant other dead. In R.Sharpley & P.R.Stone (Eds.), *The Contemporary Tourist Experience: Concepts and Consequences* (pp. 30–45). Abingdon, Oxon: Routledge

Stone, P. R. (2013). Dark Tourism, Heterotopias and Post-Apocalyptic Places: The Case of Chernobyl. In L.White & E.Frew (Eds.), *Dark Tourism and Place Identity*. (pp. 29–46). Melbourne: Routledge.

Stone, P. R. (2016) *'A commodification of death' – Dark tourism and difficult Heritage*. 'Packaging up death and the dead' for the contemporary visitor economy: A dark tourism and heritage perspective, ESRC 'Encountering Corpses' Seminar Series 2014–2017. Lancaster Castle, Lancaster, 19 October.

Stawkowski, M.E. (2016) 'I am a radioactive mutant': Emergent biological subjectivities at Kazakhstan's Semipalatinsk Nuclear Test Site. *American Ethnologist* 43(1), 144–157.

Turner, B. (1976). The organizational and interorganizational development of disasters. *Administrative Science Quarterly*, 21(3), 378–397.

United Nations Scientific Committee on the Effects of Atomic Radiation (UNSCEAR), New York, NY (United States) (2000). *Sources and effects of ionizing radiation UNSCEAR 2000 report to the General Assembly, with scientific annexes.* Volume I: Sources. United Nations (UN): UN.

Van der Veen, M. (2013). After Fukushima: revisiting Chernobyl and the collapse of the Soviet Union. *Rethinking Marxism: A Journal of Economics, Culture & Society, 25,* 121–129.

Vaughan, D. (1999). The dark side of organizations: Mistake, misconduct, and disaster. *Annual Review of Sociology, 25,* 271–305.

Weaver, C.P., Moss, R.H., Ebi K.L., Gleick, P.H., Stern, P.C., Tebaldi, C., Wilson, R.S., Arvai, J.L. (2017). Reframing climate change assessments around risk: Recommendations for the US National Climate Assessment. *Environmental Research Letters,* 2017;12, 080201. doi: 10.1088/1748-9326/aa7494.

Webb, G. R. (2018). The Cultural Turn in Disaster Research: Understanding Resilience and Vulnerability through the Lens of Culture. In: H. Rodríguez, W. Donner & J. E. Trainor (Eds.), *Handbook of Disaster Research* (pp. 109–122). Second Edition. Cham, Switzerland: Springer.

Winderman, E. (2017) Times for birth: chronic and kairotic mediated temporalities in TLC's A Baby Story. *Feminist Media Studies, 17*:3, 347–361, DOI: 10.1080/14680777.2016.1192556

WNA, (2012) Chernobyl Accident 1986. Retrieved 19 May, 2019, from the World Nuclear Association website, http://www.world-nuclear.org/info/chernobyl/inf07.html

Wu, HC & Cheng, CC. (2018) What drives supportive intentions towards a dark tourism site? *International Journal of Tourism Research 20,* 458–474. https://doi.org/10.1002/jtr.2196

Yoshida, K., Bui, H. T., & Lee, T. J. (2016). Does tourism illuminate the darkness of Hiroshima and Nagasaki? *Journal of Destination Marketing & Management, 5,* 333–340.

Young, C., & Light, D., (2016). Interrogating spaces of and for the dead as 'alternative space': cemeteries, corpses and sites of Dark Tourism. *International Review of Social Research, 6* (2), 61–72.

Magdalena Banaszkiewicz

Fun in the Power Plant. Edutainment in the Chernobyl Exclusion Zone Tourism

Abstract: In the recent years, there has been a significant rise in the popularity of tours organized to the Chernobyl Exclusion Zone (CEZ). A visit paid to the Zone usually exceeds the basic understanding of the Zone raised on stereotypes and opens a new horizon of deeper exploration of the complexity of this site. The aim of the chapter is to depict the educational potential of the tours organized to the Zone (not necessarily limiting themselves to the issues connected simply with the nuclear energy). The particular attention will be paid to the tension between education and entertaining that is considered to be a fundamental facet of visitors engagement in the intellectual process. Presentation of this topic will be based on a content analysis of the programs, participant observation and interviews with the organizers.

Keywords: heritage, entertainment, education, Chernobyl, tourism

Introduction

In the recent years, there has been a rise in the popularity of tours organized to the Chernobyl Exclusion Zone (CEZ). The Zone is a displaced area under the strict control of the Ukrainian State responsible for its security. It is not only a site of memory, a physical space symbolizing a nodal event for Ukrainian memory and identity, but also a large nature reserve, where nature has been developing practically unhindered for over 30 years. While its presence in the global popular culture (i.e. video game "S.T.A.L.K.E.R.") strongly stimulates tourism imaginaries, the continued high levels of radioactivity of some areas and "post-apocalyptic" state of material object makes the experience of visiting the Zone both risky and exiting. A visit paid to the Zone usually exceeds the basic understanding of the Zone raised on stereotypes and opens a new horizon of deeper exploration of the complexity of this site. The aim of the chapter is to depict the educational potential of the tours organized to the Zone (not necessarily limiting themselves to the issues connected simply with the nuclear energy). The particular attention will be paid to the tension between education and entertaining that is considered to be a fundamental facet of visitors' engagement in the intellectual process.

Tourism in the Chernobyl Exclusion Zone in Numbers and in Tourism Studies

The CEZ has been established in Ukraine in an area with a radius of approximately 30 kilometers from the power plant, the territory most affected by radioactive waste after the Chernobyl disaster in 1986. It is an area the population must not live, no economic activity would be carried out, and no food can be produced[1]. However, already in the 90s the first visitors started to appear in the Zone after obtaining a special permission from the Ukrainian government. The situation changed significantly in 2011 when the area was opened for official visitors under the regulations of the Ukrainian State Agency on Exclusion Zone Management. Since then, it is observed that there is a dynamic increase in the number of visitors to the CEZ: 8,000 tourists visited the Zone in 2010, almost 18,000 in 2013, and 36,000 in 2016 during the 30th anniversary of the disaster (almost 25,000 of whom were non-residents). In 2017, the number of visitors to the Zone reached 50,000 and 63 thousand in 2018. From January until the end of May 2019, the Zone was visited by as many tourists as in jubilee 2016 (almost 36 thousand)[2]. It is believed that the spring tourism boom, particularly visible for non-residents, is due to the huge popularity of the HBO series "Chernobyl", which not only gained critical acclaim, but also spectacular ratings (the fifth and final episode of the series was viewed by more than 2 million people in the US alone (Welsch, 2019) as well as audience support (on IMDb, over 280,000 users gave the show an average rating of 9.6 stars out of 10, making it the highest rated TV show on the platform (Stolworthy, 2019).

In recent years, a number of papers focusing on various aspects of visiting the CEZ have been published. One of the first researchers who focused on the process of ruination of the abandoned city of Pripayat was Paul Dobraszczyk (2010). In the consecutive year Goatcher and Brundsen (2011) presented their study on the emotional encounters of visits to the Zone and propose to use the

---

[1] This area and its existing legal order are defined in the document: *'On the legal status of the territory which was contaminated by radioactive radiation as a result of the Chernobyl disaster* of 1991' (and subsequent changes), see Про правовий режим території, що зазнала радіоактивного забрудненння внаслідок Чорнобильської катастрофи, http://zakon0.rada.gov.ua/laws/show/791%D0%B0-12, Retrieved July. 10, 2019.

[2] ЗОСТАННЯ КІЛЬКОСТІ КІЛЬКОСТІ ВІДВІДУВАЧІВ ЗОНИ ВІДЧУЖЕННЯ В РОЗРІЗІ 1 КВАРТАЛУ 2018–2019, https://cotiz.org.ua/novyny/stat1kv/, Retrieved July. 10, 2019.

term "sublime" in order to depict the special psychological state of person being in the Zone (Hannam and Yankowska, 2017). The specific character of the CEZ attracted attention of Philip Stone who, in his seminal paper published from 2013, compared the unique exclusiveness of the Zone's space to the Foucault's concept of heterotopia (Stone 2013). The environmental degradation of the CEZ served as an argument for Yankowska and Hannam (2014, p. 932) to label visits to the CEZ both as dark tourism and toxic tourism, stressing that this form of exploration "can provide a strong educational experience, raising awareness about the current environmental issues and the polluted environmental conditions around us" (Hannam and Yankowska, 2014, p. 937). However, the space of the Zone is being consequently mythologized particularly, thanks to the popular culture (Banaszkiewicz and Duda, 2019) and the visual representations easily accessible in virtuality (Banaszkiewicz and Skinner, forthcoming). Mediatization undoubtedly contributes to th treatment of the CEZ "an open-air museum of dark legends" (Afanasiev and Afanasieva, 2018, p. 38), where authenticity of physical space is the subject of performative interventions (Banaszkiewicz, 2018) but also a dissonant heritage that can stimulate intercultural dialogue (Banaszkiewicz 2020). The above studies do not exhaust the scope of issues and research problems related to the intensive process of the Zone touristification, on the one hand, and the representation of the Chernobyl disaster and the CEZ area in culture on the other. The two areas permeate each other, mutually stimulating strategies for the interpretation of the post-catastrophic heritage.

Education and Entertainment in Tourism

An attempt of holistic explanation of the phenomenon of tourism can be reduced to two concepts. The first, recognizing tourism as a "secular pilgrimage" implies that a tourist is motivated by the search of meaning, discovering authenticity and transformation of a subject that offers different experience, seeing them in the search for pleasure and entertainment, thus allowing tourists detachment from everyday worries and responsibilities (Boorstin, 1977 Pfaffenberger 1983). The concept reaches into the roots of the humanistic reflection on the condition of man, perceived as a working being. As Władysław Okoń wrote, referring to Aristotle: "Fun is a consequence of work understood as fatigue. Where is work, there must be fun, because tiredness requires rest [. . .]" (Okoń, 1995, 60). Leisure time intended even for travel stands in opposition to economic activity and, therefore, is to give pleasure impossible to get while working, identified with the duty, seriousness and responsibility. According to Johann Huizinga

(1944), the author of a classic view of man as *homo ludens* – playful creature – this "unseriousness", which is a free action, is a quintessence of fun.

However, while according to Huizinga, fun can be a deadly serious matter and is simply a property of culture that cannot be assessed as good or bad, entertainment, that is simply just fun, is a product in consumer culture, and is threat to adults getting infantile of the scale on alarming proportions. James E. Combs (2010) and Neil Postman (2000), we are entering a new phase in human history, permeated by fun to such an extent that it can be described as a world of fun. Entertainment is a distinctive form of mass culture, which, at the same time, is a consumer culture. Therefore, there is no consumer culture without entertainment. Consequently, tourists are hedonistically oriented consumers, desirous for still new experience stimulating their emotions (Bauman 1996, Salazar 2010).

Tourism based on products that give pleasure and relaxation, referred to as 3S (sun, sea, sand), has become a designate of the most popular type of mass travels, i.e., rest at resorts in warm countries. However, with the tremendous development of tourism over the last 25 years, there has occurred, which was a part of global consumption trends, the needs of tourists, and consequently change in offers. Shifting significance from product (possessions) to experience (collecting experience) (Pine & Gilmore 1999), which has become a pillar of a cathedra of consumerism of the 21st century, found its reflection in travelling (Urry & Larsen 2011). In many cases, the 3S has been replaced by 3E (entertainment, education, excitement). Moving away from mass tourism based on environmental values, and consequently turning to individual tourism based on active involvement from a tourist, as well as increasing share of tourism product of cultural tourism category, is a characteristic trend observed globally in the recent years.

Entertainment has not been abandoned, but was put in a triad along with education, i.e., cognitive element and excitement, i.e. "experience", strong and positive emotional stimulus (Robinson & Picard 2016). In relation to the increasingly common educational strategy of gamification, tourism also recognizes elements characteristic of the game world, and which are referred to as 3F (fun, friends, feedback). It is not so much fun as entertainment among friends aimed at achieving further goals, which are counted and recorded in the form of feedback (e.g. points, badges, tokens) that becomes the axis of creating tourist programs, especially for young people. Not looking far, such an approach has its roots in the scout movement, where the basis of upbringing is action. The involvement of tourists in the process of experiencing travel more than just passive gaze can be referred to as a theoretical framework for the concepts of experience economy (Pine & Gilmore 1999, Urry & Larsen 2011) that puts in the center of exchange not products or services but experiences.

Highlighting the correlation between effectiveness of education and presence of elements of entertainment in cognitive process is by no means a new phenomenon, however, growing in importance in recent decades, at heritage sites that are tailored to tourists' needs for entertainment. It is worth to remember, that not only scenarios of exhibitions or "ludic" projects translate into the effectiveness of educational process. The spectrum of motivation is its indispensable element as it guides a recipient (visitor, spectator, tourist). According to research cited by John H. Falk and Lynn D. Dierking (2010, 79), better cognitive results are achieved, when a museum visitor has a high level of motivation: "As it would be expected, individuals voicing a strong educational motivation demonstrated significantly greater learning than did those expressing a low educational motivation. However, less expected, a similar relationship was found among those individuals voicing strong entertainment motivations. These significant differences were independent of the individual's expressed educational motivations." As it turns out, a pro-entertainment attitude results in better educational outcomes. When presenting a dissonant culture, it is particularly challenging to include solutions that allow to enjoy the fun. This is due to the fact that such heritage is not subject to harmonious interpretation. Ambivalence related to its perception requires people managing the heritage to be particularly delicate and intuition driven so that not to present a one-sided, subjective and over-simplified narrative of the past, in the name of striving for making heritage experience enjoyable.

Educational Nuclear Tourism in the CEZ

Currently, the trips to CEZ are organized by various tourist entities: both Ukrainian and foreign companies, mainly using the intermediation of local tour operators. It is hard to deny that such a dynamically developing Chernobyl tourism is already more and more mass in nature, which is supported by the Agency's recent activities (e.g. simplifications in the entry procedure, introduction of electronic tickets, adjustment of infrastructure to the needs of tourists) as well as unambiguous declarations of the new central authorities of Ukraine, which perceive the Zone as a tourist attraction of great potential[3]. Due to the

---

3   From the formal point of view, tourism activity as an economic activity for many years has been prohibited in the Zone. President Volodymyr Zelenskyon 10th July 2019 during the ceremony of handing over to Ukraine the construction protecting

needs of the market, the offer of tour operators is becoming more and more diversified. It is based, of course, on one-day excursions, the main point of which is a visit to Pripyat, but what is interesting, the biggest organizers of trips to the Zone try to shape their image as active heritage stakeholders and not only commercially oriented businesses. Naturally, a narrative about "the mission" can be an effective marketing tool, but the involvement in other projects related to the dissemination of knowledge about the Chernobyl brings objective educational fruits. The CHERNOBYLwel.com office cooperates with the National Museum of Chernobyl in Kiev and organizes the Chernobyling Festival[4]. Tour operators also organize graffiti cleaning and garbage collection actions in the Zone.

Educational profile of the activity is characteristic primarily for the enterprise called The Chornobyl Tour, also known as the Chernobyl Tour operating on the market since 2008. The name of this organizer appears in two language versions – Ukrainian language "Chornobyl Tour" (official name of the company) and English language "Chernobyl Tour" (used as a domain name and in promotional materials addressed to foreign tourists). This is an interesting example of how post-colonial geopolitics translates into tourism. The Anglophone world knows Chernobyl from the Russian language version ("Chernobyl"), not the Ukrainian ("Chornobyl") or Belarusian ("Kharnobyl"). Hence, for foreign visitors, this is the name of the company. Its co-founder and a person responsible for the scientific layer of the tour, including the training of guides, is Sergii Mirnyi – one of the liquidators of the consequences of the accident at the power plant, often appearing on the international arena as an

---

the old Sarcophagus over the 4th reactor of the so-called New Ark, he signed a decree which is to be the beginning of "the transformation of the exclusion zone into one of the points of development of the new Ukraine"

https://www.president.gov.ua/en/news/glava-derzhavi-pidpisav-ukaz-shodo-rozvitku-chornobilskoyi-z-56321 During his visit, the President called for an end to the corruption and bans on tourists and for the isolation zone to be turned into a future magnet for tourists and scientists.

4 Chernobyling is a three-day festival organized for the first time in 2017 in Slavutych. The idea is to bring Chernobyl issues closer to young people through concerts, seminars, meetings with interesting people and sightseeing (and also to revitalize the cultural life of the city). As the organizers declare, the income from the event is intended for self-residents. The official language of the festival is English. The second edition of the festival in 2018 is to be held in Kiev. See Chernobyling, https://www.facebook.com/Chernobyling.festival/

expert in Chernobyl matters[5]. As S. Mirnyi argues: "Back then we fought with physical, radiation contamination, and now we eliminate a different, informational kind: contamination of human brains by misconceptions and outright myths. In the Zone, radiation contamination is largely defeated, for it has been localized and REDUCED MORE THAN MILLION TIMES as compared with the first days of the disaster. But in human minds, in their thoughts, perception and imagination it continues to persist as "deadly dangerous" – as if the cleanup was never done. This causes enormous harm to the health and life of people and whole countries. So, in order to make radiation decontamination truly efficient – as it has turned out – one needs complement it with one more, informational cleanup. And, frankly, each time, when in the end of the day I with the group leave the Zone, I feel something similar to what I felt, driving out the column of radiation recon armiks back in 1986: that after my shift a bit more people have become safer, and the world slightly different – a bit cleaner and better a place." (Chernobyl Tour – Mirnyi, https://chernobyl-tour.com/sergii_mirnyi_en.html)

Mirnyi postulates that the commercial activities of the company should be treated only as one of the forms of combating "information contamination". (Chernobyl Tour - About, https://www.chernobyl-tour.com/about_us.html), which he repeatedly emphasizes in his interviews[6]. The Chornobyl Tour includes a research department, whose work is directly supervised by Mirnyi himself. His activity consists mainly in close cooperation with the media in the field of Chernobyl issues, and recently also in lobbying the political and social environment for the inclusion of the material heritage of the Zone (mainly the selected buildings of the Pripyat) on the UNESCO World Heritage List. The educational dimension is also to be characterized by trips to CEZ organized by

---

5 Sergii Mirnyi was a commander of radiation reconnaissance platoon in Chernobyl in 1986. Accept his involvement in the activities of the Chornobyl Tour as its scientific advisor, he is a writer and scriptwriter, and an internationally known expert in the Chernobyl Disaster and mitigation of ecological-social disasters. He is an author of books "Worse than radiation" and "7 odd Chernobyl stories" (Budapest: Bogar Kiado, 2001), "Chernobyl liquidators health as a psycho-social trauma" (Budapest: Bogar Kiado, 2001) and several dozen artistic and scientific publications, presentations at international conferences.

6 The following section on Chernobyl Tour activities is based on material collected from the 2017–2019 grant field research in Kiev and CEZ, in particular the in-depth interviews with Sergii Mirnyi Yaroslav Yaroslav Yemelianenko, Ann Merrill, Svitlana Priadko conducted in February 2019.

Chernobyl Tour. For this reason, every tour guide of the Chernobyl Tour Zion is bound by labour standards that can even be compared to corporate standards. After relevant training, a candidate for a guide takes an internal exam, which only entitles him to give a tour of the company's groups. What is more, the guides are obliged to use the substantive elaboration (type of script/trip scenario) and their work is checked by other guides and ghost clients.

All these measures are aimed at maintaining a high quality of service, although they also impose certain restrictions on the guides working for the company. In addition to the "standard" one-day program, which in its scope is relatively similar to the programs of competitors, Chernobyl Tour offers a deeper exploration program that can last from one to several days and includes places less obvious (e.g. fish and rodent scientific experimental base, Yaniv railway station, Paryshiv village, meeting with selfsettlers). In addition, interested parties may also book a private tour of the Chernobyl nuclear power plant (that includes construction site of the new confinement "Arch", mockup hall at the Administrative and Service Complex (ASC-1) of the PDO, "golden corridor" extending through the building of the PDO, the control panel of the reactor, turbine hall, reactor hall, memorial to Valery Khodemchuk, buried under the ruins of the reactor and the room with the main circulation pumps).

In search of new customers, Chernobyl Tour has also created an unusual, as for Ukrainian conditions, offer of study trips of strictly educational character. It is aimed primarily at foreign visitors recruited from the academic community. In order to develop educational programs, Mirnyi has partnered with Ann Merill, a US-based specialist in organizing international educational programs, coordinating academic projects, and with extensive experience in NGO's in the socioeconomic environment of Eastern Europe. The result of this cooperation is a program of specialist thematic excursions, which can be attended mainly by foreign students, mainly from science faculties. In 2018 the Chernobyl Tour organized visits to the Zone for about 85 students and professors from the USA, UK, Azerbaijan, Norway, the Netherlands, South Korea, Germany, and Japan. Their interests ranged from engineering (including nuclear, industrial, materials science), radioecology and radiobiology, social and ecological resilience, and "dark" tourism. Interestingly, the study groups are not limited to nuclear exploration, but also present lesser-known themes such as the heritage of WWII and the Cold War.

At this point it is worth emphasizing the importance of research department in the process of creating new tourist products. A good example is the latest thematic excursion organized by Chernobyl Tour in a weekly cycle since June 2019. The "HBO Chernobyl" TOUR series is Chernobyl's direct response to the dizzying popularity of the series "Chernobyl" and aims to "revealing to the secrets and real stories of the events that occurred". (https://chernobyl-tour.com/serial_chernobyl_hbo_tour_ukraine_en.html). During the tour, both "must see" points, which did not have much meaning for the series itself, but are the Amusement Park of Prypiat with the world-famous Ferris wheel or the lunch of power plant employees in the canteen of the Chernobyl Nuclear Power Plant, as well as authentic spaces were taken into account, which are the location of the series of events (the basement of the Chernobyl nuclear power plant, in which the liquidation headquarters was located in the first days after the explosion of the fourth power unit or, the fire department from which the first firefighters left for the scene, the medical unit of the city of Pripyat, which received the first victims, the legendary bridge of Pripyat town). There will also be an "entertainment" point (a ride in an armored patrol vehicle, in which the liquidators in 1986 made a radiation reconnaissance, making the first radiation contamination maps), but also a cognition of "the real stories of people whose characters were reflected at the series" I as promised by the organizers "some of them you can even meet in person".

What is important, before the new program was added to the offer, it was preceded by analytical work on the compatibility of the series with reality and social consultations with representatives of the Kyiv magistrate, among others. At this stage, a meeting with witnesses of the 1986 events, open to the general public, was particularly important, thanks to the personal contacts of Sergei Mirnyi.[7] Of course, there is no doubt that such events contribute to the increase of media recognition of the agency itself (which, shortly after the development of the new route, organized a special study tour for media representatives). At the same time, the use of the potential of popular culture for edutainment seems to be a very effective tool, as it reaches out to people who would most likely not benefit from a more scientific offer, as in the case of educational programs.

---

7   During a meeting held on June 5, 2019, the newly opened Chernobyl Hub, a club in the courtyard of the Chernobyl Tour office in Kiev, the gathered guests watched the last episode of the HBO series together and then took part in a discussion with guests: Sergei Paryszyn, who participated in the first meeting of the crisis staff in the bunker under the power plant management building, Alexei Breus – the operator of the fourth block, who began his shift on April 26, 1986. at 7 a.m., Alexei Ananenko – participant of the diving mission in flooded rooms under the reactor and Siergiey Mirnyi. The report from the event can be viewed at the link: Учасники аварії на ЧАЕС обговорють серіал CHERNOBYL HBO, https://www.youtube.com/watch?time_continue=9&v=2DR8QqnuzZE,

Regardless of the subject matter and length of the trip, Chernobyl Tour focuses on combining elements of education and entertainment, both at the level of content and in the realization of the programs themselves: knowledge is conveyed through anecdotes, universalization, and analogies, and guiding strategies are based on intensive dialogue with the audience, performative involvement of the visitors, provoking them to reflect and interpret themselves.

Educational trips for Chernobyl Tour are only a supplement to the basic offer, which is well characterized by the aforementioned acronym 3F. In turn, the Polish organizer of the "Zero Zone", which will be presented in this chapter as the second case study, oscillates much more on the border of science than entertainment. From the formal point of view, the Zero Zone can be classified as a tourism organiser (it has a relevant legal entity), although the website repeatedly emphasizes the untypical nature of the offer, created by a group of enthusiasts, former students of the Warsaw University of Technology, who in 2007 went to Chernobyl for the first time. After establishing cooperation with the Polish Nucleonic Society, they started to organize trips regularly and are currently the oldest Polish organizer of trips to CEZ. The main pilot and guide in the Zone is Dr. Marek Rabiński, a member of the Polish Nucleonic Society and an employee of the National Research Centre in Świerk, although the trips often involve other employees of the National Centre for Nuclear Research and people professionally professionally or amateurishly interested in nuclear energy. The local guide Sergey Akulinin, a former operator of reactor turbine no. 2 in Chernobyl Nuclear Power Plant and a participant in the accident liquidation action, is a permanent collaborator of the Zero Zone on the spot. As the tour operators emphasize, referring to the "scientific" profile of the tour guides, "this does not mean that the trip is only for scientists. The trip is for everyone (with small restrictions). The participation of researchers ensures that the trip is safe for health". (Zero Zone, https://strefazero.org/index.php?id=6#). The organizers, being aware that too much "science" can be a deterrent, try to cut the program so that its entertainment elements balance the educational element.

As in the case of thematic trips organized by Chernobyl Tour, Zero Zone trips are not one-day trips – participants spend an average of 2–3 days in CEZ and the program is much more flexible than in the case of "standard" trips. In the tab presenting the specificity of trips, very strongly (both verbally and visually, as can be seen in the attached illustration), highlights the difference between the "The Zero Zone" and other tour operators consisting in greater involvement of the tourist himself (see Fig. 1):

> TO NIE JEST TYPOWA WYCIECZKA Z BIUREM PODRÓŻY.
> U nas masz głos dotyczący kształtu wyjazdu.

Fig. 1: Extract from the description of the trip from the "Zero Zone" website (screenshot), source: Zero Zone, https://strefazero.org/?id=104&ArticleId=145, date of access: 16.07.2019

The organizers strongly dissociate themselves from the stereotypical image of a tourist trip: "With us you will definitely see more and feel the atmosphere that constantly accompanies today's employees of the closed Chernobyl zone. On our tour, you are an explorer, not a tourist."

Indeed, the program of the expedition is a framework and the organizers give a lot of freedom to the participants of the trip, not only by declaring on the website that "In Pripyat the participants receive from us maps of the city and explore what they want. If the situation does not require us to do so, we do not force anyone to walk in a group". (Zero Zone, https://strefazero.org/?id=104&ArticleId=145, access: 16.07.2019) Thanks to good agreements with local guides and Agency staff, the Zero Zone actually offers more exploration opportunities. The first difference between the basic program of a one-day trip organized by a mass organizer and the program of a standard group of Zero Zone (trips) is logistics – Zero Zone groups live in Slavutych and enter the Zone area by an extraterritorial employee train. During the three days of the visit, the group explores, among others V and VI power plant block and a mink research farm, St. Ilia's Church from 1789, barge dump, cemetery of old fortresses, a monument to partisans from World War II, Yanov station, and Burakivka equipment dump. As you can see, the program definitely extends the list of attractions in relation to what a one-day tourist visits, including also those places that are not directly related to the disaster itself, but present the heritage of Polesie before the outbreak.

In addition to daily exploration of the abandoned buildings of the Pripyat on your own, there is a possibility of visiting the interior of the Chernobyl Nuclear Power Plant, although admission to it is additionally paid from April 2018. The organizers of the trips try to enter the current of popularization of knowledge at all costs. A sightseeing program combining elements of scientific cognition (such as detailed mini lectures on the functioning of power plants or nuclear power engineering) and entertainment (urbex character of cognition of the Pripyat) is the implementation of this strategy. It is important, however, that the "Zero Zone" also tries to shape its image on this basis, cooperating with the creators of YouTube (MocnyVlog, Tube Riders, Urbex History, Potato) and Polish media giants (TVN, Onet, TVP). In the spring of 2019 she organized a trip with two famous YouTubers, Krzysztof Gonciarz and the creator of "Uwaga Naukowy Bełkot" ("Attention Scientific Gibberish"), which resulted in a documentary miniseries about the Zone (Chernobyl 2019, https://www.youtube.com/playlist?list=PLRIPPC8uohcdT1a-DW7pH4lWPTELCdeov) and two episodes of popular science blog (https://www.youtube.com/watch?v=BmPry7Gr0-M). Both the forum and the fanpage on Facebook "Zero Zone" have a mixed character combining historical facts, facts about nuclear power, news from the Zone, and information about trips to the Zone.

Summary and Conclusions

The idea behind this chapter was to present an educational offer of tour operators specializing in creating trips to the Chernobyl Exclusion Zone. There is no doubt that the Zone has become a popular tourist attraction in recent years. Regardless of its formal status, the interest of tourists, not only from year to year, but also from month to month, is growing, which makes both the space of the Exclusion Zone undergo changes and the range of tourism organizers' offer changes. On the one hand, the growing range of programs is intended to respond to the heterogeneous needs of tourists, and on the other hand, the creation of new products stimulates the number and profile of visitors. In the analysis of the case studies cited above, it was attempted to demonstrate that the sightseeing programs combine elements of education and entertainment to a varying degree. Firstly, it is a consequence of general trends observed in tourism (regardless of whether they are acronym 3E or 3F, the element of fun/entertainment is inalienable), secondly, the tourist potential of the Zone itself, whose heritage is not only limited to post-catastrophic value. The fact that both factors, or both sides of the same coin, i.e., supply and demand, are taken into account allows a better understanding of the Chernobyl Exclusion Zone tourism phenomenon. The nuclear issue, although it seems to be the basic thematic axis of the trip to the Zone, is one of many threads that can be explored in its space. In-situ experience offers great opportunities for interpretation with regard to the issues of nuclear energy, so it can be assumed that programs profiled in terms of nuclear tourism will continue to be developed.

Acknowledgements

The article is a result of a project financed by the National Science Centre in Poland (no. 2016/23/D/HS3/01960).

References

Afanasiev, O.E, & Afanasieva, A.V. (2018). Museums of dark mythologies in the tourist place. In *M. Korstanje, and B. George (Eds.) Virtual Traumascapes and Exploring the Roots of Dark Tourism* (pp. 26–49). IGI Global.

Banaszkiewicz, M. (2018). *Turystyka w miejscach kłopotliwego dziedzictwa.* Kraków: Jagiellonian University Press.

Banaszkiewicz, M., & Duda, A. (2019). To be a S.T.A.L.K.E.R. On architecture, computer games and tourist experience in the CEZ. In M. Gravari-Barbas, N. Graburn, and J-F. Staszak (Eds.). *Tourism Fictions, Simulacra and Virtualities.* (pp. 197–210). London-New York: Routledge.

Banaszkiewicz, M. (2020). Tourism and Dissonance: Towards a Dialogical Approach.In M. Gravari-Barbas (Ed.), *A Research Agenda for Heritage Tourism* (pp. 119–134). Cheltenham: Edward Elgar Publ.

Bauman, Z. (1996). From Pilgrim to Tourist - or A Short History of Identity. In S. Hall and P. du Gay (Eds.). *Questions of Cultural Identity* (pp. 18–36). London.

Boorstin, D. J. (1977). *The image: A guide to pseudo-events in America.* New York: Atheneum.

Combs, J. E. (2010). *Play world: The emergence of the new ludenic age.* Westport: Praeger.

Dobraszczyk, P. (2010) Petrified Ruin: Chernobyl, Pripyat and the Death of the City. *City*, 14(4), 370–389, doi:10.1080/13604813.2010.496190.

Falk, J. H., & Dierking, L. D. (2010). *Learning from museums.* Lanham: Rowman & Littlefield.

Goatcher, J., & Brunsden, V. (2011). Chernobyl and the Sublime Tourist. *Tourist Studies*, 11(2), 2011, 115–137., doi:10.1177/1468797611424956.

Huizinga, J. (1944). *Homo ludens.* London: Routledge.

Stone, P. (2013). Dark Tourism, Heterotopias and Post-Apocalyptic Places: The Case of Chernobyl. In L. White, E. Frew (Eds.). *Dark Tourism and Place Identity* (pp. 79–93). London-New York: Routledge.

Hannam, K., & Yankovska, G. (2017). You Can't Go Home Again – Only Visit: Memory, Trauma and Tourism at Chernobyl. In S. Marshall (Ed.).

Tourism and memories of home: migrants, displaced people, exiles and diasporic communities (pp. 53–68). Bristol: Channel View Publications.

Okoń, W. (1995). *Zabawa a rzeczywistość*. Warszawa.

Pfaffenberger, B. (1983). Serious Pilgrims and Frivolous Tourists: The Chimera of Tourism in the Pilgrimages of Sri Lanka. *Annals of Tourism Research*, 10 (1), 57–74.

Pine J., & Gilmore, J. (1999) *The Experience Economy: Work is Theatre an Every Business a Stage*. Boston: Harvard Business School Press.

Postman, N. (2000). *Amusing Ourselves to Death: Public Discourse in the Age of Show Business*. London: Penguin.

Robinson, M., & Picard D. (Eds.) (2016). *Emotion in Motion: Tourism, Affect and Transformation*.London: Routledge.

Salazar, N. (2010). *Envisioning Eden: Mobilizing Imaginaries in Tourism and Beyond*. Oxford: Berghahn Books.

Stolworthy, J. (2019) Chernobyl becomes the highest-rated TV series of all time on IMDb. *The Independent*. https://www.independent.co.uk/arts-entertainment/tv/news/chernobyl-hbo-imdb-rating-breaking-bad-sopranos-game-thrones-sopranos-a8942996.html.

Urry J., & Larsen J. (2011). *The Tourist Gaze 3.0*. London: Routledge.

Yankovska, G., & Hannam, K. (2014). Dark and toxic tourism in the CEZ. *Current Issues in Tourism*, 17 (10), 929–939.

Welch, A. (2019). Animal Kingdom,' 'Archer,' and 8 other shows double in cable Live +7 ratings for June 3–9. https://web.archive.org/web/20190624182237/https://tvbythenumbers.zap2it.com/dvr-ratings/cable-live-7-ratings-june-3-9-2019/, Accessed 16.02.2021

Lina Kaminskienė

What We Find Outdoors: Discovering Nuclear Tourism Through Educational Pathways

Abstract: Outdoor education can take place in a variety of contexts, and its realisation in specific spaces can contribute to the development of educational tourism in regions that have not traditionally been classified as tourist attractions. The study revealed that leveraging outdoor education with formal and non-formal education programmes not only promotes the tourist attraction of the visited sites and places, but also creates new jobs and boosts development of museums and other cultural institutions. The decommissioning processes of the Ignalina Nuclear Power Plant have a direct impact on the social, economic, and cultural development of the Ignalina region and the city of Visaginas; therefore, the development of educational tourism has high potential here. Contextual and region-focused specific curriculum can stimulate the development of nuclear tourism in the region, taking into account the geographical location of Visaginas and the specific environment formed by the Ignalina Nuclear Power Plant. Utilising educational potential of nuclear tourism has great potential to attract learners from all over Lithuania and the neighbouring countries, mainly Latvia, Belarus and Poland. Outdoor education has a variety of ways to be implemented by utilising different pedagogical scenarios such as landscape analysis, school journeys, action research, outdoor adventure activities, cultural journalism or field studies. Outdoor education also contributes to the development of new forms of educational structures such as outdoor schools and kindergartens, forestry schools, STEM centres and others.

Keywords: outdoor education, place-based education, learning environments, educational tourism

Introduction

In this chapter, learning environment outside the school and the classroom will be analysed and linked to educational tourism and place-based education. This discussion will help to review the educational potential of areas around Visaginas city and Ignalina Nuclear Power Plant. The educational pathways of the area are currently underexplored by educational researchers and school practitioners. This chapter strives to uncover diverse educational approaches relevant to place-based education directly and indirectly enhances the development of educational tourism.

Researchers still argue whether outdoor education is a synonym to place-based education, or they are completely separate concepts. The majority of authors agree that the idea behind different names is the same, but they are just labelled differently (e.g., Knapp, 2014), and only few consider that place-based education resulted in a historical development after such movements as nature study, outdoor education and environmental education. Sarivaara and Uusiautti (2018) argue that place-based education is closely related with outdoor and environmental education. They refer to Sobel (2004) who notes that place-based education connects the classroom and the community. Place-based education is also characterised as the pedagogy of place, which incorporates concepts of experiential education, community-based education, education for sustainability and environmental education (Sobel, 2004).

Outdoor education became particularly actualised after the publication of Louv's book *Last Child in the Woods: Saving our children from nature-deficit disorder* in 2005. Louv coined the term "nature-deficit disorder" with the aim to raise awareness that the contemporary generation of children spend less and less time outdoors and this situation contradicts to the foundational principle of being human. Louv (2008) noted that staying long time indoors causes harm and negative impact on physical and psychological health, and stressed the need for restorative environments.

Historically, the idea of outdoor education is not new, dating back to the 19th-century education; however, today outdoor education has developed new characteristics, and the idea of "outdoor" might be linked to real, virtual and imaginative places. Outdoor education can be defined as pedagogical activities taking place outside the classroom or the school surroundings. Other related and synonymous terms are: learning outside the classroom, *udeskole,* outdoor adventure education, heritage education, environmental education, field studies, etc. Researcher Fägerstam (2013) notes that two traditions could be identified regarding the outdoor education: the Anglo-Saxon and the Scandinavian tradition. The former relates outdoor education with adventurous experiences that focus on team-building and the development of leadership skills (Thorburn & Allison, 2010). It is also important to note that in the Anglo-Saxon tradition outdoor education is usually, yet, not always, associated with education that takes place in special educational centres, museums, science parks, etc.

The Scandinavian tradition of outdoor education is closely linked to formal education and is implemented as part of the formal curriculum. Fägerstam (2013, p. 1) claims that in the Scandinavian tradition "the term outdoor education most often involves school-based learning outside of the classroom, in the nearby natural or cultural landscape or on school grounds, often with a cross-curricular approach". The researcher mainly refers to the works of Jordet (2007, 2010), who studied outdoor education processes in the regular Norwegian primary schools. It is worth mentioning that outdoor education in Norway has a special name "uteskole", which means outdoor school. The research performed by Jordet (2010) allowed him to develop a model of school-based outdoor learning. The key aspects of the model are that school surroundings are used as a learning arena and as a source of knowledge.

Rickinson et al. (2004) distinguish three types of outdoor activities: fieldwork and outdoor visits, outdoor adventure education, and school grounds and community-based projects. Fieldwork and outdoor visits are close to what Scandinavian researchers call school-based outdoor learning in the sense that fieldwork is usually linked to a specific curriculum (it may be geography, history, math, literature, etc.); however, it usually takes place outside the school settings and involves visits and field studies in the nature centres, parks, etc. Outdoor adventure education focuses on adventurous activities taking place most often in natural environments and involves such activities as mountain climbing, canoeing, etc. The third group of outdoor education activities is linked to school grounds and community projects that take place in nearby places, not very distant from the school.

The theoretical foundations of outdoor education are closely related to what Gruenewald (2003) defined as place-based education with the following theories of experiential learning, ecofeminism, problem-based learning, sociocultural theories of learning, Bruner's free discovery learning theory, and others. Gruenewald (2003, p. 620) claims that a number of contemporary educational strategies rely on the concept of the place: "Experiential learning, context-based learning, problem-posing education, outdoor education, environmental/ecological education, bioregional education, natural history, critical pedagogy, service learning, community-based education, Native American education—all of these approaches to education tend to include engagement with local settings."

Researchers Siskar and Theobald (2008) claim that not only the place, but also the community are the dominant concepts and approaches in contemporary education. Even though the community and the place are not identical concepts, it is most important that community cannot exist without a place. Siskar and Theobald (2008, p. 70) discuss the distinctions of the two concepts, but they both "represent a legitimate path to a much more substantive definition of what constitutes an education—or an educated person". Even though the idea of learning outdoors is not new at all, however, numerous discussions on how school curriculum could be linked and realised in various settings are still continuing. There are open questions whether outdoor learning is limited to any geographical area, whether it is a part of formal curriculum or is mainly implemented as non-formal curriculum. Moreover, is it a mono-sectorial (purely educational) or multi-sectorial issue, when discussing about educational tourism? Whether outdoor education is somehow influenced by the tendencies developed by educational centres, museums, other places of touristic attraction, or is it a vice versa process that tourism industry discovered underexplored potential of education relating it to tourism development?

Following this discussion, it would be good to return to Gruenewald's (2003) observations that a multidisciplinary analysis of the place reveals high pedagogical potential as the place may unlock a variety of historical, cultural, biological and other diversities. Linking or even accepting outdoor education as a part of formal education allows the schools to become an equal player of the local context as well as to contextualise learning in real surroundings as opposing to simulative environments developed in the class. Gruenewald's (2003) work still remains highly actualised nowadays in many countries, going through national curriculum reforms, such as the Baltic sea region countries (Lithuania, Latvia), Eastern and Central Europe (Ukraine) and Asia (Kazakhstan, Georgia, Armenia), some years earlier in Scotland (Thorburn & Allison, 2010) and Finland (the New National Core Curriculum for Basic Education, 2016). For example, Salminen et al. (2016) observe that in Finland in 2010, the statistical data showed that schoolchildren aged 10–14 (grades 4–8) were the most frequent visitors in museums and other type of exhibitions. Researchers indicate that usually school groups are focused on learning subjects from humanities, arts and natural sciences. It is very important that some significant changes might be achieved by applying changes in the national curriculum, what, for example, happened with history studies. Similarly like in Lithuania, school children start learning history as a separate subject from the 5th grade, and accordingly, national museums receive much higher numbers of visitors from this age cohort. After the changes in the national curriculum from 2016, it was decided to start history studies at an earlier age, which museums and other public institutions had to accept and adapt to the new situation.

The dimensions of place-based education, including the perceptual, sociological, ideological, political and ecological aspects, are worth deeper consideration for the countries while reviewing their school curriculum. The idea of extending the notion of pedagogy and accountability towards the place is going as a red line in Gruenewald's work, and particularly today it is obvious that modern pedagogy needs new inspirations and stimulus to respond to a very demanding new generation, preventing them from being educated in an isolated school environment. Moreover, as Cuthbertson et al. (2004) note, there are concerns that technologies put additional barriers between humans and the natural environment, which also raises new questions how technologies could be successfully exploited to bring students and environments closer.

The findings of scientific literature on outdoor education brings forth a critical analysis of current pedagogies and approaches to implement outdoor education and raises discussion on its interrelation with educational tourism. This chapter illuminates dimensions of place-based education and outdoor education and situates them in the array of contemporary educational theories and methods.

Pedagogical Approaches in Implementing Outdoor Education

Naturally, all these conceptual discussions lead to questions how outdoor education can be successfully implemented in educational institutions. The analysis of scientific literature reveals a variety of approaches and cases.

Pedagogies of place: natural history

Natural history as an educational tradition dates back to the Victorian and Edwardian England. In the USA, one of the most famous publications for teachers appeared in 1911, when Anna Botsford Comstock published the book *Handbook of Nature Study*. Natural history, as Gruenewald (2003) notes, allowed to keep close school-nature (place) connection while implementing the curriculum. Surprisingly, now in the 21st century, this connection in real implementation of the school curriculum for many teachers looks a complicated issue. Natural history is focused on deeper investigation of nature, landscape, biological and cultural diversity. It is not limited to rural areas but can be successfully implemented as pedagogical excursions in urban areas.

A team of UK researchers (Rickinson et al., 2004) analysed studies on outdoor education carried out in the period of 1993–2003. They also noted that historical development of outdoor education was well established in the early 20th century schools. One of the botanical educators was Dr Lilian Clarke who taught from 1896 to 1926, developed innovative teaching practices in the design and use of school gardens. The main principles behind the contemporary outdoor education were laid down in these "school gardens", and this type of education was aimed at promulgating the use of the "outdoor classroom"; stimulating a proactive view of learners' creating their own textbooks from "hands-on" work in the garden; recognising that teachers and learners contribute to the pace of the lesson; and documenting the teaching to share with others (Rickinson et al., 2004). The contemporary educational traditions associated with natural history are field studies, outdoor adventure activities and landscape analysis.

Landscape analysis

The landscape analysis method was developed as an educational method by the Place-based Landscape Analysis and Community Education (PLACE) programme, an innovative programme by the University of Vermont and Shelburne Farms. The main idea behind this method is to explore the region through the analysis of physical landscape, cultural landscape and ecological landscape. The method allows students to immerse in local heritage, cultures, landscapes, opportunities and experiences, using these as a foundation for the study of language arts, mathematics, social studies, science and other subjects across the curriculum. Landscape analysis is one of the approaches to implement place-based education; moreover, it also helps to develop learning through participation in service projects for the local schools and communities. The landscape analysis starts from physical analysis of the landscape, which involves not only geographical or topographical analysis of the landscape, but also urban structures, human and non-human characters. The physical, cultural and ecological aspects of the land are strongly linked and intertwined. Physical landscape analysis helps to understand the origins of flora and fauna of the region, distribution of population in the landscape. For example, Ignalina Nuclear Power Plant region in Lithuania is characterised by a big number of lakes, forests and a relatively small number of population in the area. The cultural landscape of the region always has manifold meanings, cultural artefacts and texts that need to be analysed. The cultural landscape analysis is not limited to centric objects (for example, Power Plant); it requires developing skills for identifying and analysing non-centric objects in the area and learning to unlock and understand their meaning (different living and industrial houses, new and neglected, small farms, etc.). Cultural landscape analysis focuses on a variety of objects and stimulates investigating causes why they have been constructed and what was their main purpose.

The third phase is related to ecological analysis. Ecological analysis is mainly focused to investigate interrelation of human activity with non-human structures, and thus, it strives to analyse how human beings affected and changed the regional landscape over time. Ecological analysis has many connections with environmental and sustainable education as it raises awareness about the interconnections in the regional ecosystems.

The advantage of the landscape analysis is that one may not implement it in sequential steps, however, going through all phases (performing physical, cultural and ecological landscape analysis) would definitely contribute to a whole understanding of the region by unlocking diverse natural and human activity treasures.

School journeys (excursions)

At the end of the 19th century, the so-called school excursions or school journals became popular, and the person who contributed to the spread of this concept was Catherine Dodd. She supported the idea that educators should enlarge the environment that children could experience. Dodd brought children to various excursions into rural Derbyshire. The new education approach inspired many teachers, and up to now, school journeys have been among the most common outdoor education methods. The main challenge remains that still a lot of teachers limit outdoor education to school journeys and do not sufficiently link this activity with the formal curriculum. In this way, school journeys become simply adventure activities or activities just for amusement and fun without a clear conceptual line and intentions to develop specific abilities and competences. Education scholars and practitioners (Jovaiša, 1993, Garalis, Švagždienė, Liesionienė, 2008) note that school excursions, also called educational journeys, can be of different types. Depending on the purpose, they could be classified into educational journeys with the aim to introduce to the new subjects in the curriculum; consolidating the covered topics, providing overview and complex. The researchers also clearly identify strong connections between school journeys and tourism; however, they stress that teachers and learners should invest sufficient time to prepare for a journey. The preparatory activities usually involve reading and analysing the object(s) to be visited, as well as preparation of special instruments or tools used during the journey (for example, questions to be answered and templates for data recording). School journeys can significantly contribute to strengthening learners' motivation and help to achieve the planned learning aims. During the last decades, educational programmes have been developed in museums, theatres, information centres, etc., in order to address diverse needs of learners of different age groups. School journeys have a number of characteristics that are common to field studies.

Field studies

Field studies in the UK started rising from 1943, with the establishment of the Council for the Promotion of Field Studies, and since then a network of

Fig. 1: Learning in a simulative pilot cabin at Aviation museum in Kaunas (picture from personal album of L. Kaminskienė).

field studies centres has been developed. Zaragoza and Fraser (2017) refer to Harington (2001) who differentiates formal and informal learning environments. The latter is associated with learning outside the classroom, in museums, zoos and specialised science education centres. There are various definitions of field studies. Some of them are close to the concept of school journeys (Krepel, Durrall, 1981), pointing out that field studies are school trips with an educational intent, during which students interact with the setting, displays, and exhibits to gain an experiential connection to the ideas, concepts and subject matter. Obadiora (2016) provides rather a general definition and suggests that a field trip is any teaching and learning process carried out by a group of people outside of the classroom environment. We consider that Zaragoza and Fraser (2017), Harington (2001) and Rickinson et al. (2004) provide a more precise definition of the field studies as they link field studies with a specific curriculum (most often science education) to be implemented in special education centres. Some interesting examples could be found in the Aviation museum in Kaunas (Fig. 1).

Rennie (2007), as well as Behrendt, Franklin (2014), distinguish two types of field trips (they are actually very much the same as school journeys): formal and informal. Formal field trips are usually organised to museums, centres and government agencies, and bear an initially pre-organised character. In most typical cases, these programmes are implemented by the staff of the centre or the museum, thus bringing the teacher and students into a similar position.

Moreover, the experience of such field trips is rather similar to all due to the formal programme the students go through. Researchers, however, stress that informal field trips open very wide possibilities to learning and in such a case experience is very diverse as there are no pre-arranged programmes and every student, as well as the teacher, become co-creators of the unique experience.

Outdoor adventure activities

Attarian (2003) claims that the origin of adventure programmes can be traced to organised camping, environmental and experiential education movements. Rickinson et al. (2004) notice that outdoor education for many researchers and practitioners is linked to adventurous activities such as mountaineering, climbing, orienteering and canoeing.

Attarian (2003), referring to Hale (1975), indicates that by the mid-1970s, over 190 adventure programmes were operating throughout the United States, with over half of the programmes found in college and university settings. Since then, the number of outdoor adventure programmes has been continuously growing. For example, quite recently, Allan and McKenna (2019) noted that outdoor adventure programmes have been implemented in higher education institutions to reduce students' resilience. In the UK, the outdoor adventure activities date back to the 1920s and have had significant associations with the military aims, particularly for boys, preparing for the challenges of the British Empire.

Action research

Action research, according to Gruenewald (2003), brings teachers and students to a situation which requires rethinking the existing practices and initiating specific action. Gruenewald (2003) treats action research as a democratic process that also develops a sense of ownership of the place as well as socially active personalities, not indifferent to what is happening around in the communities. The most important in this approach is that students are facilitated through the learning process that requires identification and analysis of the situation, the problem and the context, and while initiating a specific action, they experience it individually and collaboratively. In many aspects, this educational approach is very close to another educational tradition, known as service learning. Action research allows students to realise that places are also results of social constructions, and thus planning actions that may bring change or ensure conservation of the situation puts students and teachers of the school in close contacts with local communities. What is also significant in this educational process is that the role of the teacher is changed. Beames et al. (2017) states that outdoor education puts teachers and students in similar positions and opens up new collaborative possibilities and co-creative practices. Action research as one of the outdoor education approaches is flexible and open to combine other pedagogies, including natural history and cultural journalism. Thus, the teacher's role is to facilitate the reflection and investigation process, however, without usurping a dominating position.

Cultural journalism

The purpose of cultural journalism, as Gruenewald (2003) explains, is to create connections between teachers, students and the cultural life of the communities. The popular Foxfire programme has initiated schools to bring children to communities for gathering stories, interviews about local cultural life and producing journalistic pieces such as articles, journals and books.

Cultural journalism has a strong phenomenological background as it allows children to understand and discover how cultural life is developed, perceived and affected by local people; what are culture and people interactions; and how cultural traditions, cultural identities, have been developed over the time.

Graham (2007) notes that cultural journalism mediates interconnections between the learners, teachers and cultures within the community. From his point of view, cultural journalism is a powerful tool or media to analyse and understand local cultures and places that resulted through human activities. Moreover, cultural journalism does not only rely on "verbal" and communicative acts, but also involves analysis of visual culture artefacts, objects and commodities of everyday life of people. Graham (2007, p. 382) provides very interesting cases how cultural journalism could be successfully used in art education:

> *As an example, high school students in a diverse suburban community created a photographic documentation of the stories of local immigrants through studio portrait photography and interviews. The teachers introduced students to important issues of multiculturalism, social justice, and documentation through films such as Born into Brothels (2004) and El Norte (1994). The students learned the technical aspects of studio photography as well as various approaches to conducting an interview. The project took the students into places in the school and in the community where they had never gone before. The students' personal journeys were given a public audience when the photographs were exhibited in the town library, accompanied by excerpts from the interviews. Their carefully crafted work honored the experiences of people whose contributions and voices are sometimes silent.*

It is not only urban areas that potentially unlock a huge variety of material for the learners, but also natural environments with, for example, ecologically sustainable patterns from indigenous local cultures.

Outdoor education and problem-based learning

We can argue that problem-based learning is more than just a method; it could be better defined as a pedagogical strategy. Outdoor education is closely related with problem-based learning, which continues to remain one of the most popular methods and approaches of learning in various educational levels, starting from early childhood education and leading to higher education. Problem-based learning is closely interrelated with such pedagogical strategies as inquiry-based learning, service learning, sustainable education, etc. It is extremely useful for outdoor education activities, notwithstanding where it is organised: in the school settings or places outside the school environment. Problem-based learning is defined as a learning process that is organised in small groups and the purpose of which is to solve the problem through discussion (Wood, 2004); it is the process that takes place in a group collaboration in order to find out as much information as possible, by applying new knowledge to solve the problem. Problem-based learning is an interaction between the learners and the teacher, who works as a tutor, as a facilitator of the learning process. The strength of this method is that it leads learners towards self-directed learning and aims to engage them in the problem-solving process encouraging to identify knowledge gaps; this is a learner-centred learning and based on continuous discussion to solve the problem (Barell, 2007). According to Graff and Kolmos (2006), problem-based learning is effective because it covers all processes within the organisation, helps to anticipate potential threats, and enables them to address these threats timely and properly.

The goal of problem-based learning is manifold. O'Brien and Caroll (2015) present several key goals, starting from developing new skills, learning to work in teams, increasing learning motivation, and deepening problem-solving skills. Through these practices, the learning process, which involves developing critical thinking, acquiring analytical and evaluation skills, developing tolerance, accepting different approaches and perspectives, becomes a natural process.

The typical phases of problem learning involve state-of-the-art analysis, including clarification of concepts, existing knowledge and clear definition of the problem; then it leads to metacognitive processes, which are enhanced by brainstorming, group discussion, mind-mapping and other collaborative methods. After these preparatory and "warming up" phases, problem-based methodology is constructed to stimulate analysis of possible solutions of the problem, analysis of alternatives, and analysis and discussion of new knowledge that is needed to solve the problem. The cycle is finalised by meta-reflection activities and systematisation of the applied knowledge and solutions. Typical steps in the implementation of the problem-based learning could be organised in the following sequence:

- *Exploring concepts* that are unknown, incomprehensible for learners. Learning of/about the problem presented by the teacher involves identification and clarification of the main concepts so that unawareness would not impede proper understanding of the problem.
- *Finding the cause of the problem*. At this stage, the learners are faced with a problematic challenge that encourages them to raise questions and become more familiar with the problem. Learners need to make a number of hypotheses to predict the causes of the analysed problem.
- *Using the 'brainstorming' approach* – Learners use their existing knowledge to identify possible solutions to the problem. Based on their past experience, they need to identify where they lack knowledge to fully solve the problem. Possible solutions are recorded (visualised, reported, etc.) and used for further processing.
- *Second and third step review* is needed for each member of the group to analyse, systematise the results of the second and third steps to create a solution system.
- *Formulation of learning objectives*. The teacher/tutor must ensure that the learning objectives are realistic, achievable, clear, specific and relevant to the possible solution of the problem.
- *Independent work of group members*. Learners look for relevant information related to learning tasks. The information must also be reliable and practically applicable for solving the problem.
- *Systematisation of results and meta-reflection* is the last phase, which facilitates the members of the group to share the information they found while working independently. The resulting information is systematised and combined into a single solution that will be applied to the problem. After the whole process, the learners talk about their experiences, the positive and negative aspects, what they have learned, and what they have acquired to use in future situations and contexts.

When solving complex issues, it is important to look for a logical and reasoned solution to the problem, which means that there is a need for continuous analysis, which is achieved by already available knowledge. This learning is like a closed circle, forcing new knowledge to spin when a problem is solved, because when new knowledge is applied again and again, problems arise, which make the process permanent, unstoppable. It can also be noticed that problem-based learning encourages constant thinking, as each part of the process requires new ideas and thoughts; in other words, the process liberates the person and promotes cognition. The process only reaffirms that knowledge is at the heart of problem-based learning, as learners who are unable to apply knowledge cannot implement and participate in the process as full-fledged participants. Learning through problem-based learning is not only about the process, but also about the problem. The problem selection process is a responsible moment because the problem has to meet the needs and experience of the learner group. Working collaboratively on a problem may cause a variety of challenges, particularly when learners, meeting for the first time, are often distrustful and reluctant to cooperate. Therefore, it is necessary to clarify the relationships between learners and their experiences (O'Brien and Carroll, 2015). The chosen problems must be complex, potentially produce a measurable impact on the individuals and the organisation, and relevant to the entire learner group so that the learners have the basic knowledge and ability to analyse the problem; important to the learners' work and useful for their future. In other words, the problem must be qualitative, thoughtful and real, in order to become a learning stimulus and a source of new knowledge.

Problem-based learning has its advantages and disadvantages. Knowledge that is acquired during problem-based learning is not superficial as the whole learning process is focused on developing the ability to learn independently – learning is organised and driven by the learners. Problem-based learning is also useful for the development of collaborative skills – knowledge sharing, teamwork, ongoing discussion; it stimulates more active learning – learning which is not passive because it requires communication, discussion and critical analysis. This pedagogical strategy helps to improve time planning skills: problem(s) should be solved within a limited scope of time. It also improves managerial abilities – each member of the group must test the position of the leader, supervise the process, manage the time and encourage the involvement of each participant. Problem-based learning enhances lifelong learning competence – the acquired knowledge is not short-term, it is future-oriented, because the learners' knowledge covers a broad context. In conclusion, the stated benefits promote students' responsibility and a flexible approach to failure by providing opportunities to learn from ones' own mistakes.

Problem-based learning has also weaknesses related to the risk that the learners' discussion can become formal, without trying too deeply to solve the problem. Insufficient experience in this type of pedagogy may lead to the situation when a learning process become superficial or dominated by the teacher or one or more learners – other members of the group are not encouraged and involved. Yeo (2008) states that lack of experience, skills and knowledge impedes the proper implementation of the process, since problem-based learning requires both – knowledge and skills – to properly implement it. Similarly, learners themselves may become barriers to the quality of the process, as individuals may tend to dominate, control and compete, thus preventing others from entering the process and learning. Competition, if not controlled and properly monitored by the teacher, can disrupt the learning environment and prevent others from achieving positive and rewarding results.

In summarising the peculiarities of problem-based learning, one can say that it is the learning that reduces alienation among learners because all individuals are involved in the process of empowering them to become active. Constant reflection on the problem affects the person individually, which means that he/she cannot be passive during problem-based learning and thus seeks to make the process and the existing activities meaningful. Problem-based learning does not exclude the personal experience that is already available, as the key is to integrate learning into everyday activities so that individuals can be satisfied with the learning process, not just the end result. Stability also prevails in this model of learning, as individuals are equal partners in the process; there are no different roles that could cause conflict in the future. The analysis of problem-based learning also distinguished the importance of identifying knowledge gaps, as each learning session must add new knowledge to the person, important for continuous progress.

Impact of Outdoor Education: Cognitive, Affective, Social/Interpersonal and Physical/Behavioural

Numerous studies suggest that outdoor education linked with community-based education brings positive impacts and shows much higher students' engagement and motivation in the learning process. Beames et al. (2017) speaks about meaningful engagement, which is nowadays one of the main challenges for educators. There are studies showing that outdoor education contributes well to students' competences, including analytical capacities, problem solving, leadership, team working, decision-making, etc. Beames et al. (2017) claims that outdoor education might bring students to real life situations that demand psycho-motor, cognitive and socio-affective efforts to be employed, and subsequently meaningful engagement develops with each other, communities, environments, objects, etc. Similarly, Jordet's (2010) model of school-based outdoor learning clearly shows the links of bodily, sensual and cognitive processes, which take place while students are actively exploring phenomena. Moreover, social and communicative factors are extremely important, as the learning experience should be articulated and communicated, thus developing students' social capacities and bringing education to widen a community context in general. In favour of place-based education, Gruenewald (2003) and Thornburn and Allison (2010) also speak about identity development, rethinking of values, preferences, development of global and local thinking, cultural awareness, broadening understanding of power relations and regional (bioregional) dimensions.

Surface (2016) addresses to the work of Robert Marzano, who cited over forty studies from the 1970s through 2002, which proved that a positive motivation is supported when the learning process integrates and converges messages from homes, communities and schools. Coleman and Hoffer (1987) claim that in such cases students' academic performance and achievement improve and motivate their further learning. It is important to say that this results not purely from the use of different learning environments, but diverse communities as well. These findings support constructivists' ideas that outdoor education, or in a more general sense – place-based education, integrates places and communities, thus bringing students to meaningful construction and understanding of the surrounding world.

Rickinson et al. (2004) analysed research studies regarding impacts of outdoor education. They grouped these impacts into four major categories: cognitive (knowledge, understanding and other academic outcomes), affective (attitudes, values, beliefs), interpersonal/social (communication, leadership, teamwork) and physical/behavioural (physical fitness, personal behaviours, etc.). Nundy (2001, p. 4) indicated that three major benefits could be associated with fieldwork: (i) learners involved in fieldwork develop their self-confidence, there is a positive impact on long-term memory, recall, knowledge and understanding; (ii) real tasks during fieldwork contribute to the personal development of pupils, enhancing qualities of leadership, perseverance, reliability, initiative, co-operation and confidence, pupils become motivated; iii) fieldwork also reinforces links between the affective and the cognitive, each influencing the other and providing a bridge to a higher order learning. As Nundy (2001) points out, pupils who are learning outdoors and undertaking real tasks that are self-motivating learn at enhanced levels compared to pupils who learn only within a classroom context.

Many research studies indicate increased motivation of learners, for example, Zaragosa, Fraser (2017) analysed how outdoor education made impact on the motivation of learners with different English proficiency and sex. Their research shows that in most cases the learning motivation was much higher outdoors as compared to the same activities in the classroom. Another research, carried out by Cwikla, Lasalle and Wilner (2009), highlighted that the eighth grade students who raised their interest in science during field trips and other related outdoor activities were more likely to take careers in science.

Research indicates that outdoor education has positive impacts on pupils' creativity. McAnally et al. (2018) assessed the effects of an outdoor education programme with no or limited access to electronic media among 14 year-old boys. Researchers compared creative thinking, socio-emotional wellbeing and materialism with their peers attending regular classes at their normal school. Boys were assessed twice, and the results showed that boys in the special programme outperformed those in regular classes on a creative thinking task. Fig. 2 summarises the reviewed literature, which suggests that in most cases, outdoor education brings positive impacts to the holistic development of personality, developing different domains, including five main domains such as cognitive, social, affective (emotional), behavioural (physical) and personal.

In the last years, several studies have been performed in Finland, which also focused on the impacts of outdoor education and how activities out of the classroom support the development of the 21st-century skills: creativity, critical thinking, communication and collaboration (studies performed by University of Helsinki, University of Eastern Finland and other institutions). The data support evidence that learning in different environments has an impact on the development of the person's identity and metacognitive skills. Salminen et al. (2016) presume that this might be the reason why it is quite a challenge to measure individual learning outputs outside the classroom settings.

Outdoor Education and Educational Tourism

Similarly to outdoor education, the definitions of educational tourism also vary. One of the definitions, proposed by Ritchie et al. (2003), associates educational tourism with the desire to learn, and learning might be a primary or secondary motivator to travel. Educational tourism can be adult study tours, international and domestic university and school students' travel, including language schools, school excursions and exchange programmes (Ritchie et al.,

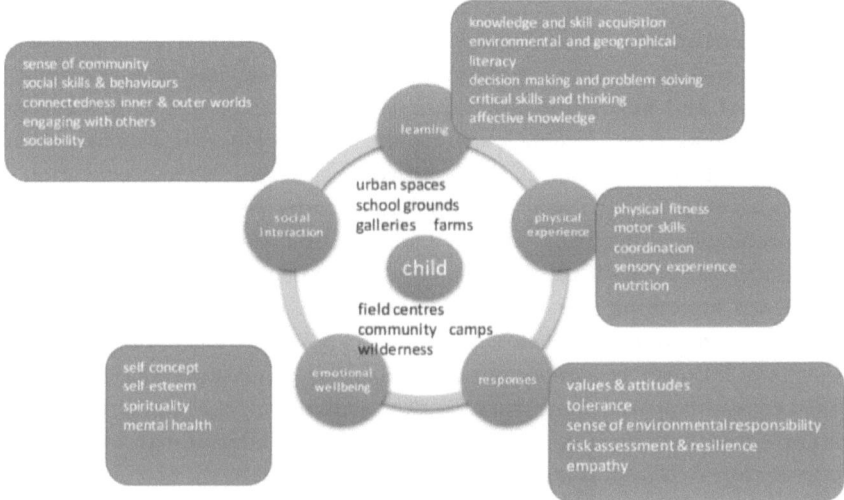

Fig. 2: Positive impacts in the five domains of child development (Kaminskienė, 2020, adapted from Malone, 2008)

2003, p. 18). Educational tourism is proposed to be analysed and understood through a segmentation process of tourism and education fields. Pitman et al. (2010, p. 220) suggest avoiding segmentation and consider educational tourism as specifically organised tours which promote an intentional and structured learning experience. Thus, according to Pitman et al. (2010), learning is a key component. Providing an overview of conceptual discussions of the definitions and components of educational tourism, McGladdery (2016) proposed a process model of educational tourism, which views educational tourism as a transformative experience. McGladdery and Lubbe (2017) argue that effective learning might occur when clearly defined and appropriate learning-stage outcomes (cognitive, affective and behavioural) of the process are developed.

A literature review of outdoor education and educational tourism helps to identify similar characteristics and even definitions of both concepts. However, taking into consideration that educational tourism involves at least two segments – education and tourism, it would not be possible to put educational tourism and outdoor education on equal sides. Outdoor education is one of the forms how place-based education is implemented, and has a very strong idea of education outside the school/classroom. Educational tourism, however, is not so much focused on activities outside or inside, but rather on the unique learning experience. Three main components link outdoor education and educational tourism: place, learning and curriculum, both formal and informal.

Educational tourism has many similarities with school journeys, excursions, field studies and field trips; however, it is a broader concept as it may also involve students' travel, language schools, etc. Historically, educational tourism has more links to education of the bourgeoisie, whereas outdoor education is very much linked with natural history and gender issues. Notwithstanding these ontological differences, outdoor education and educational tourism focus on specific real/virtual/imaginative places, where people have to travel/visit or experience something new. From this point of view, both outdoor education and educational tourism are closely linked to a broader concept of place-based education; however, we cannot put educational tourism on the same line with other educational traditions, such as experiential learning, sustainable education, contextual learning, etc., for the reason that educational tourism is a trans-sectorial concept linking education and tourism.

From the pedagogical point of view, outdoor education and educational tourism are important while providing diverse contexts of learning, and thus contributing to the realisation of specific curriculum. Outdoor education and educational tourism, as research suggest, have similar impacts on participants and result in most cases in positive changes in cognitive, affective and physical/behavioural domains. Fig. 3 illustrates links of place-based education and educational tourism that go through all dimensions, including theoretical backgrounds, educational traditions, schools and pedagogies.

For further development of educational tourism, it would be useful to refer to three outdoor educational landscape approaches proposed by Sandell and Öhman (2013): active domination, active adaptation and passive adaptation. Even though these strategies are directly linked to sustainability and environment concern, they could be potentially expanded while exploiting possibilities of urban areas, villages, museums, etc. Active domination is a strategy that treats the existing landscape and environment as "a factory" and should be adapted by using various facilities and measures by creating the necessary infrastructures or providing and expanding activities/services that optimise the best use of the place. Active adaptation is somehow similar to domination but is limited and subordinate to the landscape (time of the year, etc.). Passive adaptation implies studying and adapting the landscape, but without major interventions or changes to it.

Other strategies highlight the importance of integrating arts in outdoor education. Grimwood et al. (2018) describe that urban outdoor educational

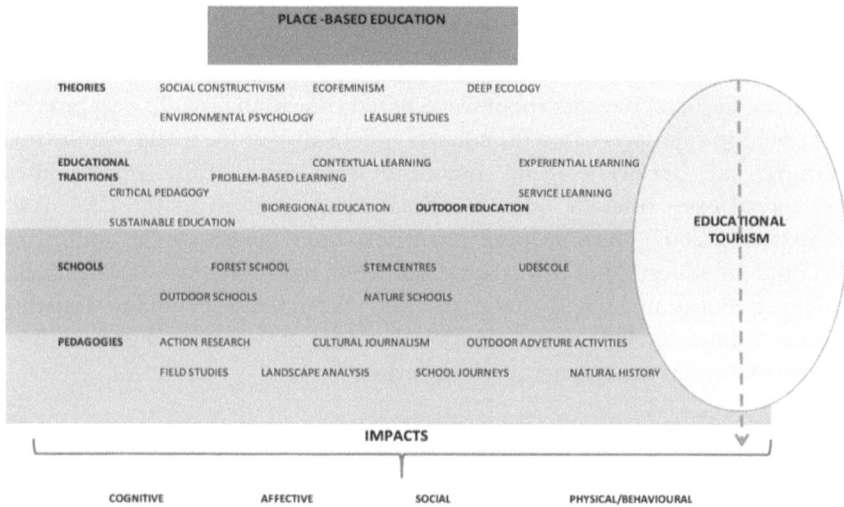

Fig. 3: Dimensions of place-based education and its links with educational tourism (Kaminskienė, 2020)

programmes become more effective when successfully combined with dramatic performances, songs and storytelling.

Changes in the national curriculum also change the profile of museums, libraries and other public institutions. As Salminen et al. (2016) note, in Finland library-school collaboration has witnessed changes from simple field-trips of school groups towards the development of libraries' pedagogical services. Accordingly, today modern libraries are places where a variety of school-oriented services has been developed. With the spread of innovative pedagogies and expansion of the concept of learning, museums have changed their approach to their groups and expanded their audience, creating the necessary infrastructure and facilities for diverse audience, including students with special needs. A significant change has occurred in communication and media used within museums, which was also very much influenced by the increased groups of students from wide age groups. As Salminen et al. (2016) indicate, museums, in addition to their traditional methods such as talking and writing, have adapted media, which uses all the senses.

Outdoor Education in Different Educational Levels and Contexts

In the last decades, the concept of place-based education brought new projects and teaching approaches into the education field all over the world. Contextual learning and service-learning practices were successfully implemented and spread over different types of education institutions in the USA, Asia (Komalasari, 2009) and European countries. There are no studies indicating that outdoor education or place-based education has limitations to certain age groups or educational levels. Studies revealing the potential of outdoor learning (Glynn, Winter 2004; Becker et al.; Monkevičienė et al., 2018) range from kindergarten, primary to tertiary level education.

Outdoor education in kindergarten

Studies show (Katz, 2010) that children learn mostly about nature, technology, engineering, while being in an environment that encourages to experience, research, experiment, collaborating with adults who raise questions, draw attention, and encourage interest. Therefore, it is important that children in pre-school facilities have the opportunity to be outdoors, in an open space with numerous challenges, objects of study and diversity in their daily activities. Children should have natural objects such as trees, shrubs, grasses and large stones, whose qualities they can learn by observing and exploring. It is also important to have real objects that can be manipulated – boards, building blocks, pipes, canvas cuttings, plates, sand, logs, cushions, as well as work tools such as spades, brooms, etc.; by manipulating these big objects by the laws of physics, they discover gravity, friction and understand structural solutions. When children are able to use different systems that include movement – wind, water, sand, balls – they can explore the laws of motion. The ability to analyse various mechanisms enables them to get acquainted with engineering solutions. Developing and expanding understanding about the natural world becomes possible by exploring the plants in the yard of the preschool, exploring the world of beetles, and caring for pets and wild animals (e.g., feeding birds in winter). Children's experience expands during their visits to spaces outside the preschool to help them discover the world around and develops their sense of responsibility towards the place they live.

Outdoor education in Lithuanian kindergartens is witnessing a renaissance. More and more pre-school institutions implement outdoor education programmes, which means that educational activities are transferred to outdoor settings, both in kindergarten and out-of-school settings. In this way, the environment is perceived not as a continuation of the internal environment, but as an impressive place for the child's experiential education, research and self-expression.

Outside the classroom, there are opportunities to get to know the environment in various ways: by researching it, finding answers to the questions raised, creating problem situations and solving them, finding relationship between causes and consequences, and most importantly – learning is realised through all the senses.

Several cases from kindergarten education present interesting examples how outdoor education contributed to the development of various competences related to pre-school curriculum. These cases were collected as part of the national research project during April–October 2018, guided by Prof. Ona Monkevičienė and the team (Monkevičienė et al., 2018). Three examples presented below demonstrate how different outdoor education is implemented in Lithuanian pre-school education institutions. One of the examples demonstrates a project in Palanga Municipality.

> *And the director offered me to take part in it, but I had to come up with the idea myself. Since our focus is on strengthening health, we are thinking about what new methods we could use. And the idea of a borderless education came up...*
> *According to the project, we visited the sea with the children, we read the story "Jūratė and Kastytis" there, we talked about this piece of literature. Before that we bought the amber for the children, we poured down those amber, and they gathered them home. From the seashore we picked pebbles, then we painted on them at the kindergarten. And throughout the spring, education was held outside the class.*
> *If there was a dew in the morning, we put on boots and we travel. We read stories outdoors and look for beetles. At first, we used plastic and paper beetles, and everyone was looking for, finding and putting in the bag. And then we named them, discussed. And after the rain we walked and found all kinds of real beetles, crows, snails. We also discussed how fluff is formed, what is dandelion milk... we are still planning to travel to the forest. And we plan to find treasure in the forest. We anticipate walking on dangerous paths, playing games in the forest. We also plan to go to the stone park. Then we want to make a stone exhibition with our parents, which is called "Let the stone speak"... So we are out in the morning after breakfast. And the morning lessons outside the class....*

This interview presents a nice combination of several types of outdoor education. We can find elements of natural history (analysing dandelion milk, snails, etc.), landscape analysis (going to the forest, to the sea), and some elements of outdoor adventure activities (walking on dangerous paths, playing games in the forest). Unfortunately, the interview does not go into detail what kind of competences were developed during these activities and what impact on cognitive, affective and physical abilities was achieved. Still, it can be identified from the interview that teachers managed to develop linguistic and literature analysis skills by going to the sea and discussing the poem, which lyrically explains how amber was created in the Baltic Sea area. Botanical knowledge is also an important part of outdoor education. Much attention is paid to physical education and health.

> *Our long-term project "Traveller to Lithuania: I know, explore, discover". We've already got well acquainted with Vilnius and that is why we went to many other places: museums, centres, and more. Now we go to Trakai, Anykščiai (bread baking, horse museum), there is such a bird village in Ignalina region. Every excursion is a stimulus for new tasks for children. Before we travel, we inquire about what we are looking for on the Internet, contacting those organizers, we already know in advance what the activities will be and we will come up with what they need to discover and learn. For example, when we were baking bread at the horse museum, getting acquainted with the ancient utensils, tasting those Lithuanian meals, kneading the bread ourselves, each shaping our own loaf, we tasted it, enjoyed the smell ... And there are so many birds, animals and even some not seen in the bird village. ... the children could feed them. ... Also, the kids could understand how the chicken are born because there was a hen in one place, sitting on the eggs, in another place with a few days of chickens, followed by a grown-up chicken ... And then we're discussing everything we saw in the group: and we have used the encyclopaedias, and we draw and discuss.*

The second case has more connection to what was defined as field studies and educational tourism. Children learn to study and observe the environment during excursions, and they become more aware of the interrelation of different phenomena. It gives children unique moments of knowledge of the nature, and creates natural conditions for observing, exploring, discovering, contemplating, summarising and experiencing everything they have learned. Excursions are not limited to the hometowns. There are interesting cases when children and teachers go to various nature monuments, to regional parks, ethnographic homesteads and so on.

> *This year we started out activities in regional parks. Education has no borders, and everything can be learned in nature .... When we arrived at the New Vilnia barrow, the teachers reminded the children how to behave in the forest, what are the basic rules of warning signs. Starting the journey through the park, each group of children received a basket, where they picked cones, acorns, leaves, etc. during the whole walk. When they climbed the barrow, the children found a stick-curse which they used throughout the trip "In the Kingdom of Trees". Each group during the walks picked one or more of the trees they liked and studied them: measured the volume of the trunk; analysed the leaves, etc. The educators told and showed how to learn the age of the trees. We brought a piece of*

> *a tree trunk and we counted the year, compared it with the trunks of other trees... The children looked at the trees from all sides, found the differences, discovered how to learn the directions of the world using the compass. The teachers explained how this affects the trunks of trees.*

The third case again presents how natural studies could be organised. Though it is mostly limited to botanical knowledge, some elements of geography (parts of the world) are also introduced. Nevertheless, these cases from pre-school education show that early age outdoor activities lack connections with communities and do not involve cultural analysis. Outdoor education activities are more focused on specific knowledge of the landscape, monuments and places; however, they lack interaction with local people. Children communicate in groups and with teachers, following pre-arranged learning programmes or plans.

Outdoor education in primary, lower and upper secondary schools

The majority of research on outdoor education (Cwikla et al., 2009; Obadiora, 2016; Zaragosa, Fraser, 2017; Grimwood etl. Al., 2018) provide cases from the primary, lower or upper secondary education level. This could be logically explained that the age group from 7 to 18 is "good clay" to apply different educational strategies in outdoor environments. Several cases from a Lithuanian gymnasium reveals that teachers are willing to bring students to different places that allow to link theoretical learning with real places and real personalities, communities.

> *After visiting B. Sruoga Museum expositions, students worked in groups on practical tasks about the writer's life and his creation, including historical and cultural conditions of that time. During the Lithuanian language lessons they analysed the work of B. Sruoga in more detail, and also prepared the project work which students presented to the gymnasium community.*

This case already presents a complex of pedagogical approaches linked to outdoor education. Literature museums are a good place where students can deepen the knowledge gained at school. Non-traditional learning can help to discover attractive activities; develop personal, social and literary competences for everyone. The museum environment creates preconditions for the creative activities of students of all kinds.

> *So we applied the idea in V. Žilinskas Art Gallery by participating in the educational program "Fun by the Greeks". Not only did the pupils get to know the Antique culture, but they also created myths and staged them. Masks, unconventional space allowed gymnasts to improvise, read and create texts. Everyone could choose roles according to their abilities. After the lesson at the V. Žilinskas Gallery, the pupils individually*

performed the practical task during the Lithuanian language lesson at school - they described, evaluated the chosen mythological personality and searched for their correspondence in the works of Lithuanian poets.

Grimwood et al. (2018) describe a case of urban outdoor education programmes. Most of these programmes took place within Toronto parks and green spaces. They included community-school and after-school programmes, as well as weeklong summer day and residential camps, targeting ages 4 to 14. The study revealed a positive attitude of educational instructors towards the urban outdoor education programmes, which shows that outdoor education is not limited only to natural landscapes.

Implementation challenges

One of the main challenges addressed by researchers is related to teachers' competency and their readiness to implement outdoor education in practice. Notwithstanding the fact that most of the schools use outdoor education as a supplementary pedagogical approach, in rather typical cases, as Maynard and Waters (2007) noted, teachers tend to use outdoor environment in a partial or limited way. It seems that teachers feel rather uncomfortable with outdoor education as it relates to the implementation of formal curriculum and consider it as part of non-formal education. As discussed in this chapter, studies show that outdoor education contributes to positive impacts on children's development.

Salminen et al. (2016) stress that in Finland teachers are taught to utilise diverse learning environments and not to concentrate on class as the only space where successful learning might be implemented. As the new national curriculum is considerably flexible and is focused on phenomenon learnings, this inspires and facilitates utilising different environments, including museums, libraries and other public institutions, which could become a very meaningful source of learning new things and contextualising them. On the national level, teachers received support through several projects which were oriented to the collaborative practices between schools and public institutions. Despite the growing number of outdoor learning activities in Finnish schools, the researchers admit that there is no official Finnish statistics on how often different learning environments outside the school environment are used.

Another challenge related to the implantation of outdoor education is discussed by Dyment et al. (2018) and relates to the pedagogical content knowledge of outdoor education and pedagogies that should/might be employed to achieve the defined goals. Researchers argue that similarly like other subjects (for example, mathematics), outdoor education has its own subject content, and teachers, as well as outdoor educators, should have competence and experience to implement outdoor education in a successful way. Dyment et al. (2018) proposed a framework of pedagogical content knowledge for outdoor education following the analogue of the framework developed for teaching mathematics.

Fägerstam (2013) identified several potentials and obstacles for outdoor education. Firstly, expectations related to implementation of outdoor education raise much higher expectations than this type of education is realised in practice by teachers. So, one of the challenges is to prepare teachers and to set realistic objectives and learning plans. Collaboration among different teachers is also considered as one of the advantages in outdoor education, and this collaboration might help to reduce boundaries among different disciplines. However, the research indicated that the raised expectations were not fully realised, and the implemented pilot outdoor education activities did not increase interdisciplinarity; besides, no stronger collaboration among teachers was observed. Fägerstam (2013) study explains this by lack of time and difficulties in planning, very limited possibilities to go further from school areas as it also requires much more time than the planned school curriculum allows to do. Notwithstanding the identified obstacles, the study shows that outdoor education enhanced participation and collaboration in the class as well as the relationship between teachers and students changed in a positive way.

Outdoor Education in the Context of Developing Educational Tourism in Visaginas and Ignalina Nuclear Power Plant

Outdoor education can take place in different contexts, but through distinctive spaces, outdoor education can contribute to the development of educational tourism in the regions that traditionally or for a long period of time have not been classified as places for tourist attraction. One of the interesting examples of educational tourism development is the city of Visaginas and the Ingalina Nuclear Power Plant. These specific spaces – the city and the area of the nuclear power plant – were developed during the Soviet period. This particular historical period and the fact that workers and specialists from different parts of the Soviet Union were involved in the construction of the power plant contributed to the result that the majority of the population in the city is Russian-speaking and of other nationalities. For more than forty years, Visaginas has become a kind of ethnic, cultural and linguistic island in Lithuania. The completion of Ignalina Nuclear Power Plant decommissioning is planned for 2038. When planning these decommissioning processes, it is important to define and ensure a further development of Visaginas city, which defined itself as a nuclear city and as a satellite of the Ignalina Nuclear Power Plant. The decommissioning processes have a direct impact on the social, economic, cultural and identity development of the Ignalina Nuclear Power Plant region – there is a need to reconceptualise and reconstruct urban and regional identities. Moreover, the city has not yet developed its economy, which has been dependent on the power plant for the last forty years. Nuclear tourism is one of the economy development strategies for the city and the region, which may adapt international experience from Europe, Japan, the US and other countries. However, nuclear tourism cannot be successfully developed without attractive concepts behind it. These concepts, as international experience shows (Mažeikienė, Gerulaitienė, 2018), have a strong link with educational goals. These educational goals may stimulate the development of nuclear tourism within the region, taking into account the geographical position of the city and Ignalina Nuclear Power Plant. The city and the plant as tourism and "educational centres" have high potential to attract pupils and students from the whole of Lithuania as well as neighbouring countries, mainly Latvia, Belarus and Poland.

The Ignalina Nuclear Power Plant was built in the area that is interesting for a variety of reasons: geographical landscape, the plant and the landscape affected by the human activity, historical perspective (the Soviet period), energy production and the role of the plant in Eastern Europe, post-Soviet countries, etc. All these aspects create an inexhaustible source for outdoor education activities.

Outdoor education, as discussed earlier, has a variety of ways how to be implemented: through school journeys, action research, problem-based learning, etc. We will discuss how Visaginas and Ignalina Power Plant context and the place can be successfully used to develop outdoor education activities on the one hand, and contribute to the development of educational tourism on the other hand.

The city of Visaginas is a very interesting place for exploration by applying such methods as cultural journalism, field studies, action research, problem-based learning and other pedagogical strategies. Cultural and ethnical uniqueness may be analysed by applying the method of cultural journalism. Learners could be assigned to analyse and record different signs in the city and try to understand what the city is "communicating" through signs, adverts, textual, visual or audio information and media. The city could also be analysed through different perspectives: historical (and then the communication of the city is important through cultural representations and archives, old newspapers and journals, documentaries that could be accessed in the library and/or other institutions); the historical perspective and story can be enriched by interviews of local people. Another perspective could be cultural; the city of Visaginas could be analysed as a multicultural place, and students could explore which acculturation strategies have been employed by local people (using time perspective yesterday and today). Cultural journalism, as described earlier in this chapter, is an extremely powerful method, which generates a vast amount of resources and information. These resources sparkle imagination and might be potentially exploited for developing tourism attractive places in the city, offering unique stories, events, etc.

One of the examples, how outdoor education can be implemented and supported with technologies, is illustrated by the following case. Emokykla (E-school), an e-learning platform, offers assignments related to Ignalina region. These assignments can be used during geography lessons for grades from 6th to 8th. Ignalina region is characterised by a spectacular landscape that was formed by the last Ice Age. The e-learning platform offers pupils to create a comic about how the site around the Ignalina Nuclear Power Plant has changed. The pupils are encouraged to describe what new components of the landscape have emerged, how they have changed the environment, and how the plant has affected the wildlife. The pupils can also virtually predict how the Ignalina region will change in twenty years after the closure of the Ignalina Nuclear Power Plant. The above-presented assignments can be done with or without visiting the place of the plan; however, deeper learning will be achieved if we combine both, e-learning and landscape analysis.

Mažeikienė and Gerulaitienė (2018) indicate that regions around nuclear power stations are also well known as nuclear tourism destinations. According to these scholars, "new forms of nuclear tourism" encourage schools to organise educational visits to power plants and the museums or tourism centres in these areas in order "to give understandable and unbiased scientific information about different topics: atoms, radiation, ionizing radiation and health, reactors, robots, physics and much more" (Mažeikienė, Gerulaitienė, 2018 p. 5674). Following these researchers, obviously, Ignalina Nuclear Power Plant is a place to enlarge knowledge in STEM subjects, including physics, chemistry and technologies. However, not less important is the fact that nuclear tourism places bear a mystified character, which stimulates imagination of tourists, willing to discover "dark" and mysterious stories related to the power plants. Thus, from the educational point of view, Ignalina Nuclear Power Plant as well as its satellite the city of Visaginas, offer a vast amount of information and almost unlimited opportunities to implement some components of the curriculum through outdoor education activities for different age groups. Obviously, this region can find its place among tourist routes in Lithuania and can be potentially integrated into a wider international network of nuclear tourism destinations.

Conclusions

Outdoor education is not a new phenomenon; however, it always deserves attention from researchers and practitioners regarding its scope, objectives and a variety of implementation ways. Historically started as natural history, outdoor education today is evident in diverse forms and pedagogical solutions: school journeys, field trips, landscape analysis, adventure activities, etc. In the last twenty to thirty years, mainly influenced by experiential pedagogy, outdoor education has proved to be a successful approach to developing learners' cognitive, affective, social and behavioural domains of different age groups.

Outdoor education and educational tourism have common stems coming from the concept of place-based education, which has been affected by a variety of theories, including ecofeminism, bioregional development, socio-cultural theories and other spheres. Still, we cannot say that educational tourism is a form of outdoor education simply for the reason that educational tourism is trans-sectorial and links education and tourism with its own objectives to attract people to specific places (be they real, imaginative or virtual) and educate them. Even though most researchers keep to the position that outdoor education, environmental education, etc. are historically changing concepts of place-based education, there are definite differences that should not be merged and fully aligned.

Notwithstanding the fact that outdoor education suggests using active pedagogies, one of the challenges remains to prepare teachers to work in creative ways while implementing outdoor education as they tend to limit outdoor education to field trips and school journeys. Researchers also identify that school raises extremely high expectations for outdoor education, yet in practice a lot should be prepared and realised step by step. Things do not change dramatically, but the schools should be ready for a systemic and gradual development of their approaches in outdoor education practices.

The final efforts should be addressed to learners – young children, teenagers and even adult learners. There is no evidence that outdoor education should be restricted to some specific age groups, even though most outdoor education practices in many countries are focused on children from 4 to 15 years. Linking educational tourism, which attracts much more adult learners, and outdoor education, which is still more related to school education, we may get very unique synergy which will not only help us to develop important skills, attitudes and behaviours of learners, but also raise awareness and responsibilities towards the places we live. Returning to Louv's coined term "nature-deficit disorder", we should think about the advantages that outdoor education and educational tourism might bring to our societies in terms of education, health, natural and cultural preservation, sustainability, and identities development in the globalised world. Outdoor education and cultural tourism should not be contrasted; on the contrary, they are developed in parallel, and our schools, education centres, museums, libraries and other institutions and actors should be ready for the change.

From the educational perspective, it is strategically important for Lithuania to develop nuclear tourism as it may open a variety of outdoor education activities combined with sustainable education, environmental education, bioregional education, service learning and others. More actively involving Lithuanian schools, kindergartens could contribute to raising awareness among young population about the variety of tourism and tourism concepts, enhancing learning of STEM, languages, history, economy, geography, physical and health education, arts and other subjects. Taking into account the experience of Scandinavian countries (Finland), more initiatives could be undertaken by the library of Visaginas by offering diverse educational programmes, cinema festivals, cultural programmes, etc. for different age groups.

References

Allan, J., & McKenna, J. (2019). Outdoor Adventure Builds Resilient Learners for Higher Education: A Quantitative Analysis of the Active Components of Positive Change. *Sports 2019, 7, 122*

Attarian, A. (2003). Trends in Outdoor Adventure Education. In R. Poff, S. Guthrie, J. Kafsky-DeGarmo, T. Stenger, & W. Taylor (Eds.), *Proceedings of the 16th International Conference on Outdoor Recreation and Education* (pp. 28–39). Bloomington, IL: Association of Outdoor Recreation and Education.

Beames, S., Humberstone, B., & Allin, L. (2017). Adventure revisited: Critically examining the concept of adventure and its relations with contemporary outdoor education and learning. *Journal of Adventure Education and Outdoor Learning, 17*(4), 275–279.

Barell, J. (2007). *Problem-based learning: An inquiry approach* (2nd ed.). Thousand Oaks, CA: Corwin Press.

Becker, Ch., Lauterbach, G., Spengler, S., Dettweiler, U., & Mess, F. (2017). Effects of Regular Classes in Outdoor Education Settings: A Systematic Review on Students' Learning, Social and Health Dimensions. *International Journal of Environmental Research and Public Health*, 14, 485; doi:10.3390/ijerph14050485

Behrendt, M., & Franklin, T. (2014). A Review of Research on School Field Trips and Their Value in Education. *International Journal of Environmental & Science Education*, 9, 235–245.

Coleman, J., & Hoffer, T (1987). *Public and Private High Schools: The Impact of Communities*. New York, NY: Basic Books, 1987.

Cuthbertson, B., Socha, T., & Potter, T. (2004). The double-edged sword: Critical reflections on traditional and modern technology in outdoor education. *Journal of Adventure Education and Outdoor Learning*, 4(2), 133–144.

Cwikla, J., Lasalle, M., & Wilner, S. (2009). My two boots...a walk through the wetlands: An annual outing for 700 middle school students. *The American Biology Teacher*, 71(5), 274–279.

Dyment, J. E., Chick, H. L., Walker, Ch. T., & Macqueen, T. P. N. (2018) Pedagogical content knowledge and the teaching of outdoor education, *Journal of Adventure Education and Outdoor Learning*, 18:4, 303–322, DOI:10.1080/14729679.2018.1451756

Fägerstam E. (2013). High school teachers' experience of the educational potential of outdoor teaching and learning, *Journal of Adventure Education & Outdoor Learning*, DOI:10.1080/14729679.2013.769887

Garalis, A., Švagždienė, B., Liesionienė, O. (2008). Ekskursijos, kaip edukacinės paslaugos, samprata ir esmė. *Ekonomika ir vadyba: aktualijos ir perspektyvos*, 4(13), 123–134.

Glynn, Sh. M., & Winter L., K. (2004). Contextual Teaching and Learning of Science in Elementary Schools. *Journal of Elementary Science Education* 16(2), 51–63.

Graff, E., & Kolmos, A. (2006). *Process of Changing to PLB. Management of Change*. Rotterdam/Taipei: Sense Publishers.

Graham, M. A. (2007) Art, Ecology and Art Education: Locating Art Education in a Critical Place-based Pedagogy. *Studies in Art Education National Art Education Association A Journal of Issues and Research, 48(4)*, 375–391.

Grimwood, B. S.R., Gordon, M., Stevens, Z. (2018). Cultivating Nature Connection: Instructor Narratives of Urban Outdoor Education. *Journal of Experiential Education* 2018, Vol. 41(2) 204–219.

Gruenewald, D. A. (2003). Foundations of Place: A Multidisciplinary Framework for Place-Conscious Education. *American Educational Research Journal* Fall 2003, Vol. 40, No. 3, pp. 619–654.

Hale, A. N. (1975). *Directory programs in outdoor adventure activities*. Mankato, MN: Outdoor Experiences, Inc.

Harington, D. G. (2001). *The development and validation of a learning environment instrument for CSIRO Science Education Centers*. Unpublished Doctor of Science Education thesis, Curtin University of Technology.

Jovaiša, L. (1993). *Pedagogikos terminai*. Kaunas, Šviesa.

Jordet, A. N. (2010). Klasserommet utenfor. Tilpasset opplaering i et utvidet laeringsrom. [The classroom outdoors. Education in an extended context]. Latvia: Cappelen Damm AS.

Jordet, A. N. (2007). *Nærmiljøet som klasserom. En undersøkelse om uteskolens didaktikk i et danningsteoretisk og erfaringspedagogisk perspektiv* [The local environment as classroom] [Doctoral Dissertation No. 80]. University of Oslo.

Katz, L. (2010, May). Stem in the early years. Paper presented at *SEED (STEM in Early Education and Development)* Conference, Cedar Falls, IOWA. Retrieved from https://ecrp.illinois.edu/beyond/seed/katz.html, Accessed August 20, 2019.

Knapp, C. E. (2014). Place-based curricular and pedagogical models. In D. Gruenewald & &A. Smith (Eds.), *Place-based education in the global age: local diversity* (pp. 5–28). Psychology Press, New York.

Komalasari, K. (2009). The Effect of Contextual Learning in Civic Education on Students' Civic Competence. *Journal of Social Sciences, 5*(4), 261–270. https://doi.org/10.3844/jssp.2009.261.270;

Kondrotienė, A., & Laučienė, S. (2011). Mokinių kūrybiškumo ugdymas diferencijuojant ir individualizuojant mokymą(si) netradicinėse edukacinėse aplinkose. *Ugdymo turinio individualizavimo ir diferencijavimo patirtis atnaujintų bendrųjų programų kontekste*. Kauno pedagogų kvalifikacijos centras. Kaunas.

Krepel, W. J., & Durral, C. R. (1981). *Field trips: A guideline for planning and conducting educational experiences*. Washington, DC: National Science Teachers Association.

Louv, R. (2008). *Last child in the woods: Saving our children from nature-deficit disorder*. Chapel Hill, NC: Algonquin Books.

Malone, K. (2008). *Every experience matters: An evidence based research report on the role of learning outside the classroom for children's whole development from birth to eighteen years*. The report commissioned by Farming and Countryside Education (FACE) in support of the UK Department of Children, School and Families Learning Outside the Classroom Manifesto.

Maynard, T., & Waters J. (2007) Learning in the outdoor environment: a missed opportunity? *Early Years, 27*(3), 255–265, DOI: 10.1080/09575140701594400

Mažeikienė, N., & Gerulaitienė, E. (2018). Educational aspects of nuclear tourism: sites, objects and museums. *EDULEARN 18 [electronic resource]: 10th international conference on education and new learning technologies*, Jul 2–4, 2018, Palma, Spain: conference proceedings, 5668–5677. doi:10.21125/edulearn.2018.1369

McAnally, H. M., Robertson, L.A., & Hancox, R. J. (2018). Effects of an Outdoor Education Programme on Creative Thinking and Well-being in Adolescent Boys. *New Zealand Journal of Educational Studies (53)*, 241–255.

McGladdery, C. A. (2016). *The relationship between international educational tourism and global learning in South African high school learners.* [Unpublished doctoral thesis]. Pretoria: University of Pretoria.

McGladdery, C. A. & Lubbe, B. E. (2017). Rethinking educational tourism: proposing a new model and future directions. *Tourism Review 72(3)*. DOI: 10.1108/TR-03-2017-0055

Monkevičienė, O., Autukevičienė, B., Kaminskienė, L., Rutkienė, A., Tandzegolskienė, I., Skerytė-Kazlauskienė, M., Monkevičius, J., Stonkuvienė, G. Vildžiūnienė, J. (2018). *Paslaugos „Tyrimo pažangi pedagoginė praktika ir pedagoginės inovacijos Lietuvos vaikų darželiuose atlikimas" ataskaita.* Retrieved from https://www.ikimokyklinis.lt/uploads/files/dir1306/dir65/dir3/3_0.php (Accessed September 8, 2019)

Nundy, S. (2001). *Raising achievement through the environment: The case for fieldwork and field centres.* Walsall, UK: NAFSO.

O'Brien, E., & Carroll, L. (2015). *A report on how problem based learning and ICT can support SMEs in Europe.* 9th Balkan Region Conference on Engineering and Business Education and 12th International Conference on Engineering and Business Education, Sibiu, Romania, October, 2019. DOI: 10.2478/cplbu-2020-0016

Obadiora, A. J. (2016). Comparative Effectiveness of Virtual Field Trip and Real Field Trip on Students' Academic Performance in Social Studies in Osun State Secondary Schools. *Mediterranean Journal of Social Sciences,* Vol. 7, (1), 467–474.

Pitman, T., Broomhall, S, McEwan, J., & Majocha, E. (2010). "Adult learning in educational tourism". *Australian Journal of Adult Learning*, Vol. 50 No. 2, 219–238.

Rennie, L. J. (2007). Learning science outside of school. In S. K. Abell & N. G. Lederman (Eds.), *Handbook of research on science education* (pp. 125–167). Mahwah: Lawrence Erlbaum Associates.

Rickinson M., Dillon J., Teamey K., Morris M., Young Choi M., Sanders D., Benefield P. (2004). *A review of Research on Outdoor Learning.* National Foundation for Educational Research and King's College London.

Ritchie, B. W., Carr, N., & Cooper, Ch. (Eds.) (2003). *Managing educational tourism.* Clevedon: Channel View Publications.

Salminen, J., Tornberg L., & Venäläinen, P. (2016). Public Institutions as Learning Environments in Finland. In H. Niemi, A. Toom, A. Kallioniemi (Eds.). *Miracle of Education: The Principles and Practices of Teaching and Learning in Finnish Schools* (pp. 253–266). Rotterdam: Sense Publishers. https://doi.org/10.1007/978-94-6300-776-4_17

Sandell, K., & Öhman, J. (2013) An educational tool for outdoor education and environmental concern. *Journal of Adventure Education & Outdoor Learning, Vol. 13, No. 1,* 36–55.

Sarivaara, E., & Uusiautti, S. (2018). Transformational elements for learning outdoors in Finland: A review of research literature. *International Journal of Research Studies in Education,* 7 (3), 73–84.

Siskar, J., & Theobald, P. (2008). The Meaning of Place and Community in Contemporary Educational Discourse. *Journal of Inquiry & Action in Education,* 1(2) 58–78.

Sobel, D. (2004). *Place-based education. Connecting classrooms and communities. Nature Literacy Series, Vol. 4.* Great Barrington: The Orion Society.

Surface, J. L. (2016). Place-based Learning: instilling a sense of wonder. *Publications of the Rural Futures Institute.* 10. http://digitalcommons.unl.edu/rfipubs/10

Thorburn, M., & Allison, P. (2010). 'Are we ready to go outdoors now? The prospects for outdoor education during a period of curriculum renewal in Scotland' *Curriculum Journal,* vol. 21, no. 1, 97–108. DOI: 10.1080/09585170903560824

Wood, E. J. (2004). Problem-Based Learning: Exploiting Knowledge of how People Learn to Promote Effective Learning. *Bioscience Education,* 3: 1, 1-12.

Yeo, K. R. (2008). How does learning (not) take place in problem-based learning activities in workplace contexts? *Human Resource Development International,* 11(3), 317–330.

Zaragoza, J. M., & Fraser, B. J. (2017). Field-study science classrooms as positive and enjoyable learning environments. *Learning Environments Research* 20, Issue 1, 1–20.

Judita Kasperiūnienė

Innovative Technological Solutions in Virtual Nuclear Education

Abstract: The need for researching nuclear learning-related processes in different countries, formal and non-formal education contexts and age groups, has been continuously growing. The plethora of didactic and technological solutions to create virtual nuclear education products often misleads educators. Our study aims to categorize and map out the existing scholarly concepts and online examples of nuclear education and virtual tours, from which to commission further reviews by identifying gaps in the research literature and virtual solutions. The findings from the review of the articles accessed in Springer Link, Scopus, Taylor & Francis, and ACM mono and multidisciplinary scholar databases and online portals, apps, and games on nuclear learning–related topics showed the difference in the concepts on nuclear education, nuclear information, and nuclear technologies for diverse scientific fields. Literature analysis has revealed a gap between modern technologies and methods of teaching and learning in formal and non-formal education contexts. Research showed that the integration of technological progress into teaching processes could be challenging. Mature educational technologies and methods may not adequately address the needs of individual learners and society. The findings, illustrating the complex nature of virtual reality solutions and virtual tour technologies applicable to online nuclear education, nuclear information, and nuclear technology studies, covering topics of empirical scholar research, online and smart solutions, allowed us to construct a framework of serious gaming for non-edutainment purposes, incorporating learning and playing tools, motivators, and educational strategies.

Our research gives insights into nuclear tourism route construction, proposing a technology acceptance model, suggesting storytelling, virtual navigation, human spatial behavior, digital and human factors, and tools and technologies research as the key topics of further empirical research and offering online tours, mobile apps, and 360 videos as technologies to facilitate virtual learning, impact learning outcomes, and raise nuclear literacy. While proposing to use serious games and tours for virtual nuclear education, we see some limitations. In nuclear education settings (formal and informal), these technologies are mainly adapted to the learner generation who accept technology. Still, some learners find it difficult to immerse in virtual activities and experience a difference between virtual and live guided tours. For educators, the biggest challenge is not only to create virtual materials on nuclear education in the text format but also, together with computer scientists and engineers, to develop virtual narrative, immersive games, simulation apps, and other types of virtual reality solutions. For nuclear scientists, the biggest challenge remains the communication of scientific information in a learner-friendly format.

Keywords: educational tools for virtual tour development, experiential gaming, geolocation technologies for education, mapping review, serious gaming for non-edutainment purposes, technology acceptance model, virtual nuclear education.

Introduction

Nuclear science is the study of atom application to various spheres of human life. It not only deals with structures, elements and forces of the nuclei; application of radioactive substances in the diagnosis and treatment of various diseases; sub-atomic processes in technology or chemical engineering, but also with nuclear safety and security; climate change; and policies, countries, and communities. Nuclear science can be studied in formal and informal nuclear education settings. Formal *nuclear education* starts in pre-primary classes and continues in schools, universities or colleges. Children start nuclear education in kindergarten through environmental lessons. Later, integrated STEM lessons develop students' *nuclear literacy* – the ability to recognize, understand, interpret, create, communicate, calculate, and use printed and written data about the atom, its properties, and applications in a variety of contexts and real-life situations. A few examples below show that in the formal education context, nuclear (or atomic) literacy is developed in a comprehensive way, integrating it into a variety of study curriculums, subjects, classes, and informal education. For example, in Hungary, "nuclear chapters of the curriculum" were integrated into school education in the last decade of the twentieth century (Toth & Marx, 1996) after the Chernobyl catastrophe. While these lessons teach atomic physics, chemistry or nature, teachers say their students are learning to observe everyday life; they learn democracy and decision-making based upon a shared understanding of information. In addition to that, students learn that nuclear education means responsibility for others and for the future (ibid.). In post-Fukushima Japan, scholars, teachers, and decision-makers believe that nuclear education (formal and informal) is very important, and the subject is actively discussed. For example, radiation lectures are provided in Japanese high schools, seeking that nuclear literacy positively transforms learners' attitudes and behaviors related to radiation in general, and disasters specifically (Tsubokura, Kitamura & Yoshida, 2018). Japanese scholars have proved that nuclear education enables young people to make better decisions about important matters in their daily lives. Another study in Turkey (Yavuz-Topaloglu, Demirhan, & Atabek-Yigit, 2019) focused on the importance of gender in examining the links between nuclear education and daily living. The results of this study showed that male respondents support nuclear power more than female, and adult females with children (mothers) are more likely to oppose nuclear power. These findings are even more important because the family in general and mothers in specific have a great influence on their children not only in formal, but also in informal education.

In the higher education context, *nuclear technology studies* are more specialized. They are mainly concentrated in fundamental, medicine or engineering faculties of universities and colleges. Although nuclear science is a relatively young science and several incidents all around the world have attracted public opinion, the interest of the young generation in the formal nuclear study has fallen (Brancucci, Flore, & von Estorff, 2014). Engineering faculties have reduced student admissions to nuclear education–related study programs (Ahn et al., 2015). Meanwhile, the first generation of nuclear experts and professionals have begun to retire, resulting in a gap between the incoming and outgoing specialists' flows (Brancucci, Flore, & von Estorff, 2014). This has led not only to a gradual shortage of skilled professionals and a greater risk of losing valuable knowledge to the nuclear community (ibid.) but also to the deteriorating public awareness of nuclear education and nuclear safety–related issues.

Informal nuclear education is a very important topic that encourages people of all ages to think, make decisions, and understand the importance of nuclear energy not only in engineering or medicine but also in everyday life. (Luk et al., 2018). Informal nuclear education happens in education laboratories, museums, exhibitions, non-formal places, and virtual spaces. In order to enhance the children's, teachers' and the community's interest in modern nuclear physics or nuclear energy, scientists search for non-traditional ways, places, environments for teaching and organizing active or immersive learning. New emerging technologies, such as virtual and augmented reality, computational dynamics, virtual laboratories, and virtual worlds have recently been more widely used not only for teaching Science, Technology, and Engineering, but also for informal science education and science communication. Virtual reality, characterized by three key elements, such as *visualization,* when the user, gamer, or learner has the ability to look around, usually with the use of a head-mounted display; *immersion,* when the person mixes imaginary and physical representation of objects; and *interactivity,* providing the degree of control over the experience, usually achieved with sensors and an input device like joysticks or keyboards (Yung & Khoo-Lattimore, 2019), is taking an increasing place in teaching and learning. The concept of immersive education can be applied to all aspects of education for different age and competency group learners: formal, informal, massive and professional training, from preschool education to life-long learning (Potkonjak et al., 2016). One of the most popular contemporary educational environments is virtual or mixed reality tours (Domingo and Bradley, 2018). Virtual tours are panoramic, virtual, augmented, or mixed reality simulations of the existing rural or urban places and environments. Technically speaking, virtual tools are collections of images accompanied by sounds or audio texts. Virtual tour development technologies have been thoroughly analyzed and described; they are constantly changing and improving (e.g. Napolitano, Scherer, & Glisic, 2018; Yung & Khoo-Lattimore, 2019; Tung et al., 2015). The educational goal of virtual tours and their relation to audiences determine not only how the virtual tour is constructed, accessible or immersive, but also what didactic messages are sent to users.

In addition to formal and informal nuclear education, *nuclear information* (informing, sharing and raising awareness of nuclei-related topics) may generate support for scientific research and technological practices; influence decision-making; inspire political, ethical and environmental thinking; and educate and strengthen communities.

In Lithuania, despite the fact that the country is no longer considered a nuclear state as the Ignalina Nuclear Power Plant was shut down more than a decade ago, public nuclear literacy for different education groups remains particularly important. By synthesizing and analyzing the existing scientific literature and online technological solutions on nuclear education and virtual tours and integrating research from various subjects (not limited to nuclear physics and environmental education, science, engineering and computer technologies), insights for the nuclear virtual route development can be created. The *aim* of this study was to categorize and map out the existing scholar concepts and online examples of nuclear education from which to commission further reviews by identifying gaps in the research literature and virtual solutions for educational material presentation. The *research questions* were: (i) how the concept of nuclear learning is presented and discussed in research literature; (ii) what topics of empirical research related to nuclear education through virtual tours dominate in scientific databases of multidisciplinary and monodisciplinary peer-reviewed literature; (iii) what topics, scenarios, and solutions are used to create freely accessible virtual tours to develop nuclear literacy.

Methodology

The mapping review (Grant & Booth, 2009) was applied to search and contextualize the research literature and internet portals that provide the possibility to travel and learn virtually. Systematic review search filters were used by summarizing empirical research articles related to nuclear education, searching information about virtual tours development and their use in teaching and learning process, focusing on the first research question (Lefebvre et al., 2017).

In the mapping review, the principles of scientific evidence-based inquiry were followed: (i) opening significant questions for empirical investigation of scientific literature and website materials; (ii) linking research to conceptual framework; (iii) using causal mapping as a method and technique (e.g. Lorenc et al., 2012; Bryson et al., 2004) that allowed direct investigation of the research question; (iv) mapping chain of reasoning (Lal, Donnelly & Shin, 2015), identifying limitations and biases, and estimating uncertainty (McMillan & Schumacher, 2010).

The mapping review was conducted in Scopus, ACM, Springer Link, and Taylor & Francis databases (2014–2018). These databases were selected seeking to find and analyze mono and multidisciplinary empirical research in social and technological sciences with a STEM perspective. The iterative stages of searching, synthesis, critical interpretation, and causal mapping were performed while researching nuclear education and science communication–related scholar topics and technological solutions of virtual tours. The following inclusion criteria were used: (i) the full article was published in scientific peer-reviewed journals in English; (ii) the article was published between 2014 and 2018; and (iii) the articles were based on an integrated approach covering nuclear education and virtual tours with the focus on STEM with a strong emphasis on citation of the selected articles. In the selection of empirical articles, preference was given to those who influenced the creation of new meta-theoretical constructs seeking theoretical saturation with respect to the framework. If the articles contained contradictory, conflicting, or underdetermined theories, the search continued seeking to purify those empirical studies, in which the variables of the chosen theory associated, correlated or reported the empirical data, making connections and preliminary substantiating theoretical statements or claims (Lorenc et al., 2012).

The construction of the research consisted of three stages inside the general iterative mapping review procedure. In the initial phase, scientific literature was identified, screened, and structured using the keywords "nuclear education" and "virtual tours", finding the key topics of empirical research. During the second phase, the ten most popular online freely accessible virtual tours were investigated and tested. Then the links between the concepts, empirical research topics, virtual tour scenarios, and technological solutions were developed, and challenges discussed.

Findings

A variety of technologies exists to provide and support virtual nuclear education. Our research focuses on *game-based learning* (e.g. Romero et al., 2016) as the method of teaching with games not specifically for virtual nuclear education. Game-based learning allows users of all ages and backgrounds to stay motivated and self-master the study curriculum effortlessly. Different types of games continuously motivate players with elements of challenge, fantasy, and curiosity (e.g. Asgari & Kaufman, 2004; Malone, 1980). While the main "official" goal of games is entertainment, scholars discuss the use of games in different learning environments and disagree over whether players learn while they are having fun. We consider that higher levels of thinking and social skills can be developed through play and learning. Creating and mastering the content means gaining facts, information and skills, and building knowledge not only from teachers, practitioners, and other experts, but also from the game environment and other gamers. Many games where users need to master their own content offer opportunities for individual or group learning. These types of games could be applied not only to informal studies, but also to formal education classrooms as additional learning materials to stimulate learner interest and to raise study motivation. Games can develop cognitive and perceptual competencies such as attention and concentration on details, characters or events; understanding of story-play, strategic thinking, problem-solving, planning, and memorizing. Games sharpen players' emotional and volitional competencies such as emotional and stress control, and endurance – the skills, critically important for social development, as well as academic performance and later life success (e.g. Hromek & Roffey, 2009). School children experiments by Goldstein and Lerner (2018) proved that in games, children could develop altruism to a stranger; comforting behaviors to someone in distress; helping behaviors to a person who needs assistance; and positive classroom social behaviors. Additionally, players develop cooperation, competition, mutual support, empathy, and moral judgment competences (Wiemeyer & Hardy, 2013).

Further in this chapter we discuss some issues concerning *serious games* – the digital games used for non-entertainment purposes not specifically in nuclear education environments. The research of empirical evidence of the impacts and outcomes of serious games, performed by Boyle and her colleagues (2016), has pointed out that the term "serious games" is becoming a new mainstream. In many cases, it is used interchangeably with *games for learning*. Serious games can be used to promote knowledge acquisition across a wide range of topics, to develop social skills and behavior change. For formal and informal nuclear education, an experiential gaming model (Kiili, 2005), based on *experiential learning theory* (Moon, 2013), flow theory (Nakamura & Csikszentmihaly, 2009) and game design (Salen, Tekinbaş & Zimmerman, 2004) can be used. In experiential learning theory, learning is described as a cycle integrating "Dewey's philosophical pragmatism, Lewin's social psychology, and Piaget's cognitive developmental genetic epistemology" (Kolb & Kolb, 2012, p. 2), and transforming active experiences into conceptual understandings and applications through reflection (Moseley et al.,2020). The four-stage cycle is observed: active experimentation, concrete experiences, reflective observation, and abstract conceptualization. Although learning could begin at any stage, all the stages need to be completed – introducing active experiencing with new concepts, models, role-playing and problem-solving, discussing, analyzing, and reflecting on live or virtual experiences (ibid.). The learner can become so involved in the game that he no longer feels the amount of time he has been playing. Time is like "disappearing" (Nakamura & Csikszentmihaly, 2009). Therefore, to keep the user not only entertained while playing, but also achieving educational goals, game developers need to collaborate with educators on the game design. *Technology acceptance model* (e.g. Marangunić & Granić, 2015) explains how learners accept and use technologies, analyzes the learners' intentions, attitudes, motivation, and beliefs concerning technologies. Game-based learning used for education scenario and game environment development consider gameplay and usability perspectives; learner technological competences and preferences; edutainment experiences; and pedagogical integration. (Fig. 1).

A virtual tour is a simulation of an existing location with the help of audio, outdoor and indoor maps, floor plans, sequential videos, or still images. These tours help in recreating a realistic representation of reality and presenting views to inaccessible areas. In nuclear education, there are some areas of the nuclear power plant restricted to non-specialist visit, or some specific situations need to be artificially constructed for information or education; therefore, virtual tours could be a solution for a broader audience. Besides, virtual tours could become an interesting alternative to fieldwork when expenses, time, or logistics are an issue for users. Nuclear education and science communication could develop not only through serious gaming techniques, but also by applying *virtual tours* as innovative and interactive tools, as well as presentations of learning materials and new knowledge, allowing users to actively immerse into a topic. In many cases, virtual tours, similar to the so-called live tours and excursions, are guided. Guiding allows the user not to be lost on a tour. *Digital guiding* could be done by using text, pictures, maps, audio, video materials, or a combination of these techniques. It includes serious game elements that motivate learners

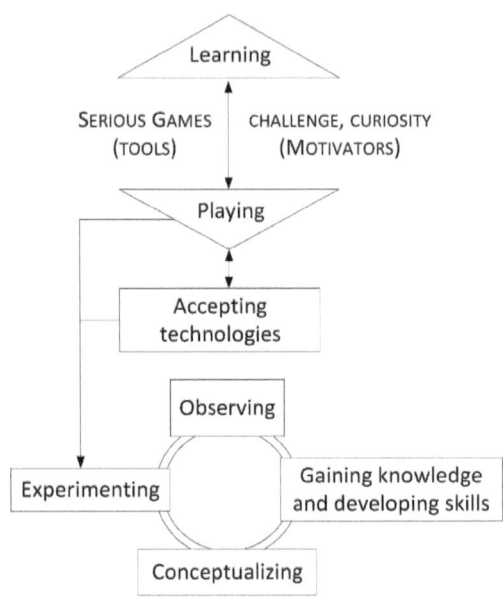

Fig. 1: The framework of serious gaming for non-edutainment purposes

and makes the overall process not only challenging, but also entertaining and immersive. Quite frequently, a person sitting in front of a computer does not even distinguish between virtual tours and digital games. Guidance features, the availability of feedback and performance reporting, and the integration of engaging and reflective capabilities enhance the overall experience, empower the learner's memories, help to interact with new knowledge, and develop practical skills (Mostafa, 2018). Through the virtual guided tours, learning, and raising awareness on nuclear tourism, the route could grow. With the help of serious games and virtual tour tools, learners are informed about science and get involved in continuous learning. During the virtual tour, they are allowed to experiment, observe, and change their environmental habits. Therefore, as with serious games, people of all ages acquire knowledge, new skills, learn how to creatively solve problems, actively experiment, reflect and observe, and conceptualize their own findings (Fig. 2).

In the next section, the research topics of the last five years (2014–2018) dominating in the empirical scientific articles in the area of nuclear education and virtual touring are identified and key themes discussed.

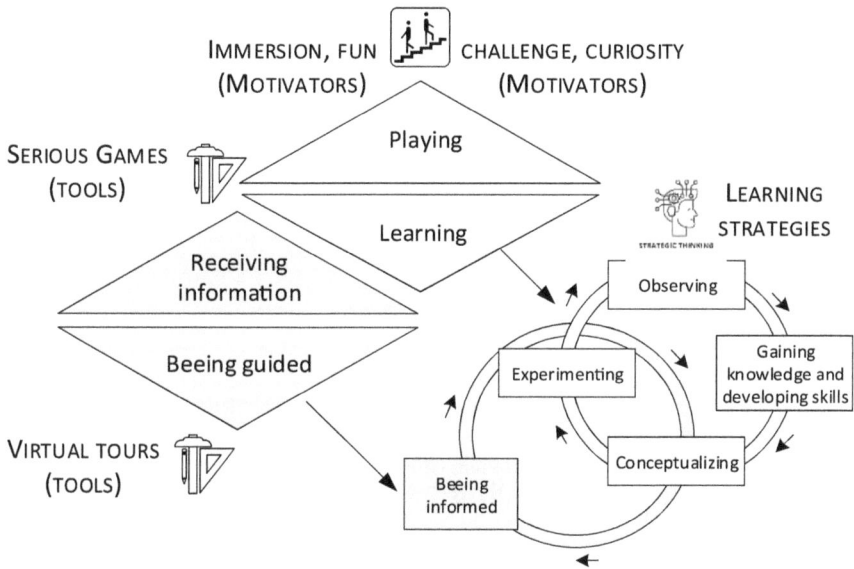

Fig. 2: Experiential gaming tools, motivators, and learning strategies for virtual nuclear education

Dominating Topics of Virtual Nuclear Education Empirical Research

In four scholar databases, 1268 empirical articles were screened (N=1268), using three main keywords: "nuclear education", "virtual tour technologies", and "virtual reality solutions". The database search exposed that articles empirically studying *nuclear education* concept are mostly published in Springer Link (n=682, 53.7 % of all researched cases) and Scopus (n=354, 27.92 % of all researched cases). In Taylor & Francis database 165 articles (n=165, 13.01 % of all researched cases) and in ACM – 67 articles (n=67, 5.28 % of all researched cases) have been found.

In Springer Link, the most popular concept was *virtual tour* (634 articles), while the *nuclear education* concept was much less popular (48 articles). In this database, *computer science* (230 articles) and *engineering* (74 articles)–related topics dominated (Fig. 2). Empirical studies on virtual tours covered technological peculiarities of user interface construction and human-computer interaction testing in artificially created environments (e.g. Zhang & Zhu, 2017; Checa, Alaguero & Bustillo, 2017); information systems application for indoor and

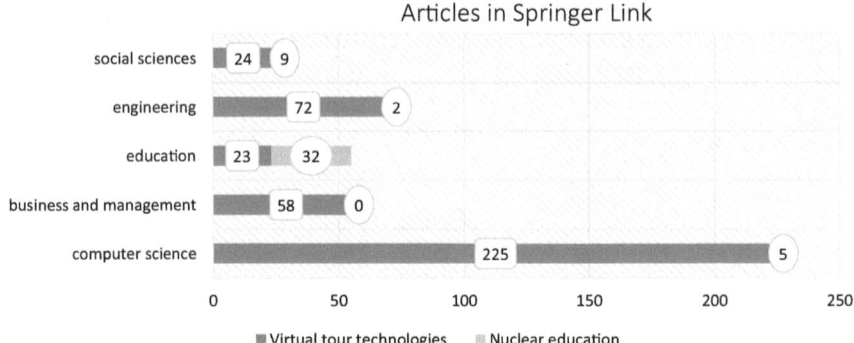

Fig. 3: The number of articles from the five most popular scientific fields researching nuclear education and virtual tours in *Springer Link* online collection of scientific, technological and medical journals, books and reference works (N=634; the number of articles researching virtual tours in the rectangle; the number of articles researching nuclear education in the oval)

outdoor museum exhibitions (e.g. Fabrizio, Chara & Brumana, 2018; Kersten, 2018), immersive web-based, panoramic, virtual, and augmented reality tours (e.g. Debailleux, Hismans & Duroisin, 2018; Bruno et al., 2016), VR games (e.g. Zhang et al., 2018; Iacono, Zolezzi & Vercelli, 2018), storytelling (e.g. Carrozzino et al., 2018; Battad & Si, 2016), etc. (Fig. 3).

In Scopus, empirical articles on *virtual reality solutions* and *virtual tour technologies* were the most popular (Fig. 4). The tourism computerization process, design of virtual tourist routes (e.g. Voronkova, 2018; Bruno et al., 2017); 3D and augmented guided tours and excursions (e.g. Lee, 2017), virtual and GPS-based navigation solutions (e.g. Wang & Chen, 2018), virtual and mobile museums, exhibitions, nature and cultural heritage, historical and wildlife preservations (e.g. Podzharaya & Sochenkova, 2018; Kersten et al., 2018), computer graphics, data visualization, 3D restorations of heritage objects (e.g. Cha et al., 2018; Castagnetti, Giannini & Rivola, 2017) were empirically researched in outdoor and outdoor educational environments, formal and non-formal learning settings. A total 65 articles empirically examined *nuclear education* and *atom engineering*–related topics. These topics covered nuclear literacy, nuclear information, and nuclear technology education. Luk et al. (2018) presented immersive virtual reality systems for nuclear literacy. Some empirical research focused on nuclear safety and raising public awareness (e.g. Wang

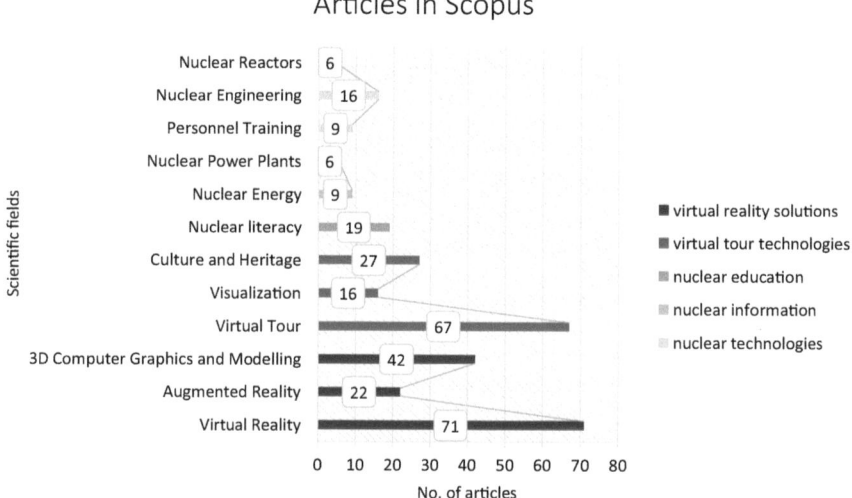

Fig. 4: Most popular article topics, researching virtual reality solutions, virtual tours and formal and informal nuclear education and training in *Scopus* peer-reviewed journals (N=372)

et al., 2017; Liu & Xia, 2014) and virtual laboratories and simulators for nuclear power plant specialist training (e.g. Yakovlev et al., 2015; Gatto et al., 2013). The authors stated that immersive learning could grab learners' attention, build an interactive educational relationship, develop a sense of belonging to nature and community, and activate life-long learning action.

In Taylor & Francis, 25 empirical articles researching *nuclear education* were found (Fig. 5). Here, studies on nuclear (atomic) literacy were published (e.g. Carson, 2018; Volpe & Kühn, 2017). The studies on virtual tours examined tours as educational phenomena. Orru, Kask and Nordlund (2019) empirically investigated social and individual motivational factors governing satisfaction with virtual nature touristic routes. These authors confirmed that a good foundation story and educational narrative may expand enthusiastic reactions and emotional responses. Virtual field trips as a technique for experiential learning in school were studied by Kenna and Potter (2018). These researchers discussed the benefits and limitations of virtual field trips to students and presented different cases of virtual field trips, stating that virtual tours are "the most viable means of accessing the world outside the classroom to incorporate experiential and authentic activities into the daily curriculum" (ibid.).

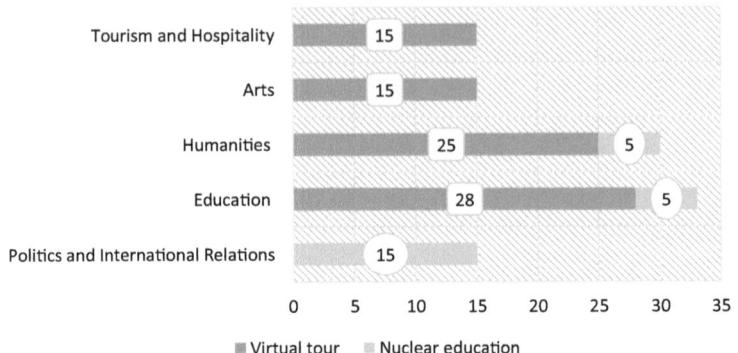

Fig. 5: Most popular article topics on a virtual tour and nuclear education-related topics in *Taylor & Francis* books and academic journals (N=165, number of articles researching virtual tours in the rectangle; the number of articles researching nuclear education in the oval)

The ACM Digital Library is the world's most exhaustive database of scholar publications and bibliographic materials covering computing and information technology. In this database, empirical articles that investigate the atomic instruction idea were not found. The most popular topics covering the virtual tour conceptual area are presented in Fig. 6. In the empirical articles, provided in the ACM database, virtual reality software and technology and virtual tutoring environments were comprehensively researched. Software and hardware systems for virtual navigation, such as 3D virtual scene generation (e.g. Wang & Chen, 2018), artificial agents as tutors (e.g. Cafaro, Vilhjálmsson, & Bickmore, 2016), and many more modern information technology–related topics, repeated in the previously mentioned databases, were contemplated. In the articles published in ACM journals, non-natural landscapes, embodiment, relational intelligence, human-like appearance, and non-verbal behavior were analyzed in artificially created environments.

The literature screening revealed three additional conceptual areas directly linked to *nuclear education* and *virtual tours* with the focus on STEM and virtual environments – *(Nuclear) Serious Games, Digital Guiding* and *Nuclear Tourism Routes*. Nuclear serious games are electronic games that have the purpose to educate, train, and change the learner's behavior through entertainment in the areas of nuclear literacy, atom physics, environmental security or related

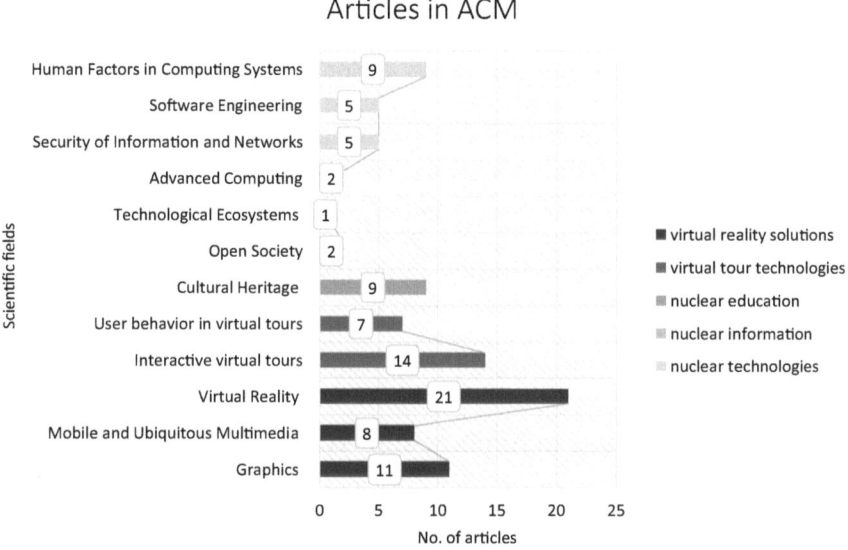

Fig. 6: Most popular article topics, researching virtual reality solutions, virtual tours and formal and informal nuclear education and training in *ACM* digital library (N=67)

subject areas by applying various problem solving, challenges, rewards, and other engagement components provided in virtual gameplay environments. Technically speaking, serious games could be computer or mobile games, simulations or interactive models, virtual environments, augmented or mixed reality and social media meeting places that provide opportunities to educate or train through responsive narrative and story, gameplay and encounters. Digital guiding is an educational activity aimed at virtually transmitting information about original objects, cultural and natural resources, constructing subjective meanings and establishing an experiential relationship, and instilling understanding and appreciation of the interpreted environment. Technically, digital guiding is done with audio or video technologies, text, or interactive communication. The nuclear tourism route is a virtual walk on specially selected areas important to nuclear energy. The learner usually constructs the nuclear tourism route himself or herself of freely chosen virtual paths, scenes, or game environments with known social, topographical, or economic accents – including images, recreation areas, and interpretive regions that reveal certain features and aspirations for nuclear literacy. In the studied scientific articles, nuclear museum education was touched very superficially. Supposedly, it is under investigation, but in this context, virtual solutions integration and game-based learning have not been sufficiently explored. Links between empirically researched conceptual areas based on researched concepts, most popular scientific fields observed in multidisciplinary and monodisciplinary peer-reviewed articles, and teaching, training, and learning challenges are presented in Tab. 1 and Tab. 2. Substantial challenges are related to the use of smart technology and the relative reluctance or inability of experts to communicate scientific knowledge in a way that is understandable to the public. For example, as modern technology solutions are constantly evolving, formal education institutions and individual learners often do not have access to the latest virtual reality devices. In addition, some specialized software is expensive.

Educational Gaming to Explore and Analyze Real-Life Issues

A variety of educational text-based materials, virtual games, and apps exist online for raising nuclear awareness and atom literacy. For example, ANSTO – one of Australia's largest public research organizations, internationally recognized players in the field of nuclear science and technology – runs a portal for business, education, and public science (ANSTO, 2019). The portal operates in the English language, thus making it accessible not only to the local reader, but also to the international audience. This portal contains general facts about nuclear science, radiation, radioisotopes, synchrotron light, and managing waste. In ANSTO, nuclear energy experts and professionals create and share educational materials. The educational and informational texts are specially adapted to different target groups: primary, secondary, tertiary education, and materials for teachers. For primary school children, ANSTO offers nuclear competitions such as *Shorebirds in Botany Bay* (raising awareness of the plight of endangered shorebirds in specific territories and local habitats that are important for shorebirds and other organisms) or *Top Coder* (mobile technology, coding, computer programming, and robotics in collaboratively environments). These educational activities stimulate thinking, develop a creative personality, and shape passionate involvement in problem-solving. For primary school children, educational activities are live and are only advertised online in ANSTO portal. For the secondary school children, workbooks and datasets are provided. These workbooks and datasets can be used in formal classes of science, chemistry, physics, or biology. Additionally, electronic workbooks can serve as required learning materials accompanying

Tab. 1: The challenges in virtual technology solutions and virtual tour technologies used within most popular scientific fields and educational settings

Concepts	Scientific fields				Challenges	Sample research
	Springer Link	Scopus	Taylor & Francis	ACM		
Virtual reality solutions		*Virtual reality*, 3D computer graphics, and modeling augmented reality		*Virtual reality*, graphics, mobile and ubiquitous multimedia	i) Although technology offers unique simulation and visualization opportunities for use in everyday life situations, urban and environmental planning; could teach a healthy lifestyle; healthier and safer living, the biggest challenge stays technology acceptance. ii) Immersion in technology activities may become unmanageable and hard to self-regulate.	Luk et al., 2018; Checa, Alaguero & Bustillo, 2017; Zhang & Zhu, 2017; Boulos et al., 2017

(continued on next page)

Tab. 1: Continued

Concepts	Scientific fields		Challenges	Sample
Virtual tour technologies	Computer science, engineering, business and management, social sciences (general), education	Virtual tour, culture and heritage, visualization	Interactive virtual tour, user behavior in virtual tours	i) Users are engaged in complex problem solving that requires coordination of multiple concepts to define (effective) solutions. ii) The tools and technologies such as virtual helmets, glasses, and smart devices could be too expensive for the individual user.	Debailleux, Hismans & Duroisin; 2018; Castagnetti, Giannini & Rivola, 2017; Battad & Si, 2016; Bruno et al., 2016; Bohlin & Brandt, 2014; Neuhofer, Buhalis & Ladkin, 2014

Tab. 2: The challenges in nuclear education, nuclear information, and nuclear technologies used within most popular scientific fields and educational settings

Concepts	Scientific fields				Challenges	Sample research
	Springer Link	Scopus	Taylor & Francis	ACM		
Nuclear education	*Education*, computer science	Nuclear literacy	Politics and international relations, *education*, humanities	Culture heritage	The gap between technology and learning methods: difficult integration of technological advances into teaching; a danger that mature educational technologies and methods might not give an adequate answer to the demands and needs of society.i)	Volpe & Kühn, 2017; Nakamura, 2016; Ahn et al., 2015; Liu & Xia, 2014
Nuclear information		Nuclear energy, nuclear power plants		An open society, technological ecosystems		Tsubokura, Kitamura & Yoshida, 2018; Carson, 2018; Wang et al., 2017; Nakamura, 2016; García-Peñalvo et al., 2015; Ahn et al., 2015
Nuclear technologies		Nuclear engineering, personnel training, nuclear reactors, waste repositories		Human factors in computing systems, the security of information and networks, software engineering, advanced computing	Lack of information and initial knowledge to understand complex nuclear solutions.	Gan & Yang, 2017; Salmani-Ghabeshi et al., 2016; Ramchurn et al., 2016

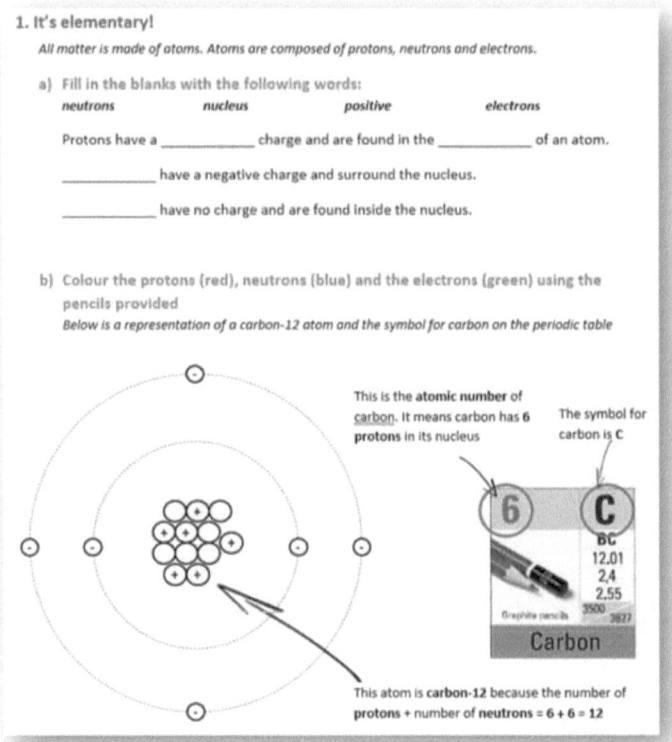

Fig. 7: An example of hands-on nuclear science workbook for secondary school children (7 to 10 years old), which could be used during a live class excursion to ANSTO (extracted from ANSTO files, 2019)

a live school excursion to the laboratories of Australia's Nuclear Science and Technology Organization. The workbooks are freely downloadable online. The exercises in the workbooks are divided into topics. Standard question types are used, such as calculated simple and multichoice, essay, description, typing, matching, gap-fill, and others. An example of the workbook is presented in Fig. 7.

In addition to workbooks and datasets, secondary school children and their teachers can virtually meet nuclear experts and participate in video-conferencing sessions. During these sessions, students plan and investigate their first physical or chemical experiments. Virtual access to high-quality radioactive sources, instruments, and scientific expertise are provided. Videoconferences last 45 minutes (a traditional lecture time) and need to be ordered in advance. During virtual sessions, students can ask questions, discuss their nuclear education-related experiments and receive expert feedback. A piece of special equipment for measuring and detection, radioactive sources and objects, and radiation shielding or similar tools are needed for experimenting. The practical training can only be done live in class under the teacher's or instructor's supervision. For tertiary education, early carrier programs are provided. Furthermore, ANSTO offers different virtual reality, mobile and online games, and apps to discover the world of nuclear science. For example, children can explore how much radioactivity it is possible to absorb in daily life, learn about health protection, the periodic table, and the atom building.

Dalton Nuclear Institute at the University of Manchester virtually shares tools, games, and information sources about nuclear energy. In their case, the educational materials are created by the University scientists and their students. Some of the tools, such as *Energy card games* or *Nuclear energy paper fortune tellers*, are offline. Others, such as *Nuclear Reactor Simulator*, are available online.

High tech educational applications, such as dynamic modeling, simulation, and 3D visualizations, are available to download from companies, experts, and individuals. For example, *Nuclear* is a 3D serious game that dynamically models an interactive atom and teaches the periodic table, *Atoms* – educational logic quiz. *Nuclear inc 2* (nuclear power plant simulator) is a serious game that not only teaches how the nuclear reaction works and how the energy of nuclear fuel is converted into electricity, but also educates how to protect yourself and your family against radiation, and explains the causes of nuclear accidents, such as Chernobyl or Fukushima. This app is available in four different languages, has a storyline, and different levels of game difficulty. Another type of application, for example, *Augmented nuclear plants*, contains educational materials in the form of text, pictures, and augmented reality models, which can be used not only informally, but also in formal education classes (Tab. 3).

The use of simulations and serious games in learning is growing. While the theoretical benefits of digital games for formal and informal teaching and learning are constantly being studied, there is still not a big choice of mobile apps that inform and develop atom literacy.

Tab. 3: The list of freely available nuclear educational apps and games

No.	Name	Target learners	Description (purpose)	Creator	Platform
Tests & quizzes					
1	How radioactive are you?	Children	Online self-evaluation test	ANSTO	Online at http://howradioactiveami.com/
2	The Brain Challenge Quiz	Children	Online quiz. Could be combined with online *Nuclear Reactor Simulator*	Dalton Nuclear Institute	Online at http://www.dalton.manchester.ac.uk/connect/learn/brain-challenge/
3	Atoms	Family	2D puzzle	Elvista Media Solutions Corp.	Android
4	Augmented nuclear plants	Formal and informal learners	An introduction to nuclear reaction, fission and fusion lesson, and assignments to students	M. Chardine	Android
Serious games					
1	Half-life hero	Children	Teaching about nuclear medicine and industrial isotopes, and their benefits to society	ANSTO	Online at https://archive.ansto.gov.au/static/halflifehero/ iOS
2	Elementals	Children, teachers	Learning the Periodic Table, supporting science education in the classroom and practicing on the go.	ANSTO	Online at https://archive.ansto.gov.au/elementals/ Android, iOS
3	Nuclear	Family	Learning about each of the elements of the periodic table by constructing a stable version of that element.	Escapist Games	iOS
4	Atom Builder	Children, teachers	Discovering the uses and properties of common isotopes, locating elements in the periodic table.	ANSTO	Online at https://archive.ansto.gov.au/static/atombuilder/

Tab. 3: Continued

No.	Name	Target learners	Description (purpose)	Creator	Platform
5	Nuclear Inc 2	Family	Serious game, teaching and training the basics of managing a nuclear reactor and a nuclear power plant in general.	Lomakin Dmitrij (ru. Ломакин Дмитрий)	Android

Virtually Enhanced Touring to Engage and Interact with New Knowledge

As stated in Yung & Khoo-Lattimore (2019), Oculus (https://www.oculus.com/), Sony (https://www.playstation.com/en-gb/explore/playstation-vr/), Samsung (http://www.samsung.com/global/galaxy/gear-vr/), Google (https://vr.google.com/), HTC (https://www.vive.com), and Microsoft (https://www.microsoft.com/en-cy/hololens) have unveiled virtual and augmented reality products to the mass market. The virtual reality tour to ANSTO's OPAL multipurpose reactor helps to discover how things happen on the atomic scale. Although virtual reality becomes more and more popular, it requires special VR helmets, glasses, and other devices. Because of that, 2D and 3D virtual tours remain popular. For example, *Nuclear Reactor Simulator* or *Nuclear Power Plant Simulator* is an "old fashioned" 2D simulator, developing nuclear literacy among various age audiences (Tab. 4).

The virtually enhanced tourism is becoming very popular, but our research has shown that there are only a few freely accessible technology-enhanced atom tourism routes. To stimulate learning, these routes not only need to have elements of experiences, but also be co-created together with teachers, instructors, and other learners. The researched concepts and technological solutions are presented in Tab. 5.

While playing serious games and traveling virtual journeys, formal and non-formal learners can enhance their nuclear education. They can learn from digital books and educational apps and self-virtually evaluate their advancement with tests, quizzes, puzzles, and similar techniques. Virtual and augmented reality tours, and online and mobile games are gaining popularity. Digital text and voice guiding help players to navigate virtual routes. Through these technological solutions, online tours and mobile apps are developed. In passive instruction-based learning (Anderson, 2008), the learner acts only as

Tab. 4: The list of popular freely available nuclear education tours and simulators

No.	Name	Target learners	Description	Creator	Platform	Online guiding	Touristic route
1	ANSTO VR	Family	VR tour inside Australia's OPAL multi-purpose reactor	ANSTO	Android, iOS	Voice, text	yes
2	Nuclear Reactor Simulator	Learners, teachers	2D nuclear reactor simulator	Dalton Nuclear Institute	Online at http://www.dalton.manchester.ac.uk/connect/learn/nrs/	Voice	no
3	Power Plant Engineering	Formal and informal classes and individual learners	Handbook of Power Plant Engineering, covering reference materials and digital book	Softonic	Android	text	no
4	Nuclear Power Plant Simulator	Family	The goal is to produce enough electricity to light up the entire city without causing a dreaded nuclear power plant meltdown.	Majik Mike Simulators	Online at http://www.nuclearpowersimulator.com/	text	no
5	Nuclear Power Plants	Family	The description of nuclear power plants from all around the world	Kirill Sidorov	Android	text	no

Tab. 5: Concepts and most popular technological solutions for virtual nuclear education

Concepts	Technological solutions	Challenges
Virtual reality solutions	Online games, game apps	It is difficult for the user to accept immersive experiences and feeling of presence; a new level of interaction achieved through all human senses
Virtual tour technologies	VR/AR tours, simulators, navigation using text, voice, maps, and GPS data	Challenges with virtual reality hardware – mobility, freedom of movement, speedy internet, data security
Nuclear education	Educational apps, online tests, online quizzes, online puzzles, digital books, online games, game apps, VR/AR tours, and simulators	Challenges with high-quality educational apps - expensive to build, quality graphics, much advertising in free apps, convincing teachers to use informal education classes, assessment, and evaluation of learning results and achievements
Nuclear information	Online tours, mobile apps, 360 and panorama videos	Some learners find it difficult to immerse into virtual activities, they experience a disconnection between online presence and live behavior
Nuclear technologies	Complex calculations, modeling, VR/AR tours	Lack of nuclear staffing able to produce virtual materials for specialists, not all scientific material is freely available on the internet

an absorber of information materials, and the teacher provides the knowledge reflected only in digital books and digital guiding using online text solutions. In our research, we concentrate on learning which takes a more active form of acquiring and accumulating knowledge. The learner himself or herself decides what information he or she needs, chooses to learn formally or informally, collects online information, constructs knowledge, and formulates the meaning of the given material. Internet and smart technologies enhance the learning experience. For example, an online self-evaluation test "How radioactive you are?" explains natural radiation (using text, images), shows practical examples (with photos, images, videos), and allows the learner to self-evaluate (using an interactive test). In this example, all the phases of Kolb's experiential learning theory (e.g. Kolb, 1984; Kolb, Boyatzis & Mainemelis, 2001) are supported. Interactivity is created by providing a variety of teaching and learning contents and enabling the learner to decide what content to choose or which path to follow. For instance, when examining augmented nuclear plants, the learner observes online materials, conceptualizes and plans the new virtual experiments, practices and self-evaluates on the net, and can make new observations. A similar search of experience was observed in technology-enhanced tourism (Neuhofer, Buhalis & Ladkin, 2014). In their analysis of technology as an enhancer of experience, these authors described two competing experiential learning scenarios. In the first scenario, technology was an integral part of the co-construction of the tourism experience. In the second scenario, technology played a complementary role in acquiring a tourist experience (ibid.). Our framework falls into the first scenario, manifesting contemporary technology as an active co-creator of nuclear education virtual experiences.

One of the biggest challenges to encourage learners to actively co-create knowledge is that of engaging students. Serious games can provide such motivating and engaging learning experiences (Kiili, 2005). In serious games, learning is explained as a cyclic process through direct immersive experience and problem-solving in the game world: permanent action and continuous practices (ibid.). For example, in ANSTO serious games *Half-life hero*, the player is a scientist who must solve real-life problems, manage the nuclear reactor, and save his country from a catastrophe. For this, a quick reaction to decision-making is necessary. The time-based element of gameplay provides a challenging, yet rewarding mechanism for players of all skill levels, while the game's quirky design appeals to kids and adults alike. In this serious game, real-life problems are presented in a fun and attractive way, which creates an immersive experience. For Kiili (2005), an experiential gaming model is based on three theories: experiential learning theory (Moon, 2013), flow theory (Nakamura & Csikszentmihalyi, 2009), and game design. In our study, the serious games part is a siding of nuclear education that is explained from experiential learning lenses (as in Kiili's), adding immersion and problem-solving.

Geolocation Technologies of Virtual Tours Development

Geolocation is a technology that uses data acquired from an individual's computer or mobile device to identify or describe the user's actual physical location (Kapoun, 2016). Geolocation technology collects two types of data. It is important to gather information about the learner or their device and the data server. Data correlation and cross-references are then performed to produce the most accurate result. (Estes, 2016). In virtual nuclear education, it is very important to explain not only the physics or benefits of the atom to man and nature but also to visually show the specific locations associated with the atom (uranium ore mines, nuclear power plant construction sites, other specific locations). Exploration of the modern nuclear world and the history of an atom can be done in two ways. Firstly, it is possible to travel and participate in a guided tour or visit science and technology museum exhibitions. When learning about an atom, this method is not always appropriate. Even dormant uranium ore mines or nuclear power plants do not usually allow visitors to come because of special security reasons. Secondly, an individual learner could virtually reach the desired place of the world, while sitting at home or school in front of a computer or smart device and exploring nuclear history, culture, peculiarities of hard-to-reach countries, places, and spaces. This method is especially useful when visiting sensitive areas or remote objects. Widespread *Google* tools, free online services, and products allow teachers and their students not only to reach atom-related sites but to create their personal virtual tours and itineraries without special programming skills. Examples of such tools are *Google Maps* (google.com/maps), *Earth* (google.com/earth), *Tour Builder* (tourbuilder.withgoogle.com), *Tour Creator* (arvr.google.com/tourcreator), and others. Teachers can use these tools in their lessons of geography, physics or history, designing teaching materials, and presenting active tasks to learners. These tools help develop not only learner's nuclear literacy, but general abilities, such as learning to learn, creativity, or communication.

Gorelick et al. (2017) studied *Google Earth* Engine "as a cloud-based platform for planetary-scale geospatial analysis that brings Google's massive computational capabilities to bear on a variety of high-impact societal issues including deforestation, drought, disaster, disease, food security, water management, climate monitoring and environmental protection" (p. 18). With *Google Earth* or *Google Maps* tools, learners can explore any place on the world map. *Earth Studio* (google.com/earth/studio) is an animation tool for Google Earth's satellite and 3D imagery. This tool help educators to record and explain lessons with animated videos and 3D pictures. Audiovisual materials could be linked to a specific location or place in the world. When developing interactive online assignments with these tools, educators can apply integrated materials to nature, humanities, or STEM lessons. Investigating how children accomplish place in everyday lives, Danby et al. (2016) showed examples how to teach preschool children geography and social interaction with *Google Earth* tool. This tool helped researchers recognizing children's competence to manipulate their social and digital worlds. Investigating to what extent the implementation of a *Google Earth*-based science curriculum increased students' understanding of nature structures, developing scientific reasoning abilities, and constructing science identity, Blank et al. (2016) discussed that students, who applied geospatial technologies in their learning, developed not only specific knowledge of earth understanding, but their science identity, and science reasoning.

Geolocation storytelling creates an interactive, and emotional, connection to learning, engaging such skills as curiosity, critical thinking, empathy, and in some instances, a call to action. In order to make educational stories easier to understand for learners of all ages, researchers recommend that it be presented in the form of static and dynamic images, not just text. Currently, the most popular forms of presenting this type of material are images, videos, infographics, and charts. By creating a variety of video materials, educators expect learners' engagement and motivation to increase. One of the tools for creating inclusive nuclear teaching materials and geolocation stories is *Tour Builder*. The *Tour Builder* was developed for informal sharing and peer-learning of adults and military veterans. Now, this tool is more widely used in formal education and informal classes. *Tour Builder* could serve as an interactive storytelling tool that connects learners to places using *Google Maps* and multimedia content. When creating an integrated educational material with *Tour Builder*, the teacher could explain the history of a specific place or power plant, tell nuclear stories with text, photos, pictures, and video materials. *Tour Builder* could be used to research the locations of famous scientific discoveries, create a tour of unusual geological features, explain how to spend summer vacation, explore the famous science and technology museums around the country, and virtually participate in physical and chemical reactions or manage nuclear processes. Teachers, sharing their ideas how to creatively use this tool, talk about using *Tour Builder* for their animal habitats and zoology classes, geographic biomes, explanation of weather and climate, social science lectures of indigenous people, and language studies. The possibilities of *Tour Builder* application are limited only by educator's imagination. Learners could use the tool to tell and share nuclear stories and personal experiences. They could link telling with a specific location on the map (Fig. 8).

Tour Creator allows educators to create 360-degree virtual tours and gamified scenes. To enhance their tours, educators could add audio recordings to scenes and link them with the specific place on Google map. The educational material created by this tool can be linked to virtual tours or exhibitions inside and outside museums and science and technology centers. Virtual educational tours could be embedded on educational website, virtual learning environment, blog, or social network. *Tour Creator* can be used creatively for analyzing any educational material and developing active tasks and assignments. For example, in a nuclear biology lesson, it is possible to create a virtual tour of the human eye – to observe the physiological and medical structure of the eye, the essential biological functions, and other topics (Fig. 9).

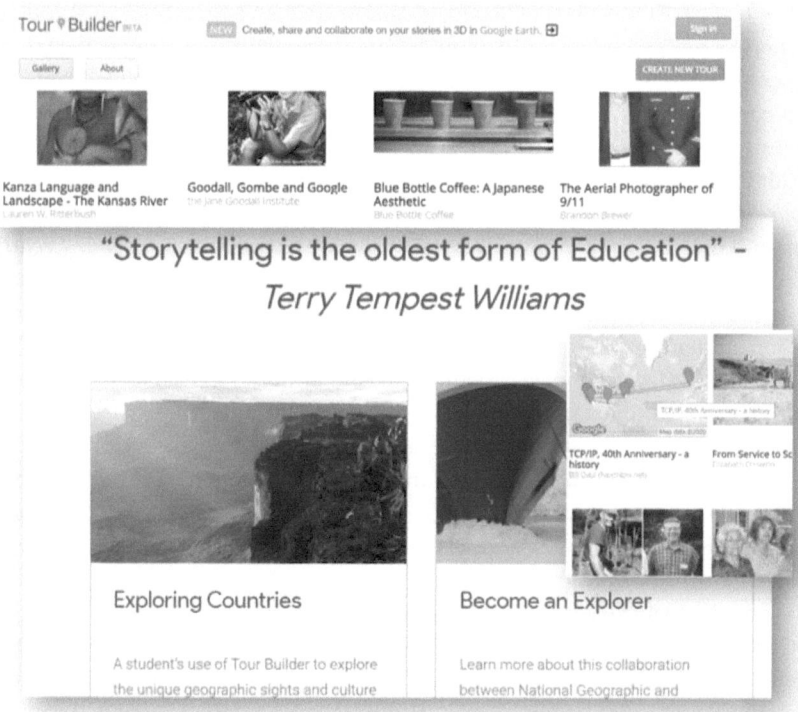

Fig. 8: *TourBuilder* story gallery

Fig. 9: Teaching and learning materials for nuclear biology classes created with the *Tour Creator*

The *Google Expeditions* (*edu.google.com/products/vr-ar/expeditions*) tool lets teachers combine virtual reality content and supporting learning material into one collection. Such an expedition or virtual tour can be installed not only on a computer but also on any smart device of a teacher or student. The tool allows teachers, along with their students, to visit virtually anywhere in the world: visit the most famous science and technology museums, observe complex or dangerous technical processes and reactions, earth and space, mountains and the ocean, and more. Teachers, who are experts in using *Tour Creation* technology say that the *Google Expeditions* app for mobile allows educators to guide tours with students following along. This application permits the teacher to keep students at the same pace while they discuss different scenes as a class. With the *Expeditions* app, classrooms have no boundaries. Davis and Schmutz (2019) presented an example, how *Google Expeditions* could be used to engage school children in the learning process. They provided examples from history, biology, anatomy, and other classes and guided teachers to use virtual reality applications in formal school settings. By presenting practical examples, Davis and Schmutz (2019) motivate educators not only to use pre-made entries in the *Google Expeditions* program, but also to create their virtual educational tours, and to make their virtual reality tours with a 360° camera.

GeoGuessr tool (*geoguessr.com*) lets individuals to create maps of places they visited and link these places to *Google Maps*. Learners could explore the places, play educational geolocation games, and perform teacher-created assignments (Fig. 10).

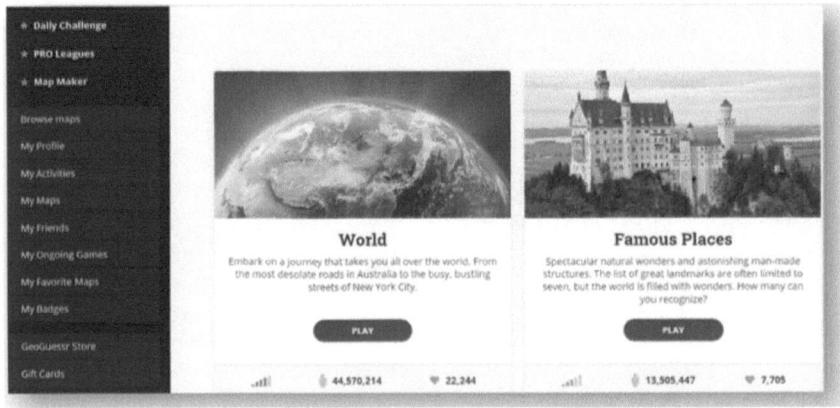

Fig. 10: *A snippet of the GeoGuessr Tool web site*

In *GeoGuessr*, learners could browse and create maps, guess the geographic coordinates of specific places, and play geolocation games. The tool can be used in history, literature, or integrated lessons. Also, the tool can be used for individual or group educational activities. Girgin (2017) presented an example of *GeoGuessr* use in geography classes. These researchers found that all learners who participated in *GeoGuessr* activities during geography lessons enjoyed game-based learning. A lot of them gave positive impressions about the game. While playing, they gained map reading skills, learned geography-related content and grew to solve problems by reasoning competence. Scholars argued that *GeoGuessr* learners reach to information and were motivated to study by themselves.

TheTrueSize (thetruesize.com), *Landlines* (lines.chromeexperiments.com) and *Time-lapse* (earthengine.google.com/timelapse) tools help you understand history, imagine the size of learner country, teach geography, national peculiarities of the country or region, and help learners to conduct geo-experiments on the area or place. The *GeoGreeting* tool (geogreeting.com) can be used for integrated language learning. With this tool, learners can send aerial imagery messages from *Google Earth* and associate foreign language learning with the history and geography of a particular country. The tool allows a learner to enter a message of up to 40 characters and email or share a link. When a learner receives such a message, he or she is associated with a particular place in the world.

Teachers can use *Space* (*google.com/maps/space*) or *AccessMars* (*accessmars.withgoogle.com*) tools to illustrate the physical and geographical structure and location of our planet in the Solar system. These tools use the latest NASA data, maps, and images, 360 videos and audio materials. In this way, students are introduced to the latest scientific achievements. Learning with tools motivate them to explore the world (Fig. 11).

These tools can be used for project-based learning. The creative use of tools in formal and informal settings stimulates students' curiosity and encourages exploration.

In this section, we have reviewed only some of the educational tools that can be used for virtual nuclear tour development. The attractiveness of tools is enhanced by the fact that they are freely accessible to all, can be used in formal and non-formal learning, and teaching with these tools requires only basic skills of internet or smart device usage. All these tools can be easily applied in virtual nuclear education.

Fig. 11: *A snippet of Space and AccessMars tools*

Discussion and Conclusions

This mapping review illustrates the complex nature of virtual reality solutions and virtual tour technologies applicable to online nuclear education, nuclear information, and nuclear technology studies covering topics of empirical scientific research, online, and smart solutions. Experiential learning tools, such as serious games and virtual tours, contribute to teaching while playing. Virtual and augmented reality gaming, going beyond the edges of the real world, occupy different areas, including education. It provides immersive experiences, an advanced level of interaction and augmented serious content. Five main senses, such as sight (visual), sound, touch (tactile feedback), smell, and taste, are activated during VR activities. The technology acceptance model and the plethora of its modifications (e.g. Lee & Lehto, 2013; Marangunić & Granic, 2015) look for an answer to the question – what causes people to accept or reject information technology? (Davis, 1989, p. 320). In our research, we join this discussion (see "accepting technologies" part in Fig. 1). The answers may be the perceived ease of use, the permanent availability, the perceived usefulness, the task-technology fit, content richness, vividness, and other dimensions. Ibrahim, Khalil, and Jaafar (2011) added to this: enjoyment, performance anticipation, and effort expectancy and gaming experience. Nuclear education tours and simulators, for example, ANSTO's VR tour inside Australia's OPAL multipurpose reactor, are an unforgettable virtual or augmented journey that allows us to explore science and make personal discoveries. The key determinants in such tours are ease of use, permanent virtual availability, immersion, fun, guiding, challenging, and raising curiosity (see Fig. 2).

The digital guides help to create virtual educational experiences by interpreting events, organizing activities, explaining places, accommodating spaces, managing time, telling stories, and co-constructing knowledge. According to Bohlin and Brandt (2014), digital guides rest on two pillars. These pillars are technology (hardware and software) and narrative (the story and the way it is composed and delivered to the learner). The first pillar – digital guiding – was significant for the virtual tours and serious games to instruct the learning process, to tell the story, navigate, explain, motivate, and encourage (Battad & Si, 2016). In the virtual tours, directions, roads, do's and don'ts, rules, instructions, helps, additional text, and audio information can be provided. In serious games and virtual tours, virtual guiding informs, helps to experiment, observes, gains knowledge and develops skills, solves educational tasks, challenges, and conceptualize (see Fig. 2). For the virtual guiding, one technical aspect that is not widely discussed in the literature is GPS functionality. In the investigated nuclear touring examples, this was not implemented, although it is becoming popular in modern tours. The second main pillar described in Bohlin and Brandt (2014) – the virtual narrative construction to inform and educate – has been minimally explored in empirical scientific articles. Although in online materials such information was not provided, either, in this case, the contacts of experts who can give more information to the teachers were identified.

Using the tools described in this section, students acquire and develop these competences: critically read, interpret cartographic and other visualizations in different media (interpretation); be aware of geographic information and its representation through GI and GIS (learning about); visually communicate geographic information (produce); describe and use examples of GI applications in daily life and in society (applying); use (freely available) GI interfaces (use); carry out own (primary) data capture (produce or gathering); be able to identify and evaluate (secondary) data (use or evaluate); examine interrelationships (analyze); extract new insight from analysis (produce); and reflect and act with knowledge (action: decision making and applying in real world) (Zwartjes & Torres, 2019).

Limitations

Since virtual and augmented reality technologies are becoming more and more popular, it is important to note that today's virtual and augmented reality solutions are still limited. Newer applications transform virtual and augmented reality into content for the virtual spaces, simulations, and 360 videos. Although technological solutions are getting more accessible to the average consumer, special glasses, hand-mounted displays, sensors, or cameras are still needed for fully immersive experiences. Virtual and augmented reality is motivating, and most of the young generation of learners have a positive attitude towards modern technologies, which are exciting, challenging, and allow to interact, create, and manipulate in virtual environments (Domingo & Bradley, 2018). However, some studies of adult learners reported the lack of AR/VR awareness because of unwillingness to accept virtual substitutes (see *Challenges* part in Tab. 1 and Tab. 4).

While proposing serious games and tours for virtual nuclear education, we can observe certain limitations. In nuclear education settings (formal and informal), these technologies are mainly adapted to the young generation of learners, having positive attitudes to educational technology – to the learners who accept technology. Research shows that not all the students, called digital natives, are competent in using technologies in educational environments. In addition, it is crucial to present the teaching material in a clear, understandable, and attractive way, considering the age and initial preparation of the learner. The biggest challenge is not only to create virtual materials on nuclear education in the text format, but also, together with computer scientists and engineers, to develop virtual narrative, immersive games, simulation apps, and other types of virtual reality solutions. The challenge for nuclear technology researchers remains the communication of scientific information in a learner-friendly format. Having this in mind, the joint forces of academic staff and technology professionals – actively involved in creating virtual learning scenarios, nuclear education, and information materials – are needed to maximize learning benefits.

References

Ahn, J., Carson, C., Jensen, M., Juraku, K., Nagasaki, S., & Tanaka, S. (2015). *Reflections on the Fukushima Daiichi nuclear accident.* Cham: Springer-Verlag GmbH.

Anderson, T. (Ed.). (2008). *The theory and practice of online learning*. Athabasca University Press.

ANSTO: Australia's Nuclear Science and Technology Organization. (26, March 2019). Retrieved from https://www.ansto.gov.au/

Asgari, M., & Kaufman, D. (2004). Relationships among computer games, fantasy, and learning. (22, March, 2021). Retrieved from https://citeseerx.ist.psu.edu/viewdoc/download?doi=10.1.1.534.355&rep=rep1&type=pdf

Battad, Z., & Si, M. (2016, September). Using multiple storylines for presenting large information networks. In *International Conference on Intelligent Virtual Agents* (pp. 141–153). Springer, Cham.

Blank, L. M., Almquist, H., Estrada, J., & Crews, J. (2016). Factors affecting student success with a Google Earth-based earth science curriculum. *Journal of Science Education and Technology, 25*(1), 77–90.

Bohlin, M., & Brandt, D. (2014). Creating tourist experiences by interpreting places using digital guides. *Journal of Heritage Tourism, 9*(1), 1–17.

Boyle, E. A., Hainey, T., Connolly, T. M., Gray, G., Earp, J., Ott, M., ... & Pereira, J. (2016). An update to the systematic literature review of empirical evidence of the impacts and outcomes of computer games and serious games. *Computers & Education, 94*, 178–192.

Boulos, M. N. K., Lu, Z., Guerrero, P., Jennett, C., & Steed, A. (2017). From urban planning and emergency training to Pokémon Go: applications of virtual reality GIS (VRGIS) and augmented reality GIS (ARGIS) in personal, public and environmental health. *International Journal of Health Geographics, 16*(7), 1-11.

Brancucci, R., Flore, M. & von Estorff, U. (2014). Post Fukushima Analysis of HR Supply and Demand. *JRC science and policy reports*. European Commission. Retrieved from https://op.europa.eu/lt/publication-detail/-/publication/cf4cfc35-067c-4851-9e5f-2541ee93d69b/language-en/format-PDF/source-114602021. Accessed at 11.02.2021

Bruno, F., Lagudi, A., Barbieri, L., Muzzupappa, M., Mangeruga, M., Pupo, F., ... & Tusa, S. (2017). Virtual diving in the underwater archaeological site of cala minnola. *The International Archives of Photogrammetry, Remote Sensing and Spatial Information Sciences, 42*, 121–126.

Bruno, F., Lagudi, A., Barbieri, L., Muzzupappa, M., Ritacco, G., Cozza, A., ... & Cario, G. (2016, October). Virtual and Augmented Reality tools to improve the exploitation of underwater archaeological sites by diver and non-diver tourists. In *Euro-Mediterranean Conference* (pp. 269–280). Springer, Cham.

Bryson, J. M., Ackermann, F., Eden, C., & Finn, C. B. (2004). *Visible thinking: Unlocking causal mapping for practical business results*. John Wiley & Sons.

Cafaro, A., Vilhjálmsson, H. H., & Bickmore, T. (2016). First Impressions in Human--Agent Virtual Encounters. *ACM Transactions on Computer-Human Interaction (TOCHI)*, *23*(4), 24.

Carrozzino, M., Colombo, M., Tecchia, F., Evangelista, C., & Bergamasco, M. (2018, June). Comparing Different Storytelling Approaches for Virtual Guides in Digital Immersive Museums. In *International Conference on Augmented Reality, Virtual Reality and Computer Graphics* (pp. 292–302). Springer, Cham.

Carson, L. (2018). Why youth and feminist activism matters: insights from anti-nuclear campaigns in practice. *Global Change, Peace & Security*, *30*(2), 261–269.

Castagnetti, C., Giannini, M., & Rivola, R. (2017). Image-based virtual tours and 3d modeling of past and current ages for the enhancement of archaeological parks: The visualversilia 3d project. In *1st International Conference on Geomatics and Restoration: Conservation of Cultural Heritage in the Digital Era, GeoRes 2017* (Vol. 42, pp. 639–645).

Cha, Y. W., Price, T., Wei, Z., Lu, X., Rewkowski, N., Chabra, R., ... & Ilie, A. (2018). Towards Fully Mobile 3D Face, Body, and Environment Capture Using Only Head-worn Cameras. *IEEE transactions on visualization and computer graphics*, *24*(11), 2993–3004.

Checa, D., Alaguero, M., & Bustillo, A. (2017, June). Industrial Heritage Seen Through the Lens of a Virtual Reality Experience. In *International Conference on Augmented Reality, Virtual Reality and Computer Graphics* (pp. 116–130). Springer, Cham.

Danby, S., Davidson, C., Ekberg, S., Breathnach, H., & Thorpe, K. (2016). 'Let's see if you can see me': making connections with Google Earth™ in a preschool classroom. *Children's Geographies*, *14*(2), 141–157.

Davis, E., & Schmutz, A. (2019). Using Google Expeditions to Introduce Virtual Reality. Online at https://digitalcommons.usu.edu/cgi/viewcontent.cgi?article=3077&context=extension_curall, Accessed 02.11.2021.

Davis, F. D. (1989). Perceived usefulness, perceived ease of use, and user acceptance of information technology. *MIS quarterly*, 319–340.

Debailleux, L., Hismans, G., & Duroisin, N. (2018). Exploring Cultural Heritage Using Virtual Reality. In *Digital Cultural Heritage* (pp. 289–303). Springer, Cham.

Domingo, J. R., & Bradley, E. G. (2018). Education student perceptions of virtual reality as a learning tool. *Journal of Educational Technology Systems*, 46(3), 329–342.

Estes, B. (2016). Geolocation—The Risk and Benefits of a Trending Technology. *ISACA Journal*, 5, 1–6.

Fabrizio, B., Chiara, S., & Brumana, R. (2018, October). A Digital Workflow for Built Heritage: From SCAN-to-BIM Process to the VR-Tour of the Basilica of Sant'Ambrogio in Milan. In *Euro-Mediterranean Conference* (pp. 334–343). Springer, Cham.

Gan, L., & Yang, S. (2017). Legal context of high level radioactive waste disposal in China and its further improvement. *Energy & Environment*, 28(4), 484–498.

García-Peñalvo, F. J., Hernández-García, Á., Conde, M. Á., Fidalgo-Blanco, Á., Sein-Echaluce, M. L., Alier, M., ... & Iglesias-Pradas, S. (2015, October). Learning services-based technological ecosystems. *Proceedings of the 3rd International Conference on Technological Ecosystems for Enhancing Multiculturality*, ACM. 467–472.

Gatto, L. B. S., Mól, A. C. A., dos Santos, I. J. L., Jorge, C. A. F., & Legey, A. P. (2013). Virtual simulation of a nuclear power plant's control room as a tool for ergonomic evaluation. *Progress in Nuclear Energy*, 64, 8–15.

Girgin, M. (2017). Use of Games in Education: GeoGuessr in Geography Course. *International Technology and Education Journal*, 1(1), 1–6.

Goldstein, T. R., & Lerner, M. D. (2018). Dramatic pretend play games uniquely improve emotional control in young children. *Developmental science*, 21(4), e12603.

Gorelick, N., Hancher, M., Dixon, M., Ilyushchenko, S., Thau, D., & Moore, R. (2017). Google Earth Engine: Planetary-scale geospatial analysis for everyone. *Remote sensing of Environment*, 202, 18–27.

Grant, M. J., & Booth, A. (2009). A typology of reviews: an analysis of 14 review types and associated methodologies. *Health Information & Libraries Journal*, 26(2), 91–108.

Hromek, R., & Roffey, S. (2009). Promoting Social and Emotional Learning With Games: "It's Fun and We Learn Things". *Simulation & Gaming*, 40(5), 626–644.

Iacono, S., Zolezzi, D., & Vercelli, G. (2018, June). Virtual Reality Arcade Game in Game-Based Learning for Cultural Heritage. In *International Conference on Augmented Reality, Virtual Reality and Computer Graphics* (pp. 383–391). Springer, Cham.

Ibrahim, R., Khalil, K., & Jaafar, A. (2011). Towards educational games acceptance model (EGAM): A revised unified theory of acceptance and use of technology (UTAUT). *International Journal of Research and Reviews in Computer Science*, *2*(3), 839-846.

Kapoun, P. (2016). Geolocation services in education outside the classroom. *International Journal of Research in E-learning IJREL*, *2*(1), 57–70.

Kenna, J. L., & Potter, S. (2018). Experiencing the world from inside the classroom: Using virtual field trips to enhance social studies instruction. *The Social Studies*, *109*(5), 265–275.

Kersten, T. P. (2018, October). 3D Models and Virtual Tours for a Museum Exhibition of Vietnamese Cultural Heritage Exhibits and Sites. In *Euro-Mediterranean Conference* (pp. 528–538). Springer, Cham.

Kersten, T. P., Tschirschwitz, F., Deggim, S., & Lindstaedt, M. (2018, October). Virtual Reality for Cultural Heritage Monuments–from 3D Data Recording to Immersive Visualisation. In *Euro-Mediterranean Conference* (pp. 74–83). Springer, Cham.

Kiili, K. (2005). Digital game-based learning: Towards an experiential gaming model. *The Internet and higher education*, *8*(1), 13–24.

Kolb, A. Y., & Kolb, D. A. (2012). Experiential learning theory. In *Encyclopedia of the Sciences of Learning* (pp. 1215–1219). Boston, MA: Springer.

Kolb, D. A., Boyatzis, R. E., & Mainemelis, C. (2001). Experiential learning theory: Previous research and new directions. *Perspectives on thinking, learning, and cognitive styles*, *1*(8), 227–247.

Kolb, D.A. (1984). *Experiential learning: experience as the source of learning and development*. Englewood Cliffs, New Jersey: Prentice-Hall.

Lal, S., Donnelly, C., & Shin, J. (2015). Digital storytelling: an innovative tool for practice, education, and research. *Occupational therapy in health care*, *29*(1), 54–62.

Lee, B. W. (2017, November). An Innovative Way of Guided Tour: A Virtual Experience of Dark Tourism. In *2017 International Conference on Information, Communication and Engineering (ICICE)* (pp. 208–210). IEEE.

Lee, D. Y., & Lehto, M. R. (2013). User acceptance of YouTube for procedural learning: An extension of the Technology Acceptance Model. *Computers & Education*, *61*, 193–208.

Lefebvre, C., Glanville, J., Beale, S., Boachie, C., Duffy, S., Fraser, C., ... & Smith, L. (2017). Assessing the performance of methodological search filters to improve the efficiency of evidence information retrieval: five literature reviews and a qualitative study. *Health technol assess*, *21*(69), 1–182.

Liu, M, & Xia, H. "Awareness of Danger: A Factor May Improve Public Acceptance to Nuclear if Included in Education." *Proceedings of the 2014 22nd International Conference on Nuclear Engineering*. Prague, Czech Republic. July 7–11, 2014. V006T12A002. ASME. https://doi.org/10.1115/ICONE22-30856

Lorenc, T., Clayton, S., Neary, D., Whitehead, M., Petticrew, M., Thomson, H., . . . & Renton, A. (2012). Crime, fear of crime, environment, and mental health and wellbeing: mapping review of theories and causal pathways. *Health & place*, 18(4), 757–765.

Luk, B. L., Lam, M. L., Chen, T. H., Zhao, J., Tsui, S. M., & Chieng, C. C. (2018, July). 3D Immersive Display Application for Nuclear Education and Public Acceptance. In *2018 26th International Conference on Nuclear Engineering*.

Malone, T. W. (1980, September). What makes things fun to learn? Heuristics for designing instructional computer games. In *Proceedings of the 3rd ACM SIGSMALL symposium and the first SIGPC symposium on Small systems* (pp. 162-169).

Marangunić, N., & Granić, A. (2015). Technology acceptance model: a literature review from 1986 to 2013. *Universal Access in the Information Society*, 14(1), 81–95.

McMillan, J. H., & Schumacher, S. (2010). *Research in Education: Evidence-Based Inquiry*. Essex: Pearson.

Moon, J. A. (2013). *A handbook of reflective and experiential learning: Theory and practice*. NY: Routledge.

Moseley, C., Summerford, H., Paschke, M., Parks, C., & Utley, J. (2020). Road to collaboration: Experiential learning theory as a framework for environmental education program development. *Applied Environmental Education & Communication*, 19(3), 238-258.

Mostafa, A. (2018). *Mediating Experiential Learning in Interactive Immersive Environments* [Unpublished doctoral thesis]. University of Calgary, Calgary.

Nakamura, J., & Csikszentmihalyi, M. (2009). Flow theory and research. In C. R. Snyder and Shane J. Lopez (Eds.), *The Oxford Handbook of Positive Psychology*, (pp. 195–206). Oxford: Oxford University press.

Nakamura, T. (2016). Training project in higher education: regional rehabilitation for safer and more secure society in Fukushima without nuclear energy. In K. Shiwaku, A. Sakurai & R. Shaw (Eds.), *Disaster Resilience of Education Systems* (pp. 131–141). Tokyo: Springer.

Napolitano, R. K., Scherer, G., & Glisic, B. (2018). Virtual tours and informational for conservation of cultural heritage sites. *Journal of Cultural Heritage*, 29, 123–129.

Neuhofer, B., Buhalis, D., & Ladkin, A. (2014). A typology of technology-enhanced tourism experiences. *International Journal of Tourism Research*, 16(4), 340–350.

Orru, K., Kask, S., & Nordlund, A. (2019). Satisfaction with virtual nature tour: the roles of the need for emotional arousal and pro-ecological motivations. *Journal of Ecotourism*, 18(3), 221-242.

Podzharaya, N. S., & Sochenkova, A. S. (2018, February). The virtual museum development with the use of intelligent and 3d technologies on the basis of the Maritime museum in Kotor. In *2018 23rd International Scientific-Professional Conference on Information Technology (IT)* (pp. 1–4). IEEE.

Potkonjak, V., Gardner, M., Callaghan, V., Mattila, P., Guetl, C., Petrović, V. M., & Jovanović, K. (2016). Virtual laboratories for education in science, technology, and engineering: A review. *Computers & Education*, 95, 309–327.

Ramchurn, S. D., Wu, F., Jiang, W., Fischer, J. E., Reece, S., Roberts, S., ... & Jennings, N. R. (2016). Human–agent collaboration for disaster response. *Autonomous Agents and Multi-Agent Systems*, 30(1), 82–111.

Romero, M., Sawchuk, K., Blat, J., Sayago, S., & Ouellet, H. (2016). *Game-Based Learning across the Lifespan*. New York: Springer Science+ Business Media,

Salen, K., Tekinbaş, K. S., & Zimmerman, E. (2004). *Rules of play: Game design fundamentals*. London: MIT press.

Salmani-Ghabeshi, S., Palomo-Marín, M. R., Bernalte, E., Rueda-Holgado, F., Miró-Rodríguez, C., Cereceda-Balic, F., ... & Pinilla-Gil, E. (2016). Spatial gradient of human health risk from exposure to trace elements and radioactive pollutants in soils at the Puchuncaví-Ventanas industrial complex, Chile. *Environmental pollution*, 218, 322–330.

Toth, E., & Marx, G. (1996, October). Nuclear literacy—Hungarian experiences. *Proceedings of the International Conference: Nuclear Option in Countries with Small and Medium Electricity Grids*, 7–9.

Tsubokura, M., Kitamura, Y., & Yoshida, M. (2018). Post-Fukushima radiation education for Japanese high school students in affected areas and its positive effects on their radiation literacy. *Journal of radiation research*, 59(suppl_2), ii65–ii74.

Tung, N. D., Barr, J., Sheppard, D. J., Elliot, D. A., Tottey, L. S., & Walsh, K. A. (2015). Spherical photography and virtual tours for presenting crime scenes and forensic evidence in New Zealand courtrooms. *Journal of forensic sciences*, 60(3), 753–758.

Volpe, T., & Kühn, U. (2017). Germany's Nuclear Education: Why a Few Elites Are Testing a Taboo. *The Washington Quarterly*, 40(3), 7–27.

Voronkova, L. P. (2018, December). Virtual Tourism: on the Way To the Digital Economy. In *IOP Conference Series: Materials Science and Engineering* (Vol. 463, No. 4, p. 042096). IOP Publishing.

Wang, A., Hong, J., Fan, R., Yu, F., Zhang, S., & Liu, Y. (2017, July). The Investigation and Analysis of Public Acceptance in the Surrounding Region of Inland Nuclear Power Plant. In *2017 25th International Conference on Nuclear Engineering* (pp. V008T12A013-V008T12A013). American Society of Mechanical Engineers.

Wang, C. S., & Chen, T. Y. (2018, February). Development of an Intuitive Virtual Navigation System for Mobile Head-Mounted Display. *Proceedings of the 4th International Conference on Virtual Reality,* ACM, 32–37.

Wiemeyer, J., & Hardy, S. (2013). Serious games and motor learning: concepts, evidence, technology. In *Serious Games and Virtual Worlds in Education, Professional Development, and Healthcare* (pp. 197–220). IGI Global.

Yakovlev, D., Pryakhin, A., Korolev, S., Shaltaeva, Y., Samotaev, N., Yushkov, E., & Avanesyan, A. (2015, July). Engineering competitive education using modern network technologies in the NRNU MEPhI. *2015 IEEE Workshop on Environmental, Energy, and Structural Monitoring Systems (EESMS) Proceedings,* IEEE. 39–43.

Yavuz-Topaloglu, M., Demirhan, E., & Atabek-Yigit, E. (2019) Is Nuclear Power Acceptance Related To Environmental Literacy In Turkey? *International Electronic Journal of Environmental Education, 9*(2), 157–173.

Yung, R., & Khoo-Lattimore, C. (2019). New realities: a systematic literature review on virtual reality and augmented reality in tourism research. *Current Issues in Tourism,* 22(17), 2056-2081.

Zhang, L., Qi, W., Zhao, K., Wang, L., Tan, X., & Jiao, L. (2018, July). VR Games and the Dissemination of Cultural Heritage. In *International Conference on Distributed, Ambient, and Pervasive Interactions* (pp. 439–451). Cham: Springer.

Zhang, Y., & Zhu, Z. (2017, June). Walk-able and stereo virtual tour based on spherical panorama matrix. In *International Conference on Augmented Reality, Virtual Reality and Computer Graphics* (pp. 50–58). Cham: Springer.

Zwartjes, L., & y Torres, M. L. D. L. (2019). Geospatial Thinking Learning Lines in Secondary Education: The GI Learner Project. In *Geospatial Technologies in Geography Education* (pp. 41–61). Cham: Springer.

Odeta Norkutė and Natalija Mažeikienė

Energy Literacy in Geography Curriculum: Redefining the Role of Nuclear Power in Changing Energy Landscapes

Abstract: The aim of this chapter is to reveal how different forms of both formal and non-formal education can be combined within the educational route of the Ignalina Nuclear Power Plant (INPP) seeking to develop the energy literacy by employing the geography curriculum. A general question raised in the chapter is how geography curriculum and teaching/learning geography could be improved by developing energy literacy (including nuclear literacy), applying the context-based learning approach and using opportunities of the outdoor learning environment (in this case, nuclear educational tourism on the site of the INPP). Seeking to achieve the aim, the data of the content analysis of geography curriculum and geography textbooks is presented. The analysis of texts on nuclear energy in the textbooks aims to identify the connection of textbook materials on nuclear energy to geographical skills and competences pointed out in comprehensive programmes and how it combines with the taxonomy levels of competences.

The research demonstrated that the national curriculum did not include direct connections to energy literacy; however, preconditions for the development of cognitive, affective and behavioural dimensions of energy literacy are created. In the textbooks, the theme of nuclear energy is mostly linked to the cognitive dimension, when basic knowledge of numbers and facts is obtained; comparisons of, e.g., volumes of nuclear fuel processing in different countries, changes in nuclear energy in different regions, and depiction of the process of recycling of nuclear waste are presented. The formation of other dimensions, i.e., behavioural and attitudinal (affective), is given little attention, and the development of these components is quite episodic.

Keywords: energy literacy, geography curriculum, energy geography, nuclear power, energy landscape, geography textbooks, The Ignalina Nuclear Power Plant, Visaginas.

Introduction

The authors of the chapter have been working on the development of the route of educational nuclear tourism in the Ignalina Nuclear Power Plant (INPP), Lithuania. The methodology of creating the educational route implies establishing links with the formal school curriculum and attracting school students and other learners as potential visitors and tourists with educational needs and interests to the site of the INPP.

The INPP and a town of Visaginas, a satellite town of the INPP encompassing its historical and social context, have been chosen as a field for creation of educational nuclear tourism and scientific investigation. The construction of the INPP started in 1974; the exploitation of the first block commenced on 31 December 1983. Back then, it was planned to be the most powerful nuclear plant in the world, with mounted four RBMK type reactors. However, after the Chernobyl catastrophe in 1986, construction of the third reactor was conserved, and the construction of the fourth block did not begin. After Lithuania restored its independence in 1990, the INPP became the most important part of the energy sector. In 1993, even 88.1 per cent of the electric energy required by the state was produced in it. At the time of Lithuania entering the NATO and the European Union, the condition in terms of safety assurance was emphasised; and, thus, the RBMK type reactors being exploited at the INPP were considered to be unsafe. Therefore, the power plant could not be exploited and, on 31 December 2009, its operation was completely terminated. The works of decommissioning of the power plant are projected until 2038.

The town was built near the INPP; staff and their family members inhabited it. At the beginning, the town was called Sniečkus (it was a surname of the First Secretary of the Communist Party of Lithuania); later (since 1992), the town was given the name "Visaginas". The nuclear plant was being exploited by specialists of nuclear energy who arrived from nuclear power plants operating all over Soviet Union. This formed an exceptional situation of the town in demographic, cultural and urbanistic terms. The town stands out by its multicultural aspect and specific linguistic milieu in Lithuania: residents representing 38 nationalities live here, majority of them speak Russian. This is a town attributed with Soviet architecture, historically formed as a mono-industrial site built for one purpose – to serve the nuclear industry, losing its major purpose after closure of the nuclear power plant and, therefore, undergoing the crisis of place identity.

The uniqueness and exceptionality of the INPP and Visaginas town provide preconditions for development of educational nuclear tourism, design an educational route combining means of formal education and educational tourism. The authors of this chapter raise questions: how to combine different forms of learning, i.e., the route of educational tourism and the process of formal education; how to make this route attractive and useful to students and teachers, as participants of formal curriculum; and how to combine cognition of nuclear energy and social, cultural, historical cognition.

In search for answers to these questions, the chapter displays the analysis of the geography curriculum content and the content of geography textbooks for school. Geography curriculum has been chosen for analysis as a school subject which potentially covers a range of nuclear energy–related topics. A general question raised in the chapter is how geography curriculum and teaching/learning geography could be improved by developing energy literacy (including nuclear literacy), applying the context-based learning approach and using opportunities of the outdoor learning environment (in this case, nuclear educational tourism on the site of the INPP).

A New Role of Geography in Teaching and Learning Energy Literacy and Promoting Education for Sustainability

There is an ongoing intensive discussion in the scientific discourse on new areas of geography research and scholarship on how to translate these new areas into new domains of geography curriculum in school. Geography becomes a central school subject in building understanding and addressing current energy issues and dilemmas. At the same time, geography curriculum becomes "a curriculum of survival" and "a curriculum of the future" aiming at promoting education for sustainability and building environmental literacy (Lambert, 2013, Butt, 2011). Additionally, geography curriculum is considered by geography researchers as the most important school subject in learning about globalisation.

Researchers of education emphasise the need for teaching energy in geography classes (Thoyre & Harrison, 2016). Energy literacy enables learners to assess energy-related decisions throughout their lives: as citizens with a "citizenship understanding" in the context of local, national and global decisions, as well as at a personal level, in the context of their daily life decisions. It is believed that "energy literacy" would help equip people to make more responsible energy-related decisions and actions (for instance, to reduce energy consumption (Van der Horst et al., 2016). To implement new energy policy decisions at the national and global levels (e.g. the introduction of renewable energy technologies, climate change measures), energy literacy becomes an educational foundation that can create social support. DeWaters et al. (2013), referring to Fowler (1976), underline the importance of energy literacy to make energy-related decisions at the individual and societal levels. According to Fowler (1976, cit. DeWaters et al. (2013)), energy literacy deals with understanding the science and technology of energy and its pervasive role in the national and world society; enables to make informed political decisions on energy-related options; and to make personal life-style decisions that are consistent with energy realities.

Thus, these different decisions that citizens have to make at different levels of the decision-making are intended to have scientific literacy (including technological and environmental literacy, scientific inquiry and problem-solving skills), but social competences and abilities, i.e., ethical aspects and citizenship understanding, are important, too.

The authors (DeWaters et al., 2013) who developed an energy literacy measurement instrument to assess energy literacy among secondary school students (forms 7 through 12) distinguished three dimensions of energy literacy: cognitive, affective (attitudes, values) and behavioural (including predispositions to behave). The cognitive element embraces contents knowledge on energy and cognitive skills. The affective element implies positive energy–related attitudes, which would allow reducing environmental impacts related to energy use, economic responsibility for using renewable resources. The behavioural component of energy literacy deals with energy-saving habits, energy-consumption patterns.

Energy Geography in the Curriculum of the Future and Survival

Many topics of energy geography are associated with environmental issues and could be transformed into education for sustainability in the geography curriculum. Authors analysing the new role of school geography call the geography "a world subject", the "curriculum of survival" and the "curriculum of the future" (Lambert, 2013, Butt, 2011). Geography is a discipline that allows us understanding the nature of globalisation, how globalisation works and what challenges and problems it creates. Geography can become a kind of an educational response to the contemporary economic and environmental crises: the global financial turmoil, the global climate change.

In contemporary geography curriculum, matters of energy are discussed in relation to global climate change by teaching about the necessity of a larger societal transition away from fossil fuels and alternative energy resources, i.e., renewable energy technologies (Thoyre & Harrison, 2016). Concerns about the environmental and economic sustainability of fossil fuel and nuclear power, energy-related environmental impacts are raised. This kind of education for sustainability seeks to educate and socialise new citizens, consumers and policymakers who become advocates of alternative energy resources. "Energy is an excellent vehicle for thinking about sustainability issues since it is imbricated in so many current environmental problems at multiple scales: climate change, air and water pollution, overconsumption, geopolitics, among others"

(Ibid., p. 33). A field of geography, "energy geography", is the "the study of energy development, transportation, markets, or use of patterns and their determinants from a spatial, regional, or resource management perspective" (Calvert, 2016, p. 104). Energy geography conceptualises energy as a social relation, and the energy mediates the human–environment relationship.

At the research level, the energy geography includes the following tasks: monitoring energy supply-chain developments; identifying place-based factors which explained observed spatial patterns of energy-sector investment; assessing environmental and economic risk, especially in the context of large-scale nuclear energy development; understanding how energy technology diffuses within and between nations; and mapping regional variations in energy production, distribution and use (Calvert, 2016). "Core topics of energy geography have traditionally included resource development, power-plant siting, land use, environmental impact assessment, energy distribution, and transport, spatial patterns of consumption, and diffusion of conservation technologies" (Solomon et al., 2004, p. 302).

Authors analysing the role of geography with its focus on globalisation (Butt, 2011) note that geography no longer emphasises geopolitics; it refers to the geo-economy as a way of revealing how global capitalism crosses borders of national and local economies. Therefore, geography allows going deeper into globalisation and patterns of global interconnectedness and interdependence. "A balanced view of globalisation must be attempted by curriculum makers, one of which takes account of its costs (for some) – unemployment, pollution, cultural change, loss of environment, resources and habitats, and poverty; but also its benefits (for others) – increasing employment opportunities, economic growth, greater exchange of goods and services, raising incomes and facilitating better access to products, services and cultures. The significance of identifying globalisation's 'winners' and 'losers', and their uneven spatial distribution, is of profound interest to geographers and geography educators." (Ibid., p. 434).

Geography, with its new focus on an analysis of global interconnectedness, turns into a discipline of cognition and potential criticism of global capitalism. Geography searches for answers to questions of how global capitalism creates inequalities, who are winners and losers of global capitalism, and how uneven spatial distribution works (Butt, 2011). Criticism of global capitalism in geography reveals general features of critical geography aiming to delineate power relations and inequalities which are produced by an uneven distribution of political, economic and social power.

At the same time, it should be noted that geography as the main school subject dealing with the understanding of globalisation in the realm of energy geography at school and in university is interconnecting and intertwined with economic geography. Energy is an important resource of economic activities and, at the same time, the energy industry is an important branch of the world and national economies. In terms of this perspective of political economy and economy geography, long-term supply and demand, regulation and pricing of energy resources are analysed.

The Changing Energy Landscape as the Main Concept of Energy Geography

Energy geography as an area of scholarship analyses the shifting of global energy landscapes with socio-technical (energy) transitions, spatial differentiation and territorially networked power relations (Bouzarovski, 2009). Such a definition would make it possible to attribute this geography to critical economic geography which analyses how political and economic power works and how power is distributed among power centres. The post-Communist states of Central and Eastern Europe (CEE) are considered as a specific energy landscape with the emergent "geographies" of energy reforms (Bouzarovski, 2009). These countries have been reforming their energy industries away from the legacies of the planned economy inherited from Communism towards a market-based energy regulation.

The authors analysing energy landscapes and their transitions (Dahlmann et al., 2017) focus on European Union energy policy processes which are aimed at the establishment of a single market for energy and the integration of renewables. Dahlmann et al. (2017) distinguish the following processes in EU24 countries between 1996 and 2013: an increase of the capacity for generating capacity (except for the Baltic region where a decline in the installed capacity is observed). In Lithuania, the energy landscape changed with the closure of nuclear power as a branch of the economy. Analysing the energy landscape of CEE, Bouzarovski (2009) provides data on the share of hydropower and nuclear energy in the total generating capacity of CEE states in 2005 (based on the data compiled from the US Energy Information Administration, International Energy Agency, and Austrian Energy Agency). At that time, a share of nuclear-generating capacity in Lithuania was more than 50 per cent. Since 2004, the first INPP reactor has been shut down; since 2009, the second reactor was closed.

Correspondingly, the energy sources in Lithuania are changing, and there is a decline in installed capacity. Other important changes in the energy landscapes in EU24 countries between 1996 and 2013: changes in patterns of fuel mixes and capacity ownership (decreasing ownership concentration and a general increase in a number of (new) owners – operators). At the same time, the Baltic region is recognised as an "energy island" in Europe with a high degree of concentration of ownership and clear path dependence due to the Soviet legacy (it is a former part of the Soviet Union's energy system) and geographically peripheral location (Dahlmann et al., 2017). In 2017, the share of renewable energy in Lithuania was 25.83 per cent. This progress in the development of renewables is partly explained by the attempt to offset the overall decrease in generating capacity related to the closure of nuclear power.

The place of nuclear energy in geography curriculum. The topic of the use of nuclear energy in the context of geography is related to the topic of education for sustainability and the global impact of the earth as well as the health and life of all humanity. Looking at the use of nuclear energy as a global phenomenon with consequences and potential harm for the environment, it is important to raise questions about the environmental impact of using nuclear energy for energy which is one of the cleaner technologies (in terms of CO_2), but the potential insecurity of the nuclear technologies and spent radioactive fuel are causing real and potentially severe damage to nature and humanity (catastrophes and disasters). As demonstrated by the Chernobyl accident, damage for nature and people has been done in both local and global contexts. These accidents and disasters as well as their negative effects create global interconnectedness. Potential harm and damage are enormous; their negative impact is hardly assessable now and in the future.

When interpreting nuclear energy as a domain of energy geography, it is important to recognise the use of nuclear energy in the global geo-economy and geopolitics. It is important to recognise how the entire global infrastructure of nuclear power is built, what countries and energy networks participate in and develop the nuclear power industry, what is an economic model of their operation and what are their geo-economic and geo-political interests. It is also important to recognise how the use of nuclear energy is incorporated into a general picture of the common energy infrastructure of countries and regions.

Energy-related topics are analysed and taught from the perspectives of critical geography. Scholars analysing energy from the point of view of critical theory highlight issues of inequality, social and environmental injustice, implications of different energy regimes for communities who have differing amounts of power in larger political economic and social systems, energy vulnerability and resilience, and the energetic political economies of power and control (Thoyre & Harrison, 2016). A critical approach to the analysis of the energy reveals the regional spatial inequalities associated with a social "energy divide" (Bouzarovski & Herrero, 2017). Energy poverty is related to deprivation and vulnerability which is experienced by citizens being unable to meet their basic domestic and household energy needs. While analysing regional inequalities in EU member countries, Lithuania together with other post-Communist states of the CEE are described as countries with the highest energy poverty levels and vulnerability of citizens in the European Union.

A New Pedagogy and Active Methods in Addressing Pedagogical Challenges and Overcoming Difficulties in Teaching Energy Geographies

Teaching energy geography is mentioned in the literature as a topic that poses a number of pedagogical challenges and difficulties (Huber, 2016). On the one hand, the subject itself is difficult due to the complexity of energy systems. Students experience a wide range of negative feelings: emotional distress, frustration, apathy, confusion, hostile defensiveness. Therefore, there is a need for effective pedagogical approaches when teaching energy geography. The search for new pedagogical approaches in energy geography is relevant to the whole subject of geography.

When discussing new trends, the need to include a new pedagogy, i.e., to build on the constructivist approach, when students construct knowledge in contexts that are meaningful to them, integrate active learning strategies, apply problem-based and inquiry-based learning, fieldwork and, at the same time, integrate new technologies in geography: digital data and imagery, new media, is emphasised (Day, 2012). On the one hand, referring to a cognitive constructivism approach, it is important to take into account the processes of constructing knowledge through accommodation and assimilation of new knowledge. Other forms of constructivism: social and pragmatic constructivism, where knowledge is constructed in interaction with other social actors (teachers, other students, community members) and in solving practical problems that are relevant to the learner, the community and society (a pragmatic approach), are also important. Although these new changes are discussed in the undergraduate teaching of physical geography, they are also important in school geography.

The constructivist approach emphasises the responsibility of learners for their learning, adjustment of learning to learners' backgrounds, skills and aspirations, collaboration among learners. Unlike passive learning, when students learn theory and learn from textbooks and examples, active learning, "learning by doing" as an inductive approach provides students with field examples, case studies or problems and it embraces classroom assignments, fieldwork and laboratory measurements, group work and student self-assessment and peer teaching. While traditional learning methods (learning in the classroom, from textbooks) are important, the following learning strategies are very useful for applying active learning in geography: problem-based (PBL), inquiry-based (IBL), experiential and service-learning (SL) through community engagement, fieldwork when the "real world" environment outside the classroom becomes a learning resource (Day, 2012).

One more important strategy of the constructivist approach to facilitate learning energy geography is a context-based approach. This concept is applied broadly in science education when one can apply scientific knowledge to the personal, social and global problems they encounter as citizens, when students can confront socio-scientific issues and when learning is based on real-world problems with an emphasis on interdisciplinary connections, where applications of science provide starting points for developing scientific ideas (Dori et al., 2018). Thus, the content of learning should be related directly to some personal or social aspect of the students' lives and represent authentic relevant issues. Students learn about certain phenomena in a specific context which covers societal, industrial and ethical aspects, people's activities and life within a community or society (ibid). According to Gilbert (2006), a context is considered as reciprocity between concepts and applications and as the social circumstances. The author distinguished four attributes of an educational context: (a) a setting, a social, spatial and temporal framework within which mental encounters with focal events are situated; (b) a behavioural environment of the encounters, the way that the task(s), related to the focal event, have been addressed is used to frame the talk that then takes place; (c) the use of a specific language, as the talk associated with the focal event that takes place; (d) a relationship to extra-situational background knowledge. The context-based approach includes "situated learning" which creates a suitable physical, social and psychological environment (Gilbert et al., 2011).

Fieldwork acquires new significance when studying the environment and when learning activities are set up outside the classroom and held in the real world. The fieldwork creates an opportunity to put theory into context. This learning strategy allows combining constructivist learning (where knowledge is constructed in real-life settings) with elements of cognitive, affective and behaviour-based learning, and experiential learning by acting and reflecting on everyday experience. Van der Horst et al. (2016) describe the experience of building energy literacy by applying fieldwork through a combination of active learning, smart meter technologies and reflection. Students asses their domestic energy use and consumption in "the field" – they monitor how electricity is used in their own homes. The fieldwork is mediated through the use of technology monitoring the consumption of electricity.

Day (2012) states that all these strategies of active learning lead to the development of various competencies: general skills, such as critical thinking, problem-solving, and subject-specific skills, such as spatial thinking, use of GIS (Geographic Information System), cartography and field methods. New approaches in teaching geography are related to IT (Information Technologies) integration: besides textbooks, electronic media and online materials, digital data, analysis software GIS and remote sensing data are increasingly used. At the same time, the role of Virtual Learning Environments (VLEs) is growing.

To overcome the learning difficulties of this subject, teachers seek to connect the everyday-ness of energy to both key concepts in geography and political questions outside the classroom (Huber, 2016). Additionally, the teaching of topics in energy geography involves the use of active methods and new pedagogy approaches applied in geography didactics.

Many energy geographers underline the general principles of constructivist and social justice perspectives in class on energy geography and highlight the use of active learning approaches in teaching, emphasising critical thinking, reflection and transformation as a pedagogical goal (Thoyre & Harrison, 2016). As noted above, energy geography belongs to a critical area of geography where development of critical thinking skills is essential. At the same time, active learning is important. Learning by doing in energy geography courses – where students complete a course-length project in active learning environments – increases student's motivation, opens the door for multiple learning styles, and enables linking theory with everyday life contexts. Through active experiential learning, students often link their experiences to the theory through reflection.

Thoyre and Harrison (2016) analyse the pedagogical experience of other authors in teaching energy geography and describe successful approaches and methods. Project-based learning activities seem to be a very promising pedagogical approach. The authors mention the work of Graybill (2016), which describes how university students are engaged in a semester-long video project on the topic of the Arctic and Urban Space. This pedagogical approach as a combination of energy geography and videography demonstrates good practice on how to employ digital technologies and digital storytelling in combination with time-intensive teamwork and work with new technologies and computer programs. According to Graybill (2016), this pedagogical practice has a dual purpose of encouraging students to become more critical consumers of knowledge production as well as providing them with a new way to create and disseminate knowledge in their futures beyond the classroom. It is achieved by engaging with multimedia knowledge on energy geographies using videography (p. 56). This videographic teaching module is an example of how to engage students and build their energy geography knowledge and skills in multimedia with the purpose of expanding the geographic knowledge for academic and non-academic audiences.

Bodzin et al. (2013) described the effectiveness of a geospatial curriculum approach to promote energy literacy at the middle level in the science education curriculum. The use of geospatial technologies (GT) as a learning technology with the focus on the concept of spatial nature of energy resources contributes to the development of knowledge on the acquisition of renewable and non-renewable resources, energy generation, storage and transport, and energy consumption and conservation. Energy, scientific and environmental literacy could be promoted by carrying out geospatial analysis and employing reasoning skills with integrating GT. The authors (Bodzin et al., 2013) delineate how GT tools (geographic information systems (GIS), virtual globes, such as Google Earth, Global Positioning System, and other related technologies) allow processing geospatial data into visualisations to facilitate problem-solving in energy learning activities. The GT could be effectively used while analysing the map view geo-referenced data locations of sustainable and non-renewable energy resource materials, proposed new power plant locations, the existing transportation infrastructure and environmental characteristics of an area. The use the GT enables students to learn about making informed decisions concerning energy resources to choose and decisions to make for a country on the supply of electricity.

Teaching About Nuclear Energy: Enhancing Energy Literacy and Scientific Literacy

Energy literacy is enhanced by energy education which is implemented in several school subjects: geography, physics, technologies and chemistry. Bartley et al. (2013) describe how energy is related to the building of multidimensional science literacy in the subject of physics. The development of these scientific and social competences is implemented through the integration of school subjects (including geography). The authors (Bartley et al., 2013) present a pedagogical scenario of lessons of physics on energy and discuss how learning about energy could be combined with building an understanding of how science and technologies influence society as well as with fostering citizenship skills. This case is an example of the development of multidimensional science literacy through library research, classroom debates and reading, synthesising, and reflecting on articles in class. Students were encouraged to work collaboratively, formulate, analyse and interpret ideas and data, to perform logical reasoning, to gather evidence. At the beginning of the lesson, the teacher gives an introduction on basic understanding of energy concepts and initiates the discussion with students on different forms of energy, renewable and non-renewable energy, transformation and conservation of energy. The students are given texts for reading in class on different types of energy; they are invited to discuss the energy use in their country (the USA), the alternative, benefits and drawbacks to using other energy sources than the main source of energy (the oil). The students are invited to make library research on alternative energy resources and discuss the findings with other students. During the library research, the students learn about solar, wind, hydroelectric and nuclear energy. The students learn about the generating capacity of nuclear energy, the process of releasing nuclear energy from uranium, the nuclear fission, and the work of reactors and nuclear power plants.

Additionally, threats of nuclear energy to people and the environment are also exposed in the literature analysis. After the students carried out their research in the library, the teacher invites students to discuss the benefits and drawbacks to using each type of energy. The teacher gives students several resources on pro and con to the use of nuclear energy, the costs and benefits of nuclear energy.

Students read articles, discuss in groups the future of nuclear energy (majority of resources suggested by the teacher are devoted to the Fukushima disaster) and produce a poster. Later, students demonstrate their poster presentation in a poster gallery walk, discuss the future of nuclear energy and write a letter to a US Senator inviting to discourage the proliferation of nuclear energy. It is noteworthy that while developing scientific literacy the students improve their citizenship competences, literacy (writing skills) and general analytic skills, when students improve their ability to "examine a topic and convey ideas, concepts, and information through the selection, organization, and analysis of relevant content" (ibid).

Energy Literacy and Energy Issues in the National Curriculum of Geography in Lithuania

In the present section, the authors, grounding on the conducted scientific overview of literature on energy and energy literacy, analyse whether the competences and content formulated in the national geography curriculum

Tab. 1: Areas of activity forming parts of the geography competency in different stages of education

Basic education level (forms 6–10)	Secondary education level (forms 11–12)
Orientation in the area and map	Orientation in the area and map
Reading of geographical information	Analysis of natural and social processes
Cognition of a region	Geographical investigation
Cognition and investigation of environment	

represent topics on energy (also including nuclear energy issues) and are directed to develop energy literacy. Major documents regulating the national curriculum in Lithuania, specifically Lithuanian General Programmes of Primary and Basic Education. Social Education (2008) and General Programmes of Secondary Education: Social Education (2010), were chosen for the analysis. The competences sought to be developed throughout the subject of geography were analysed as well as their relation to the topic of energy and energy literacy were dealt with.

The concept of specific areas of activity in the formal national curriculum would mean that these areas of activity can be treated as parts of the geography competency (Tab. 1):

Besides identification of the areas of activity, the national curriculum on geography presents recommendations for the content, i.e., related topics. It is worth noting that quite common and broad topics are listed (e.g. "Energy Industry" and "Global Economy"); therefore, later (in the stage of development of the curriculum) other developers of the curriculum (authors of textbooks, teachers preparing and selecting materials, designing education plans in schools) may put a more specific content into the frameworks of these broadly formulated topics (in our case of analysis, topics on energy). Such a broad naming and definition of the topics at the stage of implementation of the curriculum basically allow and create preconditions for the analysis of energy and energy industry as the most important branch of global and national economy, in-depth explore political economy, economic and energy geography where energy is a mediator in the human–environment relationship. Grounding on the concept of energy geography proposed by Calvert (2016, p. 104), the context of energy can be incorporated into many thematic areas, for instance, energy development, transport systems, market economy, consumption models and their impact on environment, consumption of natural resources, etc.

Seeking to reveal the theme of energy in the curriculum and to link it to the specific skills to be developed through it, the researchers revealed that the content of the competence and its levels were quite differently defined at different levels of the curriculum. The areas of activities are indicated in Lithuanian General Programmes (which later are divided into knowledge, skills and attitudes), what should be gained by school students learning geography; whereas topics linked to the mentioned parts of the geography competency are very broad and general. On the one hand, this allows teachers construct a particular content of their lessons which at that time seems relevant to them, by choosing materials from various sources (including media). Moreover, other means of implementation of the curriculum, school textbooks, provide materials which enable gaining skill set in the programmes. In this case, an assumption is drawn that quite extensive *presentation of the energy-related topic in the national curriculum creates preconditions* for development of energy literacy. Van der Horst et al. (2016) define energy literacy as abilities to make responsible decisions and take actions decreasing energy consumption at the personal (local) level; and, later on, this allows making energy-related decisions at the public (global) level, too. Grounding on the concept of energy literacy provided by DeWaters et al. (2013), geography curriculum allows perceiving the role of energy science and technologies at the national and global levels. It is underlined that the characteristics of energy literacy in the national curriculum are not specifically described and addressed.

To conduct the analysis of the evolution in the gaining of skills, the taxonomy by Bloom et al. (1956) and Anderson & Krathwohl (2001) defining the hierarchy of foreseen levels of learning skills has been employed This taxonomy was helpful in identifying how energy-related competencies of various complexity levels are introduced in different stages of education.

It is worth noting that DeWaters et al. (2013) pointed out the levels of energy literacy: cognitive, affective (attitudes, values) and behavioural (including predispositions to behave). The cognitive element embraces contents knowledge on energy and cognitive skills; the affective element implies positive energy-related attitudes, which would allow reducing environmental impacts related to energy consumption, economic responsibility for using renewable resources; and the behavioural element deals with energy-saving habits, energy consumption patterns.

When analysing the content of Lithuanian national geography education, the following questions were raised: is energy literacy sought to be developed? Are preconditions for application context-based approach in teaching about energy created and what contexts of energy consumption are defined by the national geography curriculum? How do these contexts of energy consumption lead to development of energy literacy in school students?

In terms of the area of activity "**cognition of the regions**" defining the geographical competency to be developed, there is an emphasis on the abilities to gain knowledge on geographical conditions of Lithuanian and world's regions, specific characteristics of the spatial structure, phenomena and regularities of natural and social environment. In the case of programmes for forms 11–12, this competency is developed, which is completely natural, in comparison to lower forms; the content of it is broadened and it becomes more complex because the knowledge of social processes, besides knowledge on natural processes, is included in these higher stages of education. In such a way, the title of this area of activity for forms 11–12 is broadened and named *"Analysis of Natural and Social Processes"*; moreover, it indicates that these are the abilities to gain knowledge, analyse and assess geographical conditions of Lithuanian, European and world's regions, specific characteristics of their spatial structure, phenomena and regularities of natural and societal environment.

The analysis of energy-related topics reveals a tendency of the move from a broad, more common definition of the content towards more specific forming of the content of a topic in the programmes, for instance, titles of topics in programmes for forms 5–6 underline that students must characterise continents only in general terms; whereas in senior forms more specific topics, such as *characterisation of natural, economic features of specific regions and countries, characterisation of the reclaiming of natural resources* (7–8 kl.), are introduced, which means that students will be acquainted with natural resources, including energy, of particular regions and countries, characteristics of the economy of countries will be analysed, and significance of energy for the economy and economics is mentioned.

When dealing with the topics for forms 11–12 in relation to the *gaining of knowledge on regions* as the area of activity, the situation is different because the topics are not linked to the abilities singed out in lower forms; they are rather linked to the areas of geography as a subject (natural geography, geographical cognition, social geography, regional geography); nevertheless, specific titles are found as well: "Impact of Economic Activity on Natural Environment", "Major Indicators of Country's Economy", "Energy Industry", "Industry". On the one hand, these topics are quite common; however, they refer to specific contexts. In this case, it is likely that, for instance, when analysing the "Impact of Economic Activity on Natural Environment" in relation to the context of the energy investigated by us, students should learn what impact on present-day landscape is made by fossil fuel and electricity produced from it. Then, a currently criticised and highly relevant context in the world, the CO_2 emission context, as a factor of global warming and climate change could emerge. Such approaches lead to the strategies of the decision-making mentioned by Van der Horst et al. (2016), such as renewable energy sources, other means of energy output (wind, sun, etc.), seeking to protect environment.

When analysing the area of activity "cognition of regions", the authors of the research found out that the abilities listed in the general programmes, according to development of skills in various stages of learning, from form to form (from form 5 through 12), are characteristic of a specific consistency in compliance with the levels of cognitive cognition (from knowledge to evaluation) formulated in the taxonomy designed by Bloom et al. (1956), Anderson & Krathwohl (2001).

It should be noted that within this area of activity the national geography curriculum does not reflect the level of application at all, which is underlined in scientific literature on energy literacy (Van der Horst et al., 2016); no statements linking to the level of application were found neither in formulated skills nor in titles of topics (Tab. 2).

Statements of the level of synthesis dominate (there are 8 of them), and slightly less are connected to the level of analysis (6). Relating to the theme of energy under investigation and development of energy literacy, an assumption that in the context of cognition of regions students should recognise and compare characteristics of regions in relation to energy resources, energy production, structure of economy, etc. can be drawn. However, the level of application of energy-related decisions is not reflected; therefore, a certain mismatch with the consistency in competence development is observed in this case, and there doubts on whether students are provided with opportunities to gain practical energy-related skills occur.

The analysis of the consistency in gaining the skills according to separate stages of education reveals that forms 5–6 emphasise the level of understanding because it is sufficient for students to define only several characteristic features of the continents. The description of students' skills in the framework of the national geography curriculum graduating from form 8 could be attributed to the levels of analysis and synthesis in compliance with the Bloom's taxonomy, since it is projected that students will be able to single out, compare characteristic features of separate regions. In this case, it is worth noting that the described skills were attributed to the higher level of synthesis because students learn about continents, separate regions, countries and diversity of their characteristic features exactly in this stage.

Tab. 2: Consistency in the gaining of skills and interaction of topics characterising the activity of "Cognition of Regions" in general programmes

Levels of cognitive skills	Forms 5–6 Skill	Topics	Forms 7–8 Skill	Topics	Forms 9–10 Skill	Topics	Forms 11–12 Skill	Topics
Evaluation				1		1		1
Synthesis			2	1	2	1	1	1
Analysis			1	1	1	2	1	
Application								
Comprehension	1	1		2				
Knowledge	1		1		1			

At graduation from form 10, similarly to form 9, the level of synthesis prevails (Tab. 2), since students must be able to substantiate similarities and differences of separate regions. Moreover, description of the assessment stage can be found in formulation of the topics "Find out and assess the most important political and economic alliances". In the case of forms 11–12, formulations attributed to higher levels, i.e., synthesis, analysis and assessment, dominate (Tab. 2); students achieve the highest level of cognitive skills formulated by Bloom et al. (1956) because they must be able to point out and assess characteristics and features of different regions; and, in relation to the theme of energy, the equivalents reflecting the said in the statements found in formulations of skills and topics were sought for.

To sum up, the gaining of the skills defining the activity of *cognition of regions* is characteristic of consistency because formulation of skills and topics become more complex in compliance with the stages of learning. However, the level of application is not fulfilled throughout the entire period of learning.

The theme of energy and energy literacy is not mentioned in the wording of the skills; nevertheless, the formulations of skills, such as *cognition of geographical conditions in Lithuania and the world, phenomena and regularities of natural and social environment, geographical conditions*, etc., allow integrating topics on energy resources, energy economy and economy, as well as developing energy literacy at the stage of implementation of the curriculum.

When analysing the area of activity "reading of geographical information", the researchers found out that it was defined by the following skills: reading, analysis, perception, critical assessment and interpretation of sources of geographical information, rendering of geographical information in written, visual and audial forms. It should be noted that this area is named only at the level of basic education (forms 5–10). Consistency in the gaining of skills in this area is quite interesting because major "load" is allocated to the topics formulated in the programmes for forms 9–10 (Tab. 3), where 15 topics relate to the context of energy. Formulation of topics reflecting the theme of energy covers almost all taxonomy levels, except for knowledge.

However, when comparing the interactions of topics and skills named in the programmes, it is observed that there are less skills formulated in this stage of learning, 4; and all these are linked to higher levels – synthesis and assessment (Tab. 3). When reading the programme for forms 9–10 and relating it not only to the theme of energy, it was observed that there were many topics; therefore, questions on whether such structure of the curriculum allowed achieving the anticipated levels of skills arose.

In general, when dealing with the common consistency in the gaining of skills related to the activity "reading of geographical information", it was observed that the levels of knowledge and application were reflected the least (Tab. 3), the levels of understanding (9 statements) and synthesis (11 statements) were reflected the most, which shows insufficiency of the development of application skills.

When analysing how the skills of reading geographical information are defined in the national curriculum, referring to the taxonomy levels formulated by Bloom et al. (1956) in separate stages of learning according to forms, the authors of the research point out that in this case the beginning is the level of knowledge which is recognised in the phrasing of topics; and, in relation to the development of the skill, there is a transition to the levels of comprehension and application because students graduating from form 6 must be able to recognise, use sources and information available in them. When relating to the context of energy, it could be foreseen that students will properly comprehend information on energy which is presented in various contexts or will be able to choose required texts and information available in them. Referring to the definition of energy literacy introduced by DeWaters et al. (2013), it should be understood as perception of the energy science and technologies as well as the role of their distribution at the national and global levels, how students gain knowledge on the content of energy.

The national curriculum of geography foresees that students at graduation from form 8 achieve the levels of analysis and synthesis. However, it was observed that at this stage of learning any topic related to energy (Tab. 3) could not be pointed out. It foresees that students must be able to select sources, to properly use and compare them, to analyse phenomena and processes while grounding on them and to properly generalise and render information. It is likely that when realising these objectives at the level of implementation of the curriculum (which would be a decision of the authors of textbooks or a teacher himself/herself), students could compare information on energy presented in different sources. In such case, this would be an element of energy literacy, as pointed out by DeWaters et al. (2013), the definition of dimensions which is again only information on energy selected and systematised by students.

The level of basic education ends with achieving the highest level of cognitive knowledge, assessment and evaluation, because here it defines that students should be able to select trustworthy sources and analyse, assess and render a situation of natural, social, economic and political phenomena in various regions grounding on them. It is likely that during the processes of learning students will be able to assess the influence of various energy-related contexts on different regions and their economy or development of economy. Assessment of reliability of information is treated by various authors as a feature to be developed.

When analysing topics which are formed in general programmes, the authors of the research reveal that the skills of reading geographical information are gained when following certain consistency: from a specific context, e.g. "Natural Resources and Consumption of Them", to a quite broad and complex one, such as "Dependence of People's Economic Activities on Historical (Formation of Civilisations) and Natural (Water, Climate, Minerals, Soil, Relief) Conditions of Environment", which also complies with the development of skills according to Bloom's taxonomy of cognitive skills, even though, as mentioned earlier, there is lack of consistency (Tab. 3).

Relating to the context of energy under analysis, the description of the national curriculum includes specific statements or groups of statements to be

Tab. 3: Consistency in the gaining of skills and interaction of topics characterising the activity of "reading of geographical information" in general programmes

Levels of cognitive	Forms 5–6		Forms 7–8		Forms 9–10	
Skills	Skill	Topics	Skill	Topics	Skill	Topics
Evaluation					1	2
Synthesis			3		3	5
Analysis			2	2		2
Application	1					2
Comprehension	1	1		2		4
Knowledge		1				

related to energy, sources of energy, and influence of energy on environment or people's lives. In this case, it can be anticipated that the above mentioned levels of energy literacy pointed out by DeWaters et al. (2013), cognitive, affective and behavioural, will be recognised, since diversity of topics and the skills listed in the programmes are related to all taxonomy levels.

In lower forms (forms 5–6), topics "People's Activities" and "Natural Resources and Consumption of Them" are suggested; they would meet the theme of energy. As observed, they do not name the theme of energy; nevertheless, it is likely that, within the topic "People's Activities", a more specific topic could be projected: introduction of economy activities of residents, which would cover the concept of energy economy, too. The second topic is more concrete: "Natural Resources and Consumption of Them"; its formulation clearly supposes that it projects introduction of natural resources, including energy.

In this context, it can be related to the cognitive level of energy literacy, as pointed out by DeWaters et al. (2013), because students gain specific knowledge on energy resources, reclaiming of them, production and consumption of energy.

There are five topics related to the context of energy found in forms 7–8, including two topics related to natural resources ("Urbanisation of Territories Depending on Natural (Relief, Climate, Soils), Economic (Natural Resources, Location of Industry), Social Conditions" and "Opportunities for Using Natural Resources (Renewable and Non-renewable Resources)"; three topics are related to economic activities of people ("Geography of Economy", "Dependence of People's Economic Activities on Historical (Formation of Civilisations) and Natural (Water, Climate, Minerals, Soil, Relief) Conditions of Environment", "Major Parts of the Global Economy Structure (Bioproduction Economy, Industry and Services)". From the point of view of definition of the dimensions of energy literacy (DeWaters et al., 2013), such phrasing of topics would cover the cognitive and affective levels because students gain and deepen knowledge on energy, energy economy and find out the dependence of people's economic activities on conditions of natural environment. In its turn, this supposes particular formation of attitudes and values because elements protecting environment should be identified and analysed.

In the scope of programmes for forms 9–10, even 15 topics which could be related to the context of energy were singled out. It is interesting to note that at this stage of education only one topic, as indicated in forms 7–8, should deal with natural resources and sustainable consumption of them ("Rational Consumption of Natural Resources and Importance of Protection of Biological Diversity"), whereas all other topics are related to the concept of economy and economics, and only one of them clearly indicates the topic of energy ("The Most Important Branches of Economy (*Fuel and Energy*, Metallurgy, Production of Machinery, Chemistry, etc.) and the Regions of Their Distribution in the World"); in cases of all others, the theme of energy is seemingly "hidden". Nevertheless, it can be considered that the theme of energy may be integrated, for instance, in the topic "Influence of People's Economic Activities (and What) on [Natural] Change", could project a theme on how the industry of economy changes nature; within the topic "Global Economy" and "Scheme of Production Links of Economy Branches", "The Most Important Provided Services – Production and Non-production (Social)" may project the theme of what place is taken by energy in the global economy structure and what are its relations to other branches of economy; in the topic "Situation of Lithuania in the Common System of Global Economy" one may recognise a likely projected theme of how this situation is determined by the economy sector; the topic "Economy Resources in Lithuania and the World (Natural: Flora, Fauna, Water, Earth – Its Surface and Depth; Man-Made: Work, Finances, Information, Capital)" implies a smaller topic of what energy resources created by man are; the topic "Influence of Industry on Environment (Pollution, Changing of Landscape)" has a smaller topic on what kind of effect is made on environment by mining of natural energy resources and production of energy.

When relating topics for forms 9–10 to manifestation of the dimensions of energy literacy (DeWaters et al., 2013), it can be stated that the abundance and diversity of topics presented in the general programmes completely reflect all three levels of energy literacy: cognitive, affective and behavioural. For instance, students already get acquainted with energy as a branch of industry – the concept is introduced (cognitive level); a topic on the influence of people's economic activities on environment (influence of energy industry on environment can be analysed in this case) – would meet the affective level because it leads to the formation of positive attitudes; and a topic "Influence of Industry on Environment (Pollution, Changing of Landscape)" should already clearly allow forming habits, models of energy consumption, which is related to the behavioural level (according to DeWaters et al., 2013).

To sum up the situation of gaining the skills related to the activity of *reading of geographical information* and linking it to the development of energy literacy in general programmes, it can be stated that the systematic approach is applied in the national formal curriculum, since, first of all, it recognised that students must obtain sufficiently consistent information on energy, its sources. Formulation of the skills-to-be-gained is quite consistent, encompasses higher levels of thinking, and the thematic development of the energy theme proceeds from specific topics to factual material moving towards a broader and deeper context. Linking to the concept of energy literacy, as defined by DeWaters et al. (2013), it is observed that the formal curriculum foresees development of competences in consistency because when starting from lower forms (forms 5–6) the cognitive dimension prevails; later, the cognitive dimension intertwines with the affective one; and in forms 9–10 all three dimensions (cognitive–affective–behavioural) are recognised. It can be stated that in the national educational content the characteristics of the activity of reading geographical information reveals preconditions to form skills of energy literacy.

The activity area "orientation in the area and map" at the stage of basic education (forms 5–10) is defined as skills to orient oneself in a geographical area, plans and maps of a location, to understand the objects existing in them; whereas the skills for secondary education (forms 11–12) are orientation in a diverse geographical area (local, regional and global) and cartographic drawings, to understand the objects situated there.

Having analysed the skills and consistency in gaining them in the activity of *orientation in the area and map*, a particular controversy is observed in this case, since basically throughout the entire period of learning, from forms 5 through 12, formulations of the skills in the area encompass a prevailing level of application; the transition from this level to one higher level, that of analysis is found (Tab. 4).

The topic "Maps and Plan" and learning how to use them dominate throughout the entire period of learning. This situation is illustrated by the statements underlining that students, after graduating from form 6, must be able to distinguish a plan from a map, to be able to orient in a specific location; when graduating from form 8, students must be able to use plans and maps to orient in a local or global area. Perhaps, the context slightly broadens by introducing a global and a local area; however, the emphasis on the use of plans and map remains. After graduation from form 10, the emphasis on the use of plans and maps remains, only with added one more contextual element, i.e. characterisation of natural, economic and political geographical position of objects, which in the terms of activity can be attributed to the level of comprehension because, in the aspect of complexity, students do not do anything new, only the content is supplemented. In form 12, the definition of the competency of orientation in the area and map is supplemented with the necessity for students to be able to find relations among objects in various geographical sites (local, regional, global).

The researchers draw an assumption that such approaches for formulation of the skills of "orientation in the area and map" may partly comply with the

Tab. 4: Consistency in the gaining of skills and interaction of topics characterising the activity of "orientation in the area and map" in general programmes

Levels of cognitive skills	Forms 5–6 Skill	Topics	Forms 7–8 Skill	Topics	Forms 9–10 Skill	Topics	Forms 11–12 Skill	Topics
Evaluation								
Synthesis								
Analysis			1		1		1	1
Application	2	2	2	1	2	2	1	
Comprehension								
Knowledge		1						

concept of energy geography introduced by Calvert (2016, p. 104) who has it that this is teaching on energy development, transmission, markets, consumption models and their impact on area, region or prospects of resource management. And, as Calvert (2016) adds to it, energy geography encompasses the following tasks: development of energy supply chains, identification of sites and models where investment proceeds; assessment of equipment risks on economy and environment, especially in the context of nuclear energy development; perception of how energy technologies develop among countries; and comprehension of a map, how energy is consumed in different regions and countries (production, trade etc.) (Calvert, 2016). Thus, in this case, activity of orientation in area and map grounding on the concept of energy geography can be perceived in a way so that the use of a map in processes of learning can and must lead to perception of a complex concept of energy, when not only recognition of objects depicted on a map takes place but, on the contrary, construction of maps according to various parameters (e.g. prediction of possibilities for various models in different locations and regions) which would allow students perceive energy as a complex phenomenon takes place.

Summing up, from the point of view of the definition of the competence of orientation in the area and map, it is obvious that the dominating emphasis is laid on the reading of maps and plans, finding of objects in them and revealing of their interrelations. It should be noted that, in relation to the context of energy throughout the entire period of delivery of the geography as a subject from form 5 through 12, the taxonomy levels of synthesis and assessment are not covered at all.

Then, it becomes interesting which topics lead to achieving this and how this can be linked to a chosen theme of energy. In general, in form 12, at graduation from school, students achieve the analytical level of orienting in the area and map at least.

Thus, to sum up, consistency of the gaining of skills characterising the area of activity *orientation in the area and map* in Lithuanian general education programmes is not at its strongest, since the relation of these skills to higher levels of thinking raise doubts. Nevertheless, the authors had some doubts about whether the analytical level is really achieved at least or it remains at the level of application only. If we compare with a tendency that there were not many topics related to energy and activities of orientation in the area and maps (in comparison to topics covering other competences), there are only 15 topics in total in forms from 5 through 10. As it was mentioned, a major theme from form 5 through 12 focuses on maps and their diversity, explanation of the structure of a map. The authors did not find any direct relation to energy in the programmes; nevertheless, there is a topic in the programme for forms 9–10 which could be named as "identification of natural, economic and political geographical position of objects", and it is likely that, when analysing the content and structure of maps, energy objects will also be recognised: distribution of energy resources, energy enterprises, etc. In programmes for forms 11–12, this theme is slightly extended by emphasising that specific areas of cognition will be explained while orienting in map: environment, territory, region and world, which basically "brings down" to the level of comprehension and raises doubts whether this is true.

Having revealed the topics at the energy literacy levels defined by DeWaters et al. (2013), it becomes clear that in this case the focus is only on the cognitive level because these topics pointed out, e.g., energy resources and their distribution (forms 11–12), very clearly allow understanding that students will gain the content of specific knowledge on energy resources. If assessed as a stage of secondary education, it would be too low to achieve this cognitive level only. However, when relating it to the definition of energy geography, as introduced by Calvert (2016), perhaps an assumption could be drawn that such format of presentation of skills and topics exactly allows forming a specific perception of energy; therefore, in this case, the analysis of teaching materials proves it is necessary.

The skills defining the area of activity **"cognition and investigation of environment"** at the stage of the forms 6–10 are presented as skills to carry out geographical observation and investigation of environment, to formulate hypotheses, collect data, conduct various measurements and calculations, search for solution ways, draw conclusions and assess obtained results. In forms 11–12, the name of the area of activity is made slightly more specific, "**geographical investigations**", and the skills named at the lower stage are extended by adding application of methods for cognition of geographical area and theoretical-practical modelling of situations.

Assessment of consistency of the stages of competences in terms of the Bloom's taxonomy reveals specific regression of skills (Tab. 5). If in basic education proceeding from form 5 through 10 the gaining of skills is characteristic of specific consistency, starting with skills of comprehension in form 5 and ending with evaluation in form 10, then at the stage of forms 11–12, it seems, there is regression because formulation of the skills reveal the levels of application and analysis (Tab. 5); and comparison of the topics revealed that the disclosing of the energy-related content would proceed at the levels of comprehension and analysis (skipping application).

The research reveals that formulation of the research competence is the most poorly expressed in the formal curriculum. What is the proof of it? First, skills of investigation should relate to higher levels of thinking – analysis, synthesis, assessment; however, while observing the formulation of skills or topics helping gain them within the programmes it was observed that, for instance, skills corresponding to the level of synthesis and assessment are named only in the programme for forms 9–10 (synthesis: "to render obtained results to others in various forms", evaluation: "to explain advantages and disadvantages of the conducted investigation, to feel responsibility for the results of the work carried out"). The formulation of these skills is not related to the subject of geography or a particular context under analysis; these are more statements naming general processes of investigation: rendering of results, explanation of the research results; or definition of value attitudes: demonstration of the sense of responsibility. These general skills do not reflect any specific (related to the subject of geography) content of investigation.

Perhaps, there could be a logical explanation to it: the gaining of the competence of cognition and investigation of the environment is defined at the taxonomy level of application because research is basically a practical work; however, there is lack of skills forming higher-level thinking.

Searching for connections between formulated skills to be gained and energy, it is observed that the theme of investigation is defined quite broadly, and it encompasses and could basically cover the theme of energy, too. Major topics which may relate to energy are the dominant economic aspects, e.g., "students are taught how to carry out investigation (e.g. maps; climatic zone and type; inland waters; states according to social economic indicators) and to write a result of it" (forms 7–8), "using sources of geographical information, students learn to carry out natural, social and economic investigations, to solve

Tab. 5: Consistency in the gaining of skills and interaction of topics characterising the activity of "cognition and investigation of environment" in general programmes

Levels of cognitive skills	Forms 5–6		Forms 7–8		Forms 9–10		Forms 11–12	
	Skill	Topics	Skill	Topics	Skill	Topics	Skill	Topics
Evaluation					1			
Synthesis					1			
Analysis	1		1		1		1	1
Application	1	1	2	1	2	3	1	
Comprehension	1	1				1		2
Knowledge								

problems, draw conclusions" (forms 9–10); "to characterise world's regional and specific principles (historical, natural, social, economic, political)" (forms 11–12). These examples demonstrate how investigation linked to energy could or should be oriented to quite a broad context: revealing of natural, social, economic aspects. The emphasis on the economic aspect within the energy literacy would enable, for instance, carrying out research on energy consumption in various countries and on its impact on economy etc.; however, on the other hand, if the economic aspect was dealt with only, this would narrow down the revealing of the concept of energy, since the context of energy, if assessed from the point of view of energy literacy, focuses on a much broader scope: includes energy production, environmental protection, spheres of social relations, geopolitics, geoeconomics, etc. If such specificity of revealing topics was followed, this would allow substantiating the revealing geography through its role for the future and survival, as proposed by Butt (2011) and Lambert (2013) investigating it as the geography like "a world subject", "curriculum of survival" and the "curriculum of the future". In this case, analysing the topics singled out in the formal national curriculum, the authors draw assumptions that, within the process of learning (at the stage of implementation of the curriculum), the concept of the nature of globalisation, how globalisation proceeds and what challenges and problems it creates should be emphasised. The context of energy could be analysed as the most important resource of economy, and, at the same time, reveal and investigate the industry of energy as the most important branch of global and national economy. In such a way, a necessity to analyse political economy and economy geography arises.

It should be noted that the wording of themes in the national formal curriculum is quite broad, and this creates some flexibility and an opportunity to integrate various topics, since a teacher has got sufficient freedom to put specific content into broader topics, e.g., the programme for the forms 7–8 includes the formulated topic where "students are taught to search for connections among constituent parts of the natural and social environment, to identify regional differences of them, to explain what causes determine these differences, to draw conclusions"; allows a teacher choose any context for the research work, for instance, branches of industry, and, referring to them, project the process of teaching and learning. Then, in relation to the dimensions of energy literacy (DeWaters et al., 2013), the processes of learning could link to the behavioural level, when habits of specific energy consumption and saving, models of energy consumption, etc. are being developed; implementation of the opportunities is plausible in compliance with formulation of the topics.

To sum up all skills and topics named in the national curriculum that could be related to the theme of energy and to that of how the formulation of competencies comply with the taxonomy levels pointed out by Bloom et al. (1956) and Anderson & Krathwohl (2001) it was found that in a chosen theme and areas of activities (reading of geographical information, cognition of regions, orientation in the area and map, cognition and investigation of environment) the taxonomy level of application prevailed. Nevertheless, the striving to single out and formulate higher levels, i.e., those of analysis and synthesis, is observed. Therefore, assumptions can be drawn that, in terms of description of the national curriculum, the theme of energy has potential and opportunity to be introduced and revealed within the stage of implementation of the curriculum.

If analysed in terms of separate stages of learning (from form 5 through 12), the theme of energy would be rendered the most consistently and appropriately in terms of topics in forms 7–8 and 9–10. When exploring how basic skills-to-be-developed in the educational content are formulated, a tendency that even though energy literacy is not named, still preconditions for the development of the dimensions of energy literacy (cognitive, affective and behavioural (DeWaters et al., 2013)) are created, is revealed. A systematic approach is observed in the description of the national curriculum, which enables forming skills in a chosen area because consistent transition from concrete gaining of the content of knowledge to formation of attitudes and behaviour is observed. However, seeking to completely reveal how energy literacy is developed through the school curriculum, not only national formal curriculum but also the stage of implementation of that curriculum, i.e., teaching materials, textbooks, survey of teachers and deeper investigation of the process of learning, should be analysed.

Even though the development of energy literacy in the content of Lithuanian education is not presented or specifically defined, still the description of the national curriculum renders a broad general characterisation of the theme and competencies which allow teachers choose and formulate specific topics which can be related to relevant issues, including energy.

The Concept of Nuclear Energy in Lithuanian Textbooks on Geography

The analysis of texts on nuclear energy in the textbooks seeks to identify the connection of textbook materials on nuclear energy to geographical skills and competences pointed out in comprehensive programmes and how it combines with the taxonomy levels of competences. Moreover, analysis of textbooks grounds on the concept of energy literacy and aims at finding out how the theme of nuclear energy reveals the aims of energy literacy, which, according to Flower (1976, cit. DeWaters et al. (2013)), encompasses the ability to make decisions at an individual and societal levels. In this case, it evaluates how the presentation of the nuclear energy concept in textbooks forms students' knowledge and attitudes about significance, role and impact of this type of energy at global, local and personal levels.

Referring to the energy literacy concept, school students should be able to assess significance of nuclear energy and its impact on their personal lives, living in immediate environment and global world. In such a case, grounding on statements of Van der Horst et al. (2016), characterising an energy-literate citizen, the concept of nuclear energy in geography curriculum and textbooks must be related to the goal to develop skills of decision-making at personal, regional, nationa and, global levels, which would mean that school students should gain knowledge on the impact of nuclear energy and could assess it as well as search for various decisions connected to consumption/non-consumption at various levels and in various contexts of social life. The conducted analysis of texts on nuclear energy in geography textbooks refers to the concept underlying energy development, transportation, markets or use patterns and their determinants from a spatial, regional or resource management perspective (Calvert, 2016, p. 104). Energy geography conceptualises energy as a social relation, and the energy mediates human–environment relationship. Authors investigating energy geography (Solomon et al., 2004; Calvert, 2016) single out the themes which are important to this branch of geography: monitoring energy supply-chain developments; identifying place-based factors which explained observed spatial patterns of energy-sector investment; assessing environmental and economic risk, especially in the context of large scale nuclear energy development; understanding how energy technology diffuses within and between nations; and mapping regional variations in energy production, distribution and use.

Grounding on the levels of the competence development taxonomy singled out by Bloom et al. (1956) and Anderson & Krathwohl (2001), the researchers raise a question how nuclear energy is introduced in textbooks: is this phenomenon presented as complicated and complex?

The following questions are formulated to investigate textbooks on geography:

1) What is the structure of materials within the textbooks: content, volume of texts on nuclear energy and how this content relates to the geographical skills formulated in comprehensive programmes?
2) How does the presentation of nuclear energy in geography textbooks provide preconditions for development of energy literacy?
3) How do complication and complexity of the materials rendered in textbook texts combine with the taxonomy levels of geography competences pointed out in the national curriculum?

After carrying out analysis of 32 Lithuanian geography textbooks for forms 6–12 (Bačkienė, Pundienė, Januškis, 2009a; Bačkienė, Pundienė, Januškis, 2009b; Česnavičius & Gerulaitis, 2008/2007; Česnavičius et al., 2010/2008; Česnavičius, &Valančienė, 2008; Dijokienė, 2016; Gerulaitis, et al., 2010a; Gerulaitis et al. 2010b; Kynė et al., 2008a; Kynė et al. 2008b; Kynė et al., 2016; Kynė et al., 2015; Šalna et al., 2012a; Šalna et al., 2012b; Šalna et al., 2006; Šalna et al., 2009/2005a; Šalna et al., 2009/2005b; Šalna et al., 2014; Šalna & Sapožnikovas, 2010/2006a; Šalna & Sapožnikovas, 2010/2006b; Šalna et al., 2005a; Šalna et al., 2005b; Šalna et al., 2010a; Šalna, et al. 2010b; Šalna et al., 2010c; Šalna et al., 2012; Šalna &Tuskenienė, 2009a; Šalna & Tuskenienė, 2009b; Valančienė & Česnavičius, 2008; Valančienė & Dijokienė, 2007; Varanavičienė et al., 2017) the energy-related concepts and their illustration with examples have been found: energy resources, renewable and non-renewable resources, energy economy, and social and environmental aspects of the use of energy sources and energy production. **The concept of nuclear energy in forms 7–9.** The analysis of the concept of nuclear energy was being consistently conducted starting from textbooks for form 7. The materials on the aspect under the interest of the researchers were not found in the texts of textbooks for form 6. Two textbooks by Šalna et al. (2009/2005a; 2009/2005b) for form 7 and two textbooks by Šalna and Sapožnikovas (2010/2006a, 2010/2006b) for form 8 have been analysed.

In the first instance, it was discovered that textbooks for forms 7 and 8 present materials related to nuclear energy in a way that there are dominant two areas in development of geography-related skills mentioned in the national curriculum: cognition of regions and reading of geographical information. The competence of map reading is slightly developed. Areas of different competences (acquaintance with regions, reading of geographic information, understanding of maps) intertwine because texts giving information on nuclear energy describe regions and countries of the world. Geography textbooks for form 7 introduce the following regions: Africa, Australia, Oceania, the Antarctic, South America and North America (Šalna et al., 2009/2005a.); whereas these are for form 8: Europe (Šalna & Sapožnikovas, 2010/2006a), East Europe and Russia, Asia (Šalna & Sapožnikovas, 2010/2006b), and single countries. Textbooks for form 7 include maps demonstrating sites of uranium mining in the Republic of South Africa (Šalna et al., 2009/2005a, p. 161) and Canada (Šalna et al., 2009/2005a, p. 241). Tables indicate and compare the volumes of uranium mining in various countries. For instance, Australia whose uranium mining constitutes 14 per cent, is mentioned as a country taking the second place in uranium mining in the world (Šalna et al., 2009/2005b, p. 180). However, there are no texts providing details about uranium as fossil fuel used in nuclear power plants; accordingly, such material does not substantiate introduction of these regions for students as sites where uranium is being mined (Republic of South Africa, Australia and Canada), no connections with nuclear energy. In discussion of regions and countries, the textbooks render numerical information, diagrams, indexes on the amounts of uranium mining and compares them in terms of different regions of the world. Nevertheless, our research reveals that texts of textbooks for form 7 introduce nuclear energy quite fragmentarily, which prevents from forming a generalised and whole concept for the students.

Thus, materials of textbooks for form 7 are not characteristic of a highly complex level in presenting energy; nuclear energy is presented indirectly (when talking about uranium). When comparing to the object of energy geography described by scholars (Solomon et al., 2004; Calvert, 2016), it can be stated that textbooks for this form lack a broader and consistent image of the energy sector y in relation to social, economic and political aspects of nuclear fuel mining in the discussed countries and regions. After the assessment of introduction of nuclear energy in textbooks for form 7, grounding on the concept of energy literacy, it was observed that the level of neither regional nor local energy consumption was presented: information on other regions of the world without any relation to the context of Lithuania is presented to students.

Texts for form 8 display more materials related to nuclear energy; and specific efforts to reveal characteristics of regions are observed. An explanation of pitchblende (uranic ore) as fossil which is "the most important fuel for nuclear power plants" (Šalna & Sapožnikovas, 2010/2006a) is already presented for form 8, which allows the students, differently from those of form 7, understand that "uranium" is linked to nuclear energy.

Another aspect to be related to nuclear energy in form 8 deals with the varieties of electricity production: water, thermal, renewable resources and nuclear. This information allows school students understand the diversity of energy sources and energy production industry. Specific aspects of nuclear energy discussed in textbooks for form 8 are linked to getting cognition of regions or countries, e.g., France is characterised as "having no large pools of oil or gas, though being famous for pitchblende (uranic ore) mining for many years. <...>. Currently, the country runs more than 20 such (nuclear, authors' note) power plants. They produce approximately three quarters of the entire electric power." (Šalna & Sapožnikovas, 2010/2006a, p. 84). The progress of Indian industry is connected to the development of nuclear energy: "Presently, India is rapidly developing high technologies. Nuclear power plants are being constructed all across the country." (Šalna & Sapožnikovas, 2010/2006b, p. 208). Comparing significance of natural resources in North Europe, the diagrams demonstrate different amounts of electric power produced in nuclear power plants in different countries: "In Sweden the amount is 43 per cent, in Finland it is 30.4 per cent, in Lithuania it is 77.7 per cent." (Šalna & Sapožnikovas, 2010/2006a, p. 66).

The information on nuclear power allows students start forming their understanding of energy-related regions, start understanding the spatial patterns of the energy sector, when the sector of energy comprises several different sources of energy, and nuclear energy constitutes an important part of national economies. Textbooks for the 8th form point out the countries that develop nuclear energy. These include Lithuania, too. Texts of the textbook comply with the strategic concepts of nuclear energy of the period they were written (in 2006). However, in relation to present-day global and Lithuanian strategies for energy development, these materials of the textbooks are already outdated and should be treated as historical, i.e., as a description of what was happening in the country many years ago. Seeking to fill in geography lessons with relevant content, a teacher should be assisted with materials that discuss contemporary priorities in the energy sector, for instance, the Lithuanian National Energy Independence Strategy (Lietuvos Nacionalinė energetinės nepriklausomybės strategija, 2018), which reflects the situation of state's energy sector 10 years after the closure of the nuclear power plant and underlines the development of renewable energy sources, green energy.

Over the latter 10 years, the Lithuanian energy map and strategic perception of the development of energy industry have significantly changed. Energy is no longer being produced from nuclear energy since 2009; production of biofuel, biomass and biogas as well as wind, solar and geothermal energy is being actively developed. The development of the energy sector is being carried out through the decentralisation and demonopolisation of energy production, involving residents in the sector of energy economy, installing solar panel batteries and producing geothermal energy for house heating. Emphasis on these new energy-focused priorities could contribute to the development of the energy literacy, as named by Van der Horst et al. (2016), through civic participation and entrepreneurship, when connections to both local (national, Lithuanian) and individual (electricity production in household) contexts are clear.

The most pronounced instance of acquaintance with the region and connection to the nuclear energy is presented in form 8, when Ukraine is characterised and presented through the narration about the disaster at the Chernobyl Nuclear Power Plant. The textbook includes a separate chapter "Disaster of the Chernobyl Power Plant" (Šalna & Sapožnikovas, 2010/2006b, p. 148). This chapter comprises 4 short texts about the catastrophe, its causes and consequences: "In April 1986, the world was struck by the disaster of the Chernobyl Nuclear Power Plant. Caused by severe mistakes made while conducting experiments, several explosions took place in one of the reactors; the explosions damaged the roof made of steel and concrete mass weighing 1,000 tons and breached it. Few times more radioactive substances spread in the atmosphere than in 1945, when nuclear bombs had been thrown on Hiroshima and Nagasaki." (Šalna & Sapožnikovas, 2010/2006b, p. 148). The extent of the disaster is discussed: the increase of radiation in Ukraine, Belarus, Lithuania, Germany, Switzerland, France, Italy and Nordic countries as well as concern about safety; the consequences of the disaster: and pollution of large areas with radioactive substances, impact on nature and diseases caused by radiation (Šalna & Sapožnikovas, 2010/2006b).

Moreover, the text includes 3 maps depicting the spread of radioactive clouds 2, 5 and 10 days after the disaster (Šalna & Sapožnikovas, 2010/2006b, p. 148) and a map demonstrating the spread of radioactivity on the territories of Ukraine, Belarus, Russia and Lithuania (Šalna & Sapožnikovas, 2010/2006b, p. 149). The text is illustrated with photos: the view of the nuclear power plant after the accident, desolated and neglected houses and other buildings on the site of the catastrophe, a photo of a child with disability who was born in the zone of radiation. Even though not extensive, such presentation makes quite an impact: various materials, i.e., texts, photos, maps, were used to create this narrative. The topic develops energy literacy in connection to important competences within the geography curriculum: cognition of a particular region and map reading.

The presentation of the Chernobyl-related topic extends the concept of energy in general and nuclear energy in particular, while energy is described not only as an economy part of particular regions, but also an important source of economy operation and development. The analysis of disasters in nuclear industry is a separate topic which reveals potential insecurity of nuclear energy, its harm and threat to people, nature and environment. Here, not economic but rather environmental aspects of energy use and production are disclosed. It should be noted that scientific literature and political discourse discuss the Chernobyl disaster as a turn in the entire history of the nuclear energy industry. The accident demonstrated a gigantic danger of the use of nuclear energy, changed the entire history and direction of this branch of economy. This has become the largest trigger of the anti-nuclear movement throughout the world. Since the start of the nuclear energy industry (since the 1950s of the twentieth century), it had a bright future predicted to it, and many expectations were linked to this type of energy industry (it was considered one of nature-friendly types of energy, not consuming many natural resources, not polluting atmosphere with CO_2 emission, and differently from fossil fuel-based energy).

The Chernobyl disaster has altered the perception of the nuclear energy industry and impacted its entire development. It is important that the Chernobyl disaster is characterised in terms of making effect to the country, Ukraine; also, it reveals how the radioactive fallout impacted the neighbouring countries to Ukraine and a large territory of Europe. Thus, it demonstrates how insecurity of the nuclear energy and accidents taking place expand the national limits of regions when radioactive contamination spread in a different way than energy regions do, landscapes reflected in geography maps form in a different way.

On the other hand, presentation of the region of Ukraine by mentioning the Chernobyl disaster creates a single-sided image putting it that being a region of the nuclear catastrophe is one of the most important (and the only?) features of Ukraine.

In terms of relevance and novelty of the data and information displayed in textbooks for forms 7 and 8, it can be noted that one part of the materials reflects basic information which does not change fast, renders specific knowledge on nuclear energy. For instance, the map of resources of minerals is present in the Republic of South Africa, which indicates coal, uranium, diamonds, copper, platinum, gold, iron, manganese, phosphorites (Šalna et al., 2005, p. 161). Also, a map of resources of minerals is present in Canada, demonstrating oil, gas, coal and uranium (Šalna et al., 2005a, p. 241), emphasising that uranium is a nuclear fuel. Information on the Chernobyl disaster (Šalna & Sapožnikovas, 2010/2006b, p. 148–149), nuclear power plants being constructed in India (Šalna & Sapožnikovas, 2010/2006b, p. 208) and the like, would be attributed to this group of information. Another type of materials deals with statistics and facts which become outdated fast. In this regard, the textbooks used in Lithuanian schools introduced in the current chapter are old; information of the discussed type is outdated and no longer relevant. Chapters of the textbook introducing the Lithuanian INPP where production of electricity was ceased as far back as in 2009 can be an example of such information; however, in this instance, the textbook published in 2006 informs that the power plant produces 77.7 per cent of the electric power for Lithuania (Šalna & Sapožnikovas, 2010/2006a, p. 66). Nevertheless, bearing in mind present-day topicalities, a teacher should treat the given data from a historical perspective, emphasising the specific situation prevailing at that time. Other instances reflecting outdated information could be related to presentation of particular statistical data, e.g., uranium mining in various countries (Republic of South Africa, Canada, Russia) by per cent, production of electric power in various power plants by per cent, etc.

On the one hand, a teacher delivering lessons on energy and who has to use outdated textbooks presenting the INPP as an operating power plant can emphasise the closure of it during lessons. Nevertheless, a teacher must meet the challenges concerning absence of relevant information in available textbooks: one will have either to find relevant and contemporary statistical information on the volume of energy industry (which no longer includes part of nuclear energy) or to give tasks to students to find information on what power plants and how much of energy they produce in Lithuania, countries of North Europe. This would require allocating additional time and information resources.

Discussing complexity in rendering the topic on nuclear energy one may observe that lower levels of knowledge and understanding singled out by Bloom et al. (1956) and Anderson & Krathwohl (2001) dominate in textbooks for forms 7–8. The basic information on uranium and nuclear fuel is presented in the textbooks (for form 7); however, the textbooks do not call uranium a nuclear fuel. This explanation is developed in textbooks for form 8, which could be attributed to the level of understanding. The level of complexity in presenting information increases: in the case of form 7, the nature of nuclear power plants is revealed, and in the case of form 8, knowledge is expanded by introducing statistical data – explaining what part of electricity is produced in nuclear power plants, the volumes are compared with power plants of other types. Thus, the initial formation of the image of an energy-focused region proceeds when different sources of energy in a particular country and region are described, and their connectedness is revealed.

The Concept of Nuclear Energy in Forms 9–10

It should be noted that geography textbooks for form 9 pay more attention to the problems of the general concept of energy, such as geopolitical decisions, discussion of various types of energy, and energy-related environmental issues; however, the aspects of nuclear energy industry are almost left without any direct consideration. Nuclear energy is mentioned only once, when dealing with environmental topic and discussing the process of power plant operation including the usage of water from water bodies to cool reactors down. The utilised warm water is released back to the water body, and this, in turn, causes the silting up.

Nuclear energy is presented in the greatest detail in textbooks for form 10 (Šalna et al., 2006; Valančienė & Česnavičius, 2008). From the current perspective in 2020, both textbooks include quite much outdated, irrelevant material. Nevertheless, like in the textbooks for lower forms, specific basic information allowing understanding major aspects of nuclear energy is presented.

Both textbooks for form 10 present the topic on nuclear energy by pointing out the historical development of the energy industry. The textbook by Valančienė and Česnavičius (2008, p. 24) puts it that until the 19th century firewood was a major source of energy; later on, after the 19th century, as industry underwent development, coal became a major kind of fuel; and "in ca. 1960, oil became the most important source of energy, gas started being used and, later on, nuclear energy (uranium is required to obtain it)". Further goes explanation on where electricity was being produced: "first power plants were thermal steam power plants and hydroelectric power plants. Later, they perfected, wind and nuclear power plants appeared" (Valančienė & Česnavičius, 2008, p. 44). A summary for a chapter "Resources and Energy" in Šalna et al. (2006, p. 23) reveals problems of the nuclear energy industry causing threat, such as recycling and storing of nuclear waste: "The nuclear energy industry is being developed throughout the world since the middle of the 20th century. Over the time, countries accumulated vast amounts of nuclear fuel waste. It is very expensive to recycle and store it. These substances are hazardous to human health; therefore, the humankind are facing a new problem – what to do with the nuclear fuel." (p. 23). Such approach allows drawing assumptions that students of the 10th form should find out about the development of the energy industry and economy, nuclear energy, production and recycling of nuclear fuel emerged in a specific period within that development. It is important to note that the development of nuclear energy, grounding on the text of the textbook issued in 2008, is presented not as a valuable source of energy, but also as a type of energy that causes problems (storing of nuclear fuel) to the humankind.

Historical information on the changes in consumption of energy sources and energy industry presented in both textbooks is illustrated by figures, diagrams and maps. The figure "Changes in consumption of energy resources" (Valančienė & Česnavičius, 2008, p. 23) demonstrates a historical fact stating that sources of nuclear energy appear in ca. 1965, and intensity of their consumption in 2000 comprises ca. 10 per cent of the entire energy production. The textbook by Šalna et al. (2006, p. 24) displays a figure demonstrating a curve of consumption of different types of energy over the period from 1950 to 2000. The data of the figure suggests that consumption of resources to produce nuclear energy increased twice over 50 years. A task dedicated to the analysis of this figure (Šalna et al., 2006, p. 24) suggests students investigate and comment on how consumption of fuel and energy sources changed from ancient times to the present day and to indicate the causes that resulted in that change. This topic and adjacent tasks on the historical change and dynamics of energy consumption included in the textbook reflect one of the major topics of energy geography in relation to historical changes in energy landscapes, technological and energy-related social transitions dealt with by authors analysing energy geography (Bouzarovski, 2009; Lambert, 2013, Butt, 2011). Such way of introduction of nuclear energy in textbooks reflects characteristics of geography as the "global thing" discussed by Lambert (2013) and Butt (2011) because the historical development of the nuclear energy industry throughout the world is dealt with.

In order to assess the novelty and rendering of basic knowledge of the textbook's materials, the historical approach to the development of the nuclear energy sector can be treated as important basic knowledge on the understanding of energy geography. However, it should be admitted that examples illustrating the situation of nuclear energy industry and economy (in tables, maps, diagrams) as well presented data, numbers no longer reflect the topicalities or problems of the current period. The textbooks render a concept of the change of an energy landscape; however, the "present" dealt with in the textbooks written in 2006 and 2008 does not meet the reality and present-time of 2020. Working in class, a teacher must regard the changes that took place in the area of nuclear energy sector since that time demonstrated by facts in the available textbook and must seek to present new, relevant data.

The entire materials on nuclear energy of textbooks for form 10 can be divided into several thematic areas.

Identification of the Significance of Nuclear Industry

Like in the case of form 8, the textbook by Šalna et al. (2006, p. 24) for form 10 has it that "uranium is a very important type of resources, required for producing nuclear energy"; and the textbook by Valančienė and Česnavičius (2008, p. 23) includes more information: points out that the fuel of nuclear energy is not only uranium, but also includes another element, thorium. Moreover, a table "Branches of economics" in the textbook (p. 43) presents nuclear energy industry as a constituent part of the energy economy, together with thermal, water and alternative energy.

The textbook by Šalna et al. (2006) describes advantages and disadvantages of nuclear energy quite in detail and consistently. The advantages of the use of nuclear energy pointed out in this textbook are the following: long-lasting, clean, because there is no impact of CO_2 on nature; cheap, because little fuel is required, and transportation of it is easier and cheaper; resources of fossil fuel are saved. Specific advantages are not pointed out in the textbook by Valančienė and Česnavičius (2008); however, similar emphases are made when talking about the operation of the INPP, for instance, the text explains efficiency of the nuclear energy industry, little nuclear fuel is required, there is no environmental pollution.

Drawbacks are emphasised, too: "These power plants may be dangerous and cause threat to environment" (Šalna et al., 2006, p. 32). "It is difficult and expensive to recycle radioactive substances. The sites for storing the waste are insufficient all around the world; it is dangerous to human health to transport nuclear fuel; the risks of harming people and environment is higher than the benefit. The safety systems of the power plants cannot completely prevent from severe earthquakes and terrorist acts." (Šalna et al., 2006, p. 33). Materials on the INPP provided in Valančienė and Česnavičius (2008) putting it that the problem lies in old-type reactors (even though it does not mention that the type is the RMBK, like in Chernobyl) could be an illustration to the statements of this textbook; the latter has it that, after the disaster at the Chernobyl Nuclear Power Plant, the trust in the nuclear energy system has decreased; nevertheless, it also notes that "this is the only known area of future energy industry which will be able to accommodate the energy demand" (Valančienė &

Česnavičius, 2008, p. 47). Interesting to note that, in such a way, a provision focusing on students' critical thinking in assessing nuclear energy from different perspectives, seeing both positive and negative aspects is being developed. Both textbooks for form 10 have questions intended for generalisation of the concept of nuclear energy: "What advantages and disadvantages exist in terms of nuclear energy? What is the perspective of nuclear energy industry?" (Valančienė & Česnavičius, 2008, p. 46).

However, it can be observed that the textbook provides a conclusion on the nuclear energy sector as the only known type of future energy industry, which could be treated as a clear pro-nuclear attitude and narrative supporting nuclear energy. It is important to underline that back in 2008 Lithuania still maintained a vision of a state developing the nuclear energy industry. The second block of the INPP would be closed in 2009; Lithuania maintained a hope to build a new nuclear power plant up until the referendum held in 2012, which prevented from construction of the new nuclear power plant.

Such introduction of significance of nuclear energy raises both a question and a doubt whether the topic of globality is sufficiently developed, which, according to scientists (Lambert, 2013; Butt, 2011), is important to geography as "a subject of survival" and "curriculum for the future", treated as an important purpose of geography as a subject – to raise questions about global challenges, global interconnectedness issues. As Thoyre and Harrison (2016) put it, globality is connected to problems of sustainability, questions of geopolitics, environmental crises and catastrophes. The analysis of the content of textbooks for form 10 reveals that the aspect of globality of nuclear energy industry is more emphasised not through the prism of the environmental protection topic (which would manifest as concern about insecurity of nuclear energy and effect of radiation on environment all around the world), but by underlining the global geoeconomics aspect, when nuclear energy is seen as the only type of energy than can meet the demands for energy in the future economy worldwide.

Analysing how the development of energy literacy proceeds, it is supposed that the naming of the significance of nuclear energy in textbooks, though, allows students form critical civic stance or attitudes (grounding on Van der Horst et al., 2016; Fowler, 1976, cit. DeWaters et al., 2013) because the context is presented by providing both positive (nuclear energy as cheap, making no harm to environment, effective) and negative (Chernobyl disaster, insecure equipment) aspects of the use of nuclear energy.

Moreover, grounding on the elements of energy literacy proposed by DeWaters et al. (2013), the affective element stands out, too, because students form the view (concern), stance (understanding that something should be changed) and responsibility. No doubt, the cognitive element is being developed as well, because basic knowledge on the significance of nuclear energy is obtained. From the point of view of complexity of the competence (Bloom et al., 1956; Anderson & Krathwohl, 2001), texts in textbooks for form 10 are oriented to higher levels: analysis, synthesis and evaluation because advantages and disadvantages of nuclear energy are presented; problem questions about complexity of recycling and storing of nuclear waste are raised; and tasks requiring considerations on perspectives of nuclear energy are included.

Revealing the Role of Nuclear Energy in the World

Both textbooks for form 10 include materials on nuclear energy and its significance in the world. The textbook by Šalna et al, (2006, p. 32) identifies the geopolitical aspect, i.e., "the decision to use nuclear energy greatly depends on the governmental views and public opinion" (p. 32). The reasons for nuclear industry development and countries developing it are indicated mentioning that "nuclear energy is used by the countries that lack fossil fuel" (p. 32). The textbook by Valančienė and Česnavičius (2008, p. 45) also has it that nuclear power plants are constructed on sites which lack other sources of energy.

Here, the development of energy literacy stands out when the geopolitical aspect of the energy system is emphasised (Fowler, 1976, cit. DeWaters et al., 2013; Van der Horst et al., 2016). Nuclear energy is characterised as an effective type of energy which may accommodate the needs for energy in specific regions; decisions on closure of the INPP are discussed in relation to national or global decisions.

Both textbooks introduce the countries and scope of the development of nuclear energy industry. These two point out that there are 30 countries worldwide which run nuclear power plants. Valančienė and Česnavičius (2008, p. 45) also note that the largest numbers of power plants are in the USA, France and Japan. Moreover, it states that the largest nuclear power plant in the world is the Fukushima Nuclear Power Plant (in Japan), comprising 10 operating reactors, and in total Japan "runs 16 nuclear power plants, electric energy is being produced by 52 reactors"; illustrations are provided on how nuclear energy constituted 8 per cent in 1990 and 12 per cent in 2007 (Valančienė & Česnavičius, 2008, p. 157). Here it is important to note that the textbook under analysis was written earlier than the disaster in the Fukushima Nuclear Power Plant that took place in 2011; therefore, it is obvious that this accident is not introduced in the textbooks.

The volume of production of nuclear energy is described in both textbooks. Šalna et al. (2006, p. 32) present a table "Ratio of electrical energy produced in nuclear power plants in 2004" which lists the countries running nuclear power plants in the world, percentage ratio of produced electrical energy and numbers of reactors per country. There is a discussion on the countries that run most of nuclear power plants and produce the largest parts of electrical energy there (Šalna et al., 2006, p. 46); whereas the textbook by Valančienė and Česnavičius (2008, p. 45) includes a table "Spread of nuclear power plants across countries, 2006" which indicates the numbers of reactors and their power in MW in various countries. The data is supplemented with the tasks presented in the teaching materials by Dijokienė (2016) for form 10; the said tasks ask to point out the states where electricity is produced at nuclear power plants, by choosing correct options from the list: Brazil, India, China, the Netherlands, Norway, the USA, Poland, Russia, France and the Republic of South Africa. Another task that allows consolidating the information presented in the textbooks deals with the exploration of a cartographic scheme to complete the tasks: to list the European states which currently run largest numbers of reactors; to indicate 3–5 states that produce nuclear energy; and to find the European countries which have no nuclear power plants; all insights must be substantiated.

The textbook by Šalna et al. (2006) lists the technological and economic aspects in ensuring safety, which is linked to high financial expenses. Another aspect to be noted in relation to this textbook is that the text emphasises the decrease in the development of nuclear energy industry impacted by the Chernobyl disaster.

The presentation of the significance of the nuclear energy sector at the global level in the textbooks complies with the concept of energy geography defined by Solomon et al. (2004) and Calvert (2016) because the texts partly reveal the aspects pointed out by the authors: identification of local factors explaining observed spatial patterns of investments in the energy sector and the distribution of energy technologies in the world (networks of the spread of nuclear power plants, capacities of nuclear power plants are presented); identification of environmental and economic risks by pointing out that expensive technological solutions in relation to the nuclear energy sector limit the development of this type of industry. Connecting this to the energy geography approach, the authors of the present research emphasise the significance of formation of the civil position aspect, too, since students get an opportunity to evaluate the actions linked to nuclear energy industry with regard to national (Lithuanian) and global (nuclear power plants worldwide, their capacity, expenses on technological solutions) decisions (according to Van der Horst et al., 2016).

When dealing with nuclear energy in the world, it is highly important to note the significance of novelty and relevance of the materials in the textbooks. In this case, information presented in Lithuanian textbooks is outdated in many instances because basically the distribution and capacities of nuclear power plants are changing, e.g., after the Fukushima catastrophe in 2011, the Japanese system of nuclear energy industry has significantly changed; or the change in the development of nuclear power plants in some other countries (France, Finland, etc.); therefore, a teacher working with materials of the textbook must critically evaluate such changes, select materials and data illustrating the changes.

It can be noted that, when linking to the energy literacy elements pointed out by DeWaters et al. (2013), materials on the significance of nuclear energy in the world as presented in the textbooks is oriented towards the cognitive element, and, from the point of complexity of the competence, it would meet the level of knowledge and understanding because essential information is rendered.

Revealing the Threats of Nuclear Energy

The textbook issued in 2006 (Šalna et al., 2006, p. 32) for form 10 describes the threats related to nuclear energy quite in detail, pointing out that they occur from radioactive pollution, storing of radioactive substances and that society assess this area negatively.

However, it was observed that textbooks for form 10 do not elaborate narration on the largest disaster of the nuclear power plant at Chernobyl, the most highlighted and strongest anti-nuclear narratives. Valančienė and Česnavičius (2008, p. 48) provide a photograph "Reactor of the Chernobyl Nuclear Power Plant (Ukraine)"; however, for a reader, it may be not clear from a provided image that it depicts the situation after the disaster happened. There is no text on the disaster supplementing this photograph. Moreover, this textbook only mentions the aspects of accidents in nuclear power plants at the end of the chapter dedicated to the energy industry (p. 46), presenting the volume of the Chernobyl disaster in short: "The level of radiation on the territory of the nuclear power plant reached 20–25 micro-r-units per second. This exceeded the permissible norm more than a thousand times," without providing more details. Even though these textbooks only quite fragmentarily develop the element of nuclear power plant catastrophes, still, the tasks of the teaching materials by Dijokienė (2016) include a creative project assignment on the topic dealing with the Fukushima Daiichi Nuclear Power Plant catastrophe that took place on 11 March 2011, following the earthquake. The task has it that, after the bombing of Nagasaki and Hiroshima, Japan has chosen the forms of nuclear power production, planned to have produced over 50 per cent of the energy consumed in the country at nuclear power plants by 2030. A question why did Japan make such a choice is posed to school students. To answer the question, students have to use various information sources and prepare for a discussion on why Japan needs so many nuclear power plants; moreover, they have to prepare a presentation "The Future of Energy in Japan". Such approach to the presentation of the materials reflects a connection between two largest catastrophes of nuclear power plants; a teacher using both teaching means can develop a relevant up-to-date discussion on assurance of safety of nuclear power plants, causes and consequences of catastrophes and the future prospects of nuclear energy.

One of the threats pointed out by Šalna et al. (2006) as a separate issue focuses on safety of nuclear fuel waste. It emphasises that "this is a potentially hazardous product resulting from nuclear energy activity: its concentration encompasses 98 per cent of all radioactive materials", and "it is difficult to solve issues of storing accumulated nuclear fuel waste and conserving old reactors. Currently, the nuclear fuel waste is usually stored in isolation from environment in reliably controlled reservoirs. Quite many of them are kept dug in the soil." (Šalna et al., 2006, p. 32). The text has it that the recycling of nuclear fuel is a complex and expensive technological process; therefore, only nuclear states, the UK, France, Russia and Japan, can afford it. Also, it points out that states that do not have such complex technology would prefer to pay for acceptance and recycling of the waste. The text is illustrated by a figure (p. 33) "Nuclear fuel from mining to storing waste". The text is supplemented with a task evoking a discussion by providing arguments in favour of or against the development of the nuclear energy sector (p. 32); and a task-question at the end of the chapter is dedicated to repetition of knowledge on nuclear waste requiring answer yes or no: "The storing of radioactive waste does not cause any big problems." (Šalna et al., 2006, p. 46).

The identification of threats allows forming environmental literacy, problem-solving skills highlighted by Van der Horst et al. (2016) related to the development of energy literacy because the textbooks present texts on danger posed by nuclear waste, conservation of old reactors. Relating to the elements of energy literacy proposed by DeWaters et al. (2013), materials of the textbook on threats and difficulties are to be linked to the affective element because statements on radioactive pollution, dangers and further situation of old reactors allow students understand and form, on the one hand, concern, anxiety about the future of the humankind; on the other hand, it may form their responsible attitude towards consumption, critical thinking and decide on their choices (e.g. which type of energy is more suitable for use: renewable or nuclear?).

From the point of view of the complexity of the competence (Bloom et al., 1956; Anderson & Krathwohl, 2001), materials of the text, in opinion of the authors of the present paper, comply with the levels of analysis or synthesis because quite complex problems on technological solutions which can be made to recycle and store nuclear waste are presented; problems revealing distrust of society in the nuclear energy system are mentioned. Such approach in presenting the materials allows a teacher arrange various discussions, debates which would help students gain skills of critical thinking.

Review of the Situation of Nuclear Energy in Lithuania

Both discussed textbooks quite consistently present the situation of nuclear industry in Lithuania. Texts emphasise the situation of the time when the textbooks were being prepared (in 2006). However, the textbooks include specific links or considerations on likely perspectives, even though they are not very clear or firm. In such a way, an actual political, economic situation of that time when Lithuania was preparing for the closure of the second block in 2009 and there were no clear visions concerning the future of nuclear energy is reflected. At that time, opportunities of construction of a new nuclear power plant jointly with Latvia and Estonia were being discussed. When delivering nuclear energy topics during geography lessons nowadays, it is clear that the projected prospects did not come true when the referendum of 2012 determined the refusal of constructing a new nuclear power plant.

Nuclear energy in Lithuania is discussed in the following aspects: by introducing the concept of sources of energy, indicating the consumed amount of nuclear fuel to produce electrical power, "in the energy balance for 1980–2005 it constituted 34–37 per cent" [comparing to other sources of energy] (Šalna et al. 2006, p. 38); the type of energy is discussed by pointing out that nuclear the energy system also operates complementing other types sources: "In Lithuania, the largest part of electrical energy is produced by power plants of three types: hydroelectric, thermal and nuclear" (Šalna et al., 2006, p. 45). The volume of this type of energy industry is illustrated by a diagram (Šalna et al., 2006, p. 39) "Production of electrical energy in Lithuania" which demonstrates the types of power plants and changes in production of electricity in 2003 and 2005; whereas the textbook by Valančienė and Česnavičius (2008) presents corresponding numbers demonstrating that the INPP is one of the most powerful power plants in the world. The textbook by Šalna et al. (2006, p. 38) comprises a map "Lithuanian system of energy" which depicts all types of power plants in Lithuania related to production of energy that were operating at the time of writing the textbook under discussion: the nuclear power plant, thermal power plants, hydroelectric power plants, oil pipelines, gas pipes, oil refinery and oil processing plants. Such rendering of the material shows that the rendering of nuclear energy industry in line with other types of energy facilitates students' better understanding of the map depicting the energy system and the structure of the energy economy.

Moreover, both analysed geography textbooks for form 10 also present the prospects of the INPP linked to geopolitical and economic aspects. First, the textbook by Šalna et al. (2006) points out that reactors of the INPP are analogous to those in the Chernobyl Nuclear Power Plant; their "time of exploitation has already expired; therefore, there is a grounded consideration that the power plant poses threat" (Šalna et al., 2006, p. 39), which would be linked to technological and environmental aspects. However, both textbooks provide details supporting the view that obligations undertaken before entering the European Union are one of the major causes to close the INPP. In such a way, the geopolitical factor is underlined as one of significant aspects in terms of the closure of the INPP. Moreover, perspectives of the development of nuclear energy industry are highlighted by stating that Lithuania, jointly with Estonia and Latvia, "have come to an agreement to jointly build a new, modern nuclear power plant in Lithuania" (Šalna et al., 2006, p. 39).

The economic aspect of the closure of the INPP is emphasised by the textbook by Valančienė and Česnavičius (2008), putting it that this might disturb the structure and prices of energy sources because too "little of local renewable resources are used" (Valančienė & Česnavičius, 2008, p. 49), the funding of the closure of the INPP from the EU funds allocated for mitigating negative financial, technical and social effects is pointed out. In this case, it would be appropriate to discuss on the closure of the INPP as a complex process in technological and geo-economic as well as geo-political and social aspects; therefore, evaluations of this process cannot be unilateral.

Materials provided by Šalna et al. (2006) on nuclear energy in Lithuania and the INPP are illustrated with a photograph "The first block of the Ignalina Nuclear Power Plant" (p. 39) and a text of the news agency ELTA "An Electrical Bridge from Lithuania to the West" published in 2005 on the foreseen solution of the problems of electricity supply after the closure of the INPP: "Lithuania and Poland will put efforts to implement the project of the electrical bridge to the West by 2009, when the closure of the second block of the Ignalina Nuclear Power Plant is anticipated" (Šalna et al., 2006, p. 40).

To consolidate knowledge on the situation of the nuclear energy sector in Lithuania, 4 questions are given in the textbook by Šalna et al. (2006, p. 46): "How has the production of electrical power changed in Lithuania after the closure of the first block of the Ignalina Nuclear Power Plant? Predict how the system of energy industry in Lithuania will change after the closure of the INPP. What positive aspects and negative effects will the closure of this giant of economy have? Which electric power plants produce most of the energy in Lithuania? And a statement requiring an answer whether "yes" or "no": Very much of electric energy is produced at the Ignalina Nuclear Power Plant." The prognostic elements on the situation of the nuclear energy sector are substantiated in tasks of more recent teaching materials by Dijokienė (2016): school students are asked to characterise the currently existing energy sector in Lithuania and, by using the cartographic scheme, to present a prediction on the changes of the energy sector by 2020. The scheme includes the operating (in 2016) energy systems, and the role of a nuclear power plant is linked to the withdrawal from the common system. However, the materials (by Dijokienė) published in 2016 point out the aspect of further plans of nuclear energy sector developments to be discussed, as it presents a task to prepare a presentation for peers on strategic projects of Lithuania, including a plan to construct a new nuclear power plant in 2018–2020. Basically, these materials are still being used in schools; therefore, teacher's critical point of view as well as knowing and informing of students about the decisions made back in 2012 to not build the new nuclear power plant are highly important.

As observed, texts included in the textbooks for form 10 lack references to quite a unique social aspect of the nuclear energy industry in Lithuania, such as consequences of the closure of the INPP and construction of a new INPP for the nuclear town Visaginas, a satellite of the INPP, and its community.

To sum up the connections of the texts in geography textbooks for form 10 to the skills of the geographical competence being developed, it is clearly seen that the development of the skills of reading geographical information dominates because the texts render various, quite consistent information, data and concepts. Moderate orientation to the skills of cognition and investigation of environment is found, when informative materials of the texts are illustrated with tables and diagrams. Skills of orientation in area and map are slightly developed when materials of the texts are illustrated with maps and photographs. Cognition of other regions (except Lithuania) is developed very episodically, when information on nuclear energy industry in some countries of the world is presented; nevertheless, this is more of an informative kind of materials. Lithuania is a major region receiving most of the attention while discussing the situation of the nuclear energy sector in Lithuania. Tasks presented in the texts are more dedicated to reproduction of knowledge, even though particular elements of the development of analytical skills can be recognised, too. Texts on the situation of the nuclear energy sector in Lithuania correspond to the theme of energy geography according to Bouzarovski (2009), when the change of energy landscape is identified in connection to social-technical solutions, spatial differentiation and territorial network where specific relations of power manifest.

In terms of assessment of introductory materials in the textbooks for form 10 in the aspect of energy literacy dimensions (DeWaters et al., 2013), it can be stated that cognitive and behavioural elements prevail because presentation of the historical development of the nuclear energy industry renders basic knowledge, and the aspect of knowledge on the technological development forms attitudes towards future behaviour in the aspect of consumption of this type of energy.

In relation to the development of energy literacy skills characterised by Van der Horst et al. (2016) and Fowler (1976, cit. DeWaters et al. (2013)), it can be observed that materials on nuclear energy in the textbooks for form 10 allow developing civic attitudes and gain analytical skills, when students are given tasks inviting to predict the situation after the closure of the INPP, change of the energy economy, and perceive this type of energy as manifestation of science and technology as well as of its role in both national and global area. In such a way, students discover the historical development of the nuclear energy industry, which allows understanding the effect of this area of energy on societal processes: increase of energy production, booming of industry, and, however, an issue of the danger of nuclear waste is raised.

Regarding the complexity of energy literacy (according to Bloom et al., 1956; Anderson & Krathwohl, 2001), it is observed that the materials on nuclear energy of Lithuania are characteristic of complexity and complication; therefore, these would be attributed to the levels of synthesis and evaluation. Such conclusions are drawn by the researchers of this case because even though presentation of informative materials (tables, numbers) is more focused on the development of analytical skills in relation to why particular processes happened (closure of the nuclear power plant, threat of the reactor etc.), the following questions project expression of profound insights on likely changes in energy industry, predicting the situation after the closure of the INPP.

Obviously, the aspect of outdated information cannot be left unnoticed because the materials presented in the textbooks deal with topicalities that were relevant in 2006, e.g., agreements to build a new nuclear power plant, which at the present moment are no longer significant because they will not be implemented, though. Thus, teachers must very carefully analyse the materials themselves and present what may be relevant in the contemporary time, e.g., point of view towards presently being built Astravets Nuclear Power Plant in Belarus and issues of electricity supply from this power plant.

The Concept of Nuclear Energy in Forms 11–12

When revealing the concept of nuclear energy, the analysis of two geography textbooks: Česnavičius and Gerulaitis (2008/2007) "Bendroji geografija" (Lith. General Geography) and Česnavičius et al. (2010/2008) "Regioninė geografija" (Lith. Regional Geography), has been carried out. The analysis of the content of these textbooks pointed out that several new aspects, such as urban and military, were added to the context of nuclear energy and a topic of technological progress was moderately developed.

Quite a large part of materials on nuclear energy displayed in the textbooks repeats what was presented for lower forms, especially form 10, including small additions. Statements about production of electric power given in the textbooks could be identified as repeated materials: "Majority of electricity is produced in thermal, nuclear and hydroelectric power plants" (Česnavičius & Gerulaitis, 2007, p. 188); nuclear energy started being used half a century ago, and power plants are built on sites which are poor in other energy resources; uranium, more rarely thorium, is used for production of fuel (p. 188); electrical energy comprises: thermal, nuclear, hydroelectric and alternative energy (p. 184); nuclear power plants, having little effect on environment in terms of pollution, are efficient; the utilisation of nuclear fuel waste and danger of nuclear fuel in terms of likely explosions are mentioned.

The repeated materials in the textbook by Česnavičius and Gerulaitis (2008/2007) can be those dealing with the information on the INPP because, like for form 10, it indicates that these reactors are among the most powerful ones in the world; but, differently from form 10, their type, RBMK-1500, is mentioned (p. 191). It is stated that they are considered to be unreliable. It is mentioned that the INPP was closed; however, no causes are revealed (like in form 10, due to the obligations undertaken when entering the EU, old model (insecurity of the Chernobyl type reactor). The tasks included in the exercise book by Šalna et al. (2012) moderately expand the materials presented in the textbook texts because there is a set of tasks on nuclear energy worldwide and particularly in Europe, displaying 6 texts from information publications on the nuclear energy sector situations in Germany, Japan and the EU; the safety issues in relation to nuclear power plants in various countries of the world; the development of the nuclear energy industry in China; and the technological use of a nucleus in space industry. Students are asked questions requiring submission of arguments: for instance, why are nuclear power plants being massively closed in Japan and other countries since 2011; how will this impact national economies; what was the impulse for the change in the nuclear energy sector since 1989; and why do some countries, despite emerging threats, continue developing the nuclear energy industry.

When analysing both textbooks for forms 11–12, it is observed that they both could supplement each other due to differences in their contents. The textbook "Regioninė geografija" (Lith. Regional Geography) by Česnavičius et al. (2010/2008) focuses on specific characteristics of regions and countries; therefore, one can find materials on nuclear energy industry of separate countries or regions, Great Britain, China, USA, Brazil and Republic of South Africa, in it; nevertheless, the information is not highly detailed, is presented in different volumes, e.g. nuclear energy industry of Great Britain is introduced following the principle of historical consistency: when the first nuclear power plant was built, how volumes of energy produced in nuclear power plants changed, what are the sources of the fuel, how many reactors operate. The nuclear energy in Japan is presented by relating it to the aspect of militaristic use of nuclear energy, telling that the first experience of the state in terms of nuclear energy was undergone as one of the largest tragedies, the dropping of nuclear bombs on Hiroshima and Nagasaki. Nuclear industry of other states or regions, Asia, USA, China, Brazil, also Japan, is presented in this textbook by displaying numbers or even less: presented maps demonstrate references to the sites of uranium mines or locations of power plants. Materials on nuclear power plants in Japan, the largest nuclear plant in Japan, Fukushima, containing 10 operating reactors (p. 188) presented in the textbook by Česnavičius and Gerulaitis (2008/2007) could supplement the materials of the earlier-mentioned textbook resulting in a more comprehensive view of nuclear industry in Japan.

Such character of materials on nuclear energy presented in the textbooks allows the authors of the research draw connections to the characteristics of global understanding of energy and energy literacy defined by Solomon et al. (2004) and Calvert (2016): this information allows students understand the changes in the energy supply-chain, how energy technology diffuses within and between nations; and mapping regional variations in energy production, distribution and use. Also the textbooks include the aspects which were not introduced in earlier forms or only presented in brief, not further developing: aspects of urbanistic, militaristic, and scientific and technological progress.

Social and Urban Aspects of Nuclear Energy

One of the aspects introduced in the textbook by Česnavičius and Gerulaitis (2008/2007) deals with the connection between nuclear energy industry and foundation of settlements. One of the social aspects linked to the effect of the nuclear energy industry is included in the exercise book (Gerulaitis & Bačkienė, 2009) in the form of a task asking students to classify the causes of forced and voluntary migration of residents, while indicating the example of an environmental disaster, the explosion of the Chernobyl Nuclear Power Plant. The foundation of settlements was only mentioned in lower forms; the town of Visaginas was identified; however, more details were not provided. The first more comprehensive discussion on Visaginas as a satellite town constructed to serve the nuclear power plant is presented in the textbook for forms 11–12 (Česnavičius & Gerulaitis, 2007). Since the textbook by Česnavičius and Gerulaitis (2008/2007) designed for form 11–12 pays much attention to geo-political and geo-economic aspects, it also manifests when talking about problems of Visaginas, the town of the INPP: "4 reactors could be constructed in the power plant (2 were built). After the changes in the political and economic conditions, a part of the planned town construction remained unfinished." (Česnavičius & Gerulaitis, 2007, p. 57). In this textbook, the topicalities of the closure of the INPP are related to economic-financial aspects, as it puts it: "Early closure of the Ignalina NPP without having required funding from the EU and other Western states and international financial institutions would be an unbearable burden for the national economy to carry" (Česnavičius & Gerulaitis, 2007, p. 191). The financial means for the closure of the INPP are not directly allocated to the power plant only, but also they are dedicated to the restructuring of the economy sector of energy while developing renewable energy resources. The exercise book by Šalna et al. (2012, p. 26) includes a task to express one's own opinion concerning the necessity of the project of a new nuclear power plant in Visaginas.

Also, the text mentions the funding for the solutions of social problems occurring in Visaginas after the closure of the INPP. The text of the textbook (Česnavičius & Gerulaitis, 2007, p. 191) has it that, after the closure of the INPP, the nuclear energy industry system in Lithuania is planned to be developed in cooperation with companies of other countries – France, Russia, USA and Canada. Also, a laconic emphasis is put on the prospects of town development, when assessing the INPP as a larger employer in the town, and after the loss of it, occurring social problems; moreover, it underlines that specific changes in performance are foreseen: "Now different activities, not related to the nuclear power plant, are being developed, small businesses and trade prevail in Visaginas. Several "scenarios" for the future of Visaginas town have been designed." (Česnavičius & Gerulaitis, 2007, p. 57).

The presented social and urban aspects of nuclear energy are linked by the authors of the research to the formation of civic attitudes pointed out by Van der Horst et al. (2016), Fowler (1976, cit. DeWaters et al. (2013), when it aims at evaluation of the decisions made in relation to energy at national and global levels. Nevertheless, the introduction of the social situation in Visaginas lacks comprehensiveness. Texts on the volume of construction of the INPP, expansion of the town, political decisions concerning the closure of the INPP and further prospects allow students form quite a comprehensive view and understanding of nuclear industry as a strategic factor supporting national economy, that construction of the nuclear power plant is also related to infrastructural, social decisions (town, maintenance scheme); however, this technology becomes outdated and, therefore, political decisions concerning the closure or construction of a new power plant are being made. The case of Lithuania demonstrates that there can be a closure scenario which essentially changes both national energy system and social structure of a single town.

In terms of the level of complexity of the competence (following Bloom et al., 1956, Anderson & Krathwohl (2001)), the authors attribute this arrangement of the contents in the textbooks to the striving to develop the competences of the synthesis level because quite complex questions encompassing different aspects of nuclear industry are raised, e.g., construction of the nuclear power plant, closure of it and solution of social problems; closure of the nuclear power plant and collaboration with other states; and closure of the nuclear power plant and economic issues of the funding. Of course, assessment of the presented outdated facts or data should not be forgotten. Facts on the capacity of the INPP can be treated as basic information, for instance, when talking about types of nuclear power plants, capacity of reactors. However, when analysing the social problems of Visaginas town at the present moment (more than 10 years after the publishing of the textbook), one may face difficulties, since there is no available sufficient information on what is taking place in the town now.

Militaristic Aspects of the Use of Nuclear Energy

Basically, a militaristic aspect is a completely new one linked to the nuclear context, occurring in the textbooks for forms 11–12. The text by Česnavičius and Gerulaitis (2008/2007) points out that "nuclear technologies created nuclear weapons, a huge threat to the humankind, too" (p. 165). Nevertheless, this aspect is not developed in detail and comprehensively because the presented facts are quite fragmentary and do not help to create a sophisticated and complex concept of nuclear weapons, their effect and geopolitical aspects. First, the aspect of nuclear armament is introduced through a narration about the nuclear bombs that were dropped on Nagasaki and Hiroshima during the US war with Japan in 1945 (Česnavičius & Gerulaitis, 2007, p. 135). Explanations elaborate that nuclear armament causes tension in society, even though after the Cold War the situation became more moderate because the Treaty of the Non-proliferation of Nuclear Weapons was signed, and some countries (Kazakhstan, Belarus, Ukraine, the Republic of South Africa) introduced nuclear disarmament. Seeking to illustrate information on nuclear weapons, a table demonstrating which 8 states dispose nuclear weapon is given (Česnavičius & Gerulaitis, 2007, p. 137).

One of the topics deals with the strengthening of geo-economic and geo-political influence of China through the military power: "The Chinese possess weapons of mass destruction, carry out nuclear weapon testing" (Česnavičius & Gerulaitis, 2007, p. 135). At the end of the text on nuclear armament (p. 137), a glossary is given, including an explanation of the concept "nuclear club" (Russ. ядерный клуб), an unofficial joint title of the states creating and possessing nuclear weapons.

It is important to note that this discussion of nuclear armament across the world is not related to nuclear energy industry, as given in the textbooks. The authors of this chapter hold the opinion that a critical analysis of nuclear energy should encompass the ability to recognise the connections between military and peaceful use of the nucleus. From the historical perspective, the peaceful nuclear energy in the USA, Russia derives from arms industry. On the other hand, setting up the infrastructure of nuclear energy facilities in separate countries may create preconditions for nuclear weapons industry.

From the point of energy literacy, the militaristic aspect of nuclear energy presented in the textbooks can be linked to the understanding of energy as science and technology as well as of its role in the national and global space put by Fowler (1976, cit. DeWaters et al. (2013)) in connection to geopolitical aspects, too. It was observed that the materials present another important aspect, the concept of nuclear weapons, by indicating the danger they may cause. The authors of the research observe that the materials could be more detailed, consistent and attractive to students; however, even such presentation of information like in the textbooks under investigation has obvious significance because the data on states having nuclear weapon is introduced, countries that develop these technologies of armament are identified. Connecting with the elements of energy literacy proposed by DeWaters et al. (2013), the militaristic nuclear aspects found in the textbook texts would correspond to the development of the affective element revealing the formation of pacific, anti-militarist attitudes, responsible behaviour because not only material on the dropping of the nuclear bomb on Japan is presented, but also questions on the nuclear armament of the present time are raised.

Even though the textbooks put major emphasis on the dropping of nuclear bombs on Japan, still a teacher delivering the curriculum could link to contemporary reality of armament, e.g., ballistic missiles testing in North Korea and programmes of uranium enrichment in Iran, which would enable connecting the concept of nuclear energy with the statements of Thoyre and Harrison (2016) having that energy can be a means to consider the questions of environmental sustainability and security in the global world.

Scientific and Technological Progress and Nuclear Energy

The use of nuclear energy as a result of technological and scientific progress is another aspect to be singled out in the textbooks for forms 11–12. In the textbook by Česnavičius and Gerulaitis (2008/2007), nuclear energy sector is linked to the development of science and technology by stating that this branch of economy is open to science (p. 195); the table presented on p. 165, "OECD classification of industry sectors which are open to science (adapted)" indicates that nuclear energy is attributed to the area of industry of high technologies of moderate complexity, and it is grounded with a statement about inventions in the area of physical sciences intended for perfection of the energy sector (p. 165). Statements in the exercise book (Gerulaitis, Bačkienė, 2009) dealing with the invention of the nuclear weapon and the launching of a nuclear reactor are related to the aspect of scientific and technological progress changing the geopolitical map because there is a task to classify various events according to importance in a particular period of time and to write about significance of these events. A statement putting that the society confidence in nuclear energy after the Chernobyl disaster decreased, however, and new technological solutions allow creating new and reliable nuclear reactors which are the fundamentals of the future power plants can be linked to the technological progress of nuclear energy sector: "Currently, this is the only known area of future energy industry which can meet the energy demand" (Česnavičius & Gerulaitis, 2007, p. 189).

It can be emphasised that the importance of the technologies of modern energy industry is underlined, though it is not elaborated in detail; nevertheless, an idea about modernisation of nuclear energy sector, inventions in the physical science, is raised, and nuclear industry is identified as the only area warranting the increasing needs for energy. This material reveals an approach expressed by Fowler (1976, cit. DeWaters et al. (2013)) on energy literacy, having it that an understanding of energy as science and technology as well as of its role in the national and global space is needed. The topic of globality is dealt with by introducing technological aspects of nuclear energy and stating that this type of energy will remain the most important in ensuring the energy needs throughout the world. This allows students design a global whole view and perceive that nuclear energy is important to the entire world.

Generalisation on the Explanation of Nuclear Energy in the Textbooks for Forms 10–12

One can recognise that this presentation is highly purposively oriented to the geo-economic aspect because economic matters are emphasised the most: extraction of nuclear (uranium) raw materials, volumes of production of nuclear energy, number of power plants, states that produce nuclear energy. This corresponds with the object of energy geography, the perception of how global infrastructure of nuclear energy sector is created, which countries and how they develop this energy industry. This aspect is highly expressed in the textbooks through presenting data on nuclear energy industry, describing examples. Texts of the textbooks reveal significance of the use of nuclear energy in the common infrastructure of separate countries or regions comparing the volumes of consumption of nuclear energy with other types of energy, volumes of nuclear fuel processing with other kinds of fuel. However, having evaluated the formation of energy geography as a holistic image of energy while discussing economic models of various countries, their geo-political and geo-economic interests in the area of nuclear energy, this aspect is elaborated quite superficially.

Less attention is paid to the consumption of nuclear energy, as a global phenomenon causing the environmental effect worldwide. In this approach, materials of the textbooks introduce the contradictions of the phenomenon of nuclear energy. They underline that nuclear energy is clean, pollution-free in terms of CO_2 emission and efficient because low consumption of fuel is required, the only one which can meet high demand of consumption. On the other hand, some statements deal with the threats posed by this type of energy, such as consequences of accidents and catastrophes, occurring insecurity caused by aging technologies of reactors, recycling and storing of waste fuel.

When dealing with development of energy literacy, one can put it that the textbooks for forms 10–12 display the cognitive and behavioural elements because the materials on nuclear energy are presented through comparison of various data, pointing out topicalities, which results in encouraging students analyse, assess and evaluate; make decisions; and form attitudes and responsible behaviour. If emphasising the complexity of the competence of energy literacy, it could be stated that materials of the textbooks for forms 10–12 are oriented to higher, more complex, i.e., analysis and synthesis, levels of evaluation.

No doubt, the aspect of outdated textbooks should be pointed out once again, which results in treating part of the information as historical materials only, no longer relevant in the present time. Due to this reason, teachers face quite a complicated task to select modern topicalities, data and information which would be suitable, reliable and important today.

Conclusions

In the course of the research, while carrying out the analysis of scientific literature, the thematic field of energy literacy and energy geography was pointed out. Grounding on these concepts, the authors analysed the national curriculum of geography and geography textbooks. The national geography curriculum emphasise competences of cognition of regions, reading of geographical information, maps, orientation in area and scientific research competences. These competences are related to the development of energy literacy.

The theme of energy geography (in this case, exploring nuclear energy in depth) is elaborated in the textbooks by indicating the competence of cognition to be developed as projected in the national curriculum. The textbooks introduce nuclear energy in various regions of the world to students: where uranium is processed, what countries and regions develop nuclear energy industry, and what is the role of nuclear energy sector. The development of the theme on nuclear energy in the analysed geography textbooks involves major aspects of energy geography: economic, geopolitical, environmental and social. However, most of the attention is paid to the economic aspect: the diversity of energy sources, traditional and new sources of energy, energy production in the world and particular regions are discussed. Regions and countries of the world where mining of uranium take place, nuclear power plants operate are discussed.

The textbooks form the concept of energy regions helping students understand the spatial patterns of the energy sector, when the energy economy comprises several different sources of energy and nuclear energy industry constitutes an important part of national economies. Nuclear energy is discussed in connection with other types of energy (hydro-electrical, thermal, renewable resources) and, in such a way, students can form a more general concept of the energy economy.

The textbooks discuss the development and change of nuclear sector as a branch of industry. The diversity of energy landscape, technological and social changes related to energy is presented. However, a new, contemporary approach to the nuclear energy, when it is no longer treated as the only prospective field of energy use since more attention in global economy and political discourse is focused on renewable sources of energy, green economy as the field of the future, is not revealed.

The textbooks introduce outdated information, since Lithuania no longer produces nuclear energy; therefore, this source of energy is not part of the field of present-day and future energy system. These significant changes in the energy landscape and political provisions are reflected in a major document on energy development, the Lithuanian National Energy Independence Strategy (2018), emphasising the modern environmental, sustainable aspects, such as production of power and electricity from renewable resources (biofuel, sun, biomass), engagement of citizens in the area of energy production and consumption, which is directly connected to stimulation of energy-focused participation and development of energy literacy.

Another area that presents nuclear energy deals with the environmental aspect emphasising potential insecurity of nuclear energy, damage and threats of it to people, nature and environment. This topic is developed by presenting nuclear accidents (Chernobyl disaster), radioactive pollution, potential insecurity and problems of storing of radioactive waste. Comparing the volume allocated to the discussion of this topic with the volume related to the economic aspect, it is observed that the environmental aspect is given far less attention than the economic introduction of nuclear energy.

Moreover, the textbooks also present the social aspect of nuclear energy when dealing with connections between nuclear energy industry and foundation of settlements. However, this aspect is developed much less in comparison to the economic one.

The analysis of the textbooks revealed growing complexity of the development of geographic competences when dealing with nuclear energy (Bloom et al., 1956, Anderson & Krathwohl, 2001) in the aspect of the cognitive levels. In the national curriculum, the theme on energy is presented following the principle of consistently growing complexity, through transition from the level of knowledge and understanding in forms 7–8 to the development of higher skills (analysis, synthesis) in forms 9–12. Materials of the textbooks, as element of implementation of the curriculum, reveal that the aspect of energy is also developed quite consistently, through all levels, starting with knowledge in lower forms and moving to more complex levels in senior forms. When talking about nuclear energy in form 7, uranium is introduced; and in more senior forms uranium is explained as nuclear fuel; later, threat to environment, geopolitical processes of development of energy industry and nuclear energy sector are discussed (forms 11–12).

The research demonstrated that the national curriculum did not include direct connections to energy literacy; however, preconditions for the development of cognitive, affective and behavioural dimensions of energy literacy are created. In the textbooks, the theme of nuclear energy is mostly linked to the cognitive dimension, when basic knowledge of numbers and facts is obtained, comparisons are presented, e.g. volumes of nuclear fuel processing in different countries, changes in nuclear energy in different regions, and depiction of the process of recycling of nuclear waste. The formation of other dimensions, i.e., behavioural and attitudinal (affective), is given little attention, and the development of these components is quite episodic.

When implementing the geography curriculum and using the geography textbook to deliver the topic on energy, teachers face basic problems: it is not clear how the theme of energy meets the development of energy literacy, textbooks display outdated factual information, it is difficult to trace the consistency in the formation of the competences. To make the process of development of energy literacy consistent, grounding on the findings of this investigation, it is recommended to link the formal curriculum to non-formal activities, which could be implemented by employing the resources of educational tourism. The educational route of nuclear tourism in the INPP and Visaginas town being designed by the researchers that was mentioned at the beginning of the chapter can facilitate teachers gaining the most relevant information on the situation of energy industry (including nuclear), complexly forming the cognitive, affective and behavioural dimensions in energy literacy.

The analysis of the geography curriculum and textbooks helped the researchers to better perceive how it was possible to bring the content of the nuclear tourism route, as non-formal education, closer to formal education (in this case, the geography curriculum). Within the route, the information on the INPP may be introduced as a moment of the historical development of energy sector in Lithuania, i.e., by presenting that when the INPP operated it was a very significant part in the development of the entire economy of Lithuania; however, the economic, political and social changes that took place after the closure of the power plant made a significant impact on the context of energy system and the development of the whole country. In this case, the route may represent the situation of the closure of the INPP, the changes, current energy sources-focused map (underlining that nuclear energy is no longer produced and other types of energy are becoming more important) and available prospects for the development of energy sector. Knowledge on the INPP can be connected to the concept of the changing energy landscape.

The route could include a relevant revelation of the social aspect of energy production and use because exactly in real environment (Visaginas town) one can present and perceive the way how nuclear industry influenced social processes: foundation of mono-industrial settlements in hinterlands, participation of top level specialists of nuclear energy sector, functioning of social and cultural infrastructure. The economic, demographic, social environment of Visaginas also allows understanding how social processes take place after the closure of the nuclear power plant.

The analysis of the textbooks on geography revealed that aiming at a comprehensive concept of the phenomenon (nuclear energy) materials must be characteristic of consistently growing complexity; therefore, the projection of its presentation could ground on the logic available in the textbooks: by rendering the content growing from specific very simple concepts to more complex levels. In such a case, teachers would find the possibilities to choose: to deliver the nuclear topic consistently, starting from lower forms (forms 6–7), or choosing the logic of growing complexity of content and consistency (e.g. in form 9, starting from elementary knowledge and facts, consistently moving to the most complex geo-political, geo-economic evaluation). In such a way, it appears important to render and choose the learning strategies. Seeking to render simple information and knowledge, methods of work in groups, discussions, studying and discussing of sources can be employed both in classroom and field within the tourism route; whereas when moving to the development of more complex skills, learning strategies encompassing several skills are necessary to apply when using a virtual tourist route or visiting Visaginas. In such a case, in both formal education lessons and in the settings of non-formal outdoor education, the following methods are recommended: digital storytelling, use of the GIS, problem-based learning, project design and narrative play strategies. When applying these methods, students could perform investigations of natural and social environment employing the GIS; present relevant projects on energy consumption and environmental protection; solve relevant issues of town development (project design); and carry out investigations of a demographic situation, place and cultural identity of the town and its residents (digital storytelling, narrative play). Thus, purposeful interaction of formal and non-formal education would appear ensuring complexity of the development of the geographic competence because the process of teaching would proceed in real-world environment and by applying interactive methods.

References

Anderson, L. W., & Krathwohl, D. (Eds.). (2001). *A Taxonomy for Learning, Teaching, and Assessing: A Revision of Bloom's Taxonomy of Educational Objectives*. New York: Longman.

Bartley, E., Brown, P. L., Concannon, J. P., & Stumpe, L. (2013). What's there to debate about nuclear energy? Promoting multidimensional science literacy by implementing STS strategies. *Science Activities: Classroom Projects and Curriculum Ideas*, 50:2, 41–48, DOI: 10.1080/00368121.2013.779227

Bloom, B. S. (Ed.). Engelhart, M. D., Furst, E. J., Hill, W. H., Krathwohl, D. R. (1956). *Taxonomy of Educational Objectives, Handbook I: The Cognitive Domain*. New York: David McKay Co Inc.

Bodzin, A. M., Fu, Q., Peffer, T. E., & Kulo, V. (2013). Developing energy literacy in US middle-level students using the geospatial curriculum approach. *International Journal of Science Education*, 35:9, 1561–1589.

Bouzarovski, S. (2009). East Central Europe's changing energy landscapes: a place for geography. *AREA*, 2009, Vol. 41 No. 4. *Royal Geographical Society* (with The Institute of British Geographers, 452–463.

Bouzarovski, S., & Tirado Herrero, S. (2017). The energy divide: Integrating energy transitions, regional inequalities and poverty trends in the European Union. *European Urban and Regional Studies*, 24(1), 69–86. https://doi.org/10.1177/0969776415596449

Butt, G. (2011) Globalisation, geography education and the curriculum: what are the challenges for curriculum makers in geography? *Curriculum Journal*, 22:3, 423–438.

Calvert, K. (2016). From 'energy geography' to 'energy geographies': Perspectives on a fertile academic borderland. *Progress in Human Geography* 2016, Vol. 40(1), 105–125.

Dahlmann, F., Kolk, A., & Lindeque, J. (2017). Emerging energy geographies: Scaling and spatial divergence in European electricity generation capacity. *European Urban and Regional Studies*, Vol. 24(4), 381–404.

Day, T. (2012). Undergraduate teaching and learning in physical geography. *Progress in Physical Geography* 36(3), 305–332.

DeWaters, J., Qaqish, B., Graham, M., & Powers, S. (2013). Designing an energy literacy questionnaire for middle and high school youth. *The Journal of Environmental Education*, 44:1, 56–78, DOI: 10.1080/00958964.2012.682615

Dori, Y. J., Avargil, Sh., Kohen, Z., & Saar, L. (2018). Context-based learning and metacognitive prompts for enhancing scientific text comprehension. *International Journal of Science Education*, 2018, Vol. 40, No. 10, 1198–1220, https://doi.org/10.1080/09500693.2018.1470351

Fowler, J. M. (1976). Energy, education, and the "wolf" criers. *The Science Teacher*, 43(3), 25–32.

Gilbert, J. K. (2006). On the nature of "context" in chemical education. *International Journal of Science Education*, 28:9, 957–976, DOI: 10.1080/09500690600702470

Gilbert, J. K., Bulte, A. M. W., & Pilot, A. (2011). Concept development and transfer in context-based science education. *International Journal of Science Education*, 33:6, 817–837, DOI: 10.1080/09500693.2010.493185

Graybill, J. K. (2016). Teaching energy geographies via videography. *Journal of Geography in Higher Education*, 40:1, 55–66, DOI: 10.1080/03098265.2015.1089474

Huber, M. (2016). Teaching energy geography? It's complicated. *Journal of Geography in Higher Education*, 40:1, 77–83, DOI: 10.1080/03098265.2015.1089476

Lambert, D. (2013). Geography in school and a curriculum of survival. *Theory and Research in Education* 11(1), 85–98.

Solomon, BD, Pasqualetti, MJ, & Luchsinger, DA (2004). Energy geography. In: Gaile G. & Willmott C. (Eds,), *Geography in America at the Dawn of the 21st Century* (pp. 302–313). Oxford: Oxford University Press.

Thoyre, A., & Harrison, C. (2016). Introduction: teaching energy geographies. *Journal of Geography in Higher Education*, 40:1, 31–38, DOI:10.1080/03098265.2016.1132539

Van der Horst, D., Harrison, C., Staddon, S., & Wood, G. (2016). Improving energy literacy through student-led fieldwork – at Home. *Journal of Geography in Higher Education*, 2016, Vol. 40, No. 1, 67–76, http://dx.doi.org/10.1080/03098265.2015.1089477

Textbooks and other sources:

Bačkienė, R., Pundienė, V., Januškis, V. (2009a). Geografija. 7 kl. 1-oji kn. (serija "Šok"). [Geography. For Form 8. Part 1. Set "Jump")]. Šviesa, Kaunas.

Bačkienė, R., Pundienė, V., Januškis, V. (2009b). Geografija. 7 kl. 2-oji kn. (serija "Šok"). [Geography. For Form 8. Part 1. Set "Šok")]. Šviesa, Kaunas.

Česnavičius, D., Gerulaitis, V. (2008/2007). Bendroji geografija. Vadovėlis 11–12 kl. Trečioji knyga [General Geography. Textbook for Forms 11–12. Book 3]. Šviesa, Kaunas.

Česnavičius, D., Gerulaitis, V., Kynė, G. (2010/2008). Regioninė geografija. Vadovėlis 11–12 klasei [Regional Geography. Textbook for Forms 11–12]. Šviesa, Kaunas.

Česnavičius, D., Valančienė, E. (2008). Lietuva. Europa. Pasaulis. 10 kl. [Lithuania. Europe. World. For Form 9]. Šviesa, Kaunas.

Dijokienė, S. (2016). Geografija. Pasaulio ir Lietuvos ūkis. Mokymosi medžiaga 10 klasei [Geography. World and Lithuanian Economy. Study Materials for Form 10]. Ugda.

Gerulaitis, Š., Bačkienė, R. (2009). Bendroji ir regioninė geografija 11–12 klasei. Antrasis pratybų sąsiuvinis [General and Regional Geography for Forms 11–12. Exercise Book 2]. Šviesa, Kaunas.

Gerulaitis, Š., Kynė, G., Varanavičienė, G., Krušinskienė, R., Tamošiūnienė, D. (2010a). Geografija. 8 kl. 1-oji kn. (serija "Šok"). [Geography. For Form 8. Part 1. Set "Jump")]. Šviesa, Kaunas.

Gerulaitis, Š., Kynė, G., Varanavičienė, G., Krušinskienė, R., Tamošiūnienė, D. (2010b). Geografija. 8 kl. 2-oji kn. (serija "Šok"). [Geography. For Form 8. Part 2. Set "Jump")]. Šviesa, Kaunas.

Kynė, G., Kirilovas, V., Mikulienė, I., Žiūrienė. S. (2008a). Mūsų daug – pasaulis vienas. Geografijos vadovėlis 6 kl. 1-oji kn. (serija "Šok"). [Mūsų daug – pasaulis vienas. Geography Textbook for Form 6. Part 2 (Set "Jump")]. Šviesa, Kaunas.

Kynė, G., Kirilovas, V., Mikulienė, I., Žiūrienė. S. (2008b). Mūsų daug – pasaulis vienas. Geografijos vadovėlis 6 kl. 2-oji kn. (serija "Šok"). [Mūsų daug – pasaulis vienas. Geography Textbook for Form 6. Part 2 Set "Jump")].]. Šviesa, Kaunas.

Kynė, G., Ubartas, T., Šabanovas, S., Barauskienė, L. (2016). Geografija 7 kl. (serija "Atrask") [Geography for Form 7. (Set "Discover")]. Šviesa, Kaunas

Kynė, G., Varanavičienė, G., Railienė, L., Ubartas, T., Kamarauskaitė, A. (2015). Geografija. 6 kl. (serija "Atrask"). [Geography. For Form 6. (Set "Discover")]. Šviesa, Kaunas.

Nacionalinė energetinės nepriklausomybės strategija [National Energy Independence Strategy] (2018).

Pradinio ir pagrindinio ugdymo bendrosios programos. Socialinis ugdymas (2008) [General Programmes of Primary and Basic Education. Social Education] https://www.smm.lt/uploads/documents/svietimas/ugdymo-programos/6_Socialinis-ugdymas.pdf

Šalna, R. Baleišis, E., Baubinas, R., Daugirdas, V. (2012a). Žemė. Geografijos vadovėlis. 1 d. 9 kl. [Earth. Geography Textbook for Form 9. Part 1]. Briedis.
Šalna, R. Baleišis, E., Baubinas, R., Daugirdas, V. (2012b). Žemė. Geografijos vadovėlis. 2 d. 9 kl. [Earth. Geography Textbook for Form 9. Part 2]. Briedis.
Šalna, R., Baubinas, R., Mačiulytė, J., Padriezas, V., Tuskenienė, V., Žolynas, M. (2006). Žemė. Geografijos vadovėlis 10 klasei [Earth. Geography Textbook for Form 10]. Briedis.
Šalna, R., Čepaitytė, G., Sapožnikovas, G. (2009/2005a). Žemė. Geografija. 1 d. 7 kl. [Earth. Geography Textbook for Form 7. Part 1]. Briedis.
Šalna, R., Čepaitytė, G., Sapožnikovas, G. (2009/2005b). Žemė. Geografija. 2 d. 7 kl. [Earth. Geography Textbook for Form 7. Part 2]. Briedis.
Šalna, R., Mačiulytė, J., Padriezas, V., Pakamorienė, A., Sapožnikovas, G., Tuskenienė, V., Žolynas. M. (2014). Žemė. Geografija. 10 kl. [Earth. Geography Textbook for Form 10]. Briedis.
Šalna, R., Sapožnikovas, G. (2010/2006a). Žemė. Geografijos vadovėlis 8 klasei. I dalis [Earth. Geography Textbook for Form 8. Part 1]. Briedis.
Šalna, R., Sapožnikovas, G. (2010/2006b). Žemė. Geografijos vadovėlis 8 klasei. II dalis [Earth. Geography Textbook for Form 8. Part 2]. Briedis.
Šalna, R., Sapožnikovas, G., Čepaitytė, G. (2005a). Žemė. Geografijos vadovėlis 7 klasei. I dalis [Earth. Geography Textbook for Form 7. Part 1]. Briedis.
Šalna, R., Sapožnikovas, G., Čepaitytė, G. (2005b). Žemė. Geografijos vadovėlis 7 klasei. II dalis [Earth. Geography Textbook for Form 7. Part 2]. Briedis.
Šalna, R., Sapožnikovas, G., Motiejuitė, G., Šiumeta, M., Šalna, R. (2010a). Gamtinė geografija. Geografijos vadovėlis. 11–12 kl. (serija "Gaublys"). [Natural Geography. Geography Textbook for Forms 11–12. (Set "Globe")]. Didakta.
Šalna, R., Sapožnikovas, G., Motiejuitė, G. (2010b). Žemė. Geografija. 1 d. 8 kl. [Earth. Geography for Form 8. Part 1]. Briedis.
Šalna, R., Sapožnikovas, G., Motiejuitė, G. (2010c). Žemė. Geografija. 2 d. 8 kl. [Earth. Geography for Form 8. Part 1]. Briedis.
Šalna, R., Šalna, R. Šiumeta, M. (2012). Gaublys. Visuomeninė geografija. Geografijos pratybų sąsiuvinis 11–12 klasei II dalis. [Globe. Social Geography. Geography Exercise Book for Forms 11–12. Part 2] Didakta.
Šalna, R., Tuskenienė, V. (2009a) Žemė. Geografijos vadovėlis. 1 d. 6 kl. [Earth. Geography Textbook for Form 6. Part 1]. Briedis.
Šalna, R., Tuskenienė, V. (2009b) Žemė. Geografijos vadovėlis. 2 d. 6 kl. [Earth. Geography Textbook for Form 8. Part 2].

Valančienė, E., Česnavičius, D. (2008). Lietuva. Europa. Pasaulis. Geografija. Vadovėlis 10 klasei [Lithuania. Europe. World. Geography. Textbook for Form 10]. Šviesa, Kaunas.

Valančienė E., Dijokienė S. (2007). Lietuva. Europa. Pasaulis. Geografija. 9 kl. [Lithuania. Europe. World. Geography for Form 9]. Šviesa, Kaunas

Varanavičienė, G., Krušinskienė, R., Šabanovas, S., Sakalauskienė, I. (2017). Geografija 8 klasei (serija "Atrask") [Geography for Form 8 (set "Discover"). Šviesa, Kaunas.

Vidurinio ugdymo bendrosios programos: socialinis ugdymas. (2010). Vidurinio ugdymo bendrųjų programų 5 priedas [General Programmes of Secondary Education: Social Education. Annex 5 to the General Curricula of Secondary Education]. https://www.smm.lt/uploads/documents/svietimas/ugdymo-programos/vidurinis-ugdymas/Socialinis_ugdymas_5_priedas.pdf

List of Figures

Ilona Tandzegolskienė
Revisiting Educational Potential of the Industrial Heritage Tourism: Ruhr Area in Germany and Ignalina Power Plant Region in Lithuania

Fig. 1: Process and analysis of industrial heritage construction (compiled by I. Tandzegolskienė). 31
Fig. 2: Model of educational tourism (McGladdery & Lubbe, 2017, p. 321). 34
Fig. 3: Process of learning (McLeod, 2017). 35
Fig. 4: A learning process model of educational tourism (adapted from McGladdery & Lubbe, 2017, p. 324). 36
Fig. 5: Elements of analysing sustainable and long-term heritage including educational tourism (adapted from Murphy and Boyle, 2005; Vangas-Sánchez et al., 2009; Storm, 2014; Sharma, 2015; Loures, 2016). 39
Fig. 6: Methodological Diagram (Capelo et al., 2011; Loures et al., 2017). . 40
Fig. 7: Route of industrial heritage (author of the photographs I. Tandzegolskienė). 42
Fig. 8: The closed Duisburg plant is dominated by the concept of industrial nature (author of the photographs I. Tandzegolskienė). . 44
Fig. 9: Aesthetic value of Zollern Colliery architectural monument (author of the photograph I. Tandzegolskienė). 47

Linara Dovydaitytė
The Pedagogy of Dissonant Heritage: Soviet Industry in Museums and Textbooks

Fig. 1: Model of Ignalina NPP at the Energy and Technology Museum, Vilnius, 2018. (Photograph by L. Dovydaitytė). 79
Fig. 2: Fragment of the exhibition "Made in Vilnius" at the Energy and Technology Museum, Vilnius, 2018. (Photograph by L. Dovydaitytė). 80
Fig. 3: The models of two repositories for radioactive waste at the INPP Visitor Centre, 2018. (Photograph by L. Dovydaitytė). 83
Fig. 4: A view of the permanent display at Visaginas Museum, Visaginas, 2019. (Photograph by L. Dovydaitytė). 89
Fig. 5: Fragment of the history textbook by Laužikas et al. (2008). 96

Fig. 6: Fragment of the history textbook by Gečas et al. (2001). 100
Fig. 7: Fragment of the history textbook by Petreikis et al. (2014). 102

Ineta Dabašinskienė
Language Transformation in Post-Industrial Landscape: a Case Study of Atomic City Visaginas
Fig. 1: The bilingual sign (Lithuanian-Russian) on a stone marking the establishment of the town in 1975 ("The town for people of a nuclear power plant will be built here, August, 1975"). (Photograph by I. Dabašinskienė). 115
Fig. 2: Graffiti in the residential area of Visaginas ("Stop NATO. No war"). (Photograph by I. Dabašinskienė). 120
Fig. 3: Bilingual signs of Russian-Lithuanian or Russian-English in Visaginas in private domains. (Photograph by I. Dabašinskienė). 124
Fig. 4: The poster inside the Culture center. Most of the informants declared their positive feelings to Visaginas, their hometown. (Photographs by Ineta Dabašinskienė). 128

Eglė Gerulaitienė and Natalija Mažeikienė
Energy Tourism at Nuclear Power Plants: Between Educational Mission and Retention of "Safety Myth"
Fig. 1: Interrelationships of energy tourism and other types of special interest tourism (adapted from Frew & Shaw, 1995) by Frantál & Urbánková (2017). 138
Fig. 2: Torness Nuclear Power Plant and Visitor Center (Photographs by E. Gerulaitienė). 150
Fig. 3: Interactive exhibition in the Visitor Centre (Photograph by E. Gerulaitienė). 155
Fig. 4: Interactive displays and exhibits to familiarize visitors with nuclear energy (Photograph by E. Gerulatienė). 156
Fig. 5: Interactive educational exhibit – measurement of radiation level (Photograph by E. Gerulaitienė). 157
Fig. 6: Statistics of INPP visitors in 2018–2019 (based on data by INPP Communication Unit). 161
Fig. 7: Excursion in Ignalina Nuclear Power plant: EDUATOM project researchers Ilona Tandezegolskienė and Eglė Gerulaitienė 165
Fig. 8: INPP Information Centre (Photograph by E. Gerulaitienė). 167
Fig. 9: Topics reflected in Torness and Ignalina Information Centres and Nuclear Power Plants. 168

Natalija Mažeikienė and Eglė Gerulaitienė
Chernobyl Museum as an Educational Site: Transforming "Dark Tourists" Into Responsible Citizens and Knowledgeable Learners
Fig. 1:   A Dark Tourism Spectrum: Perceived product features of dark tourism within a 'Darkest–Lightest' Framework of Supply by Stone (2006). 181
Fig. 2:   Road signs depicting abandoned and uninhabited towns after the Chernobyl disaster (Photograph by E. Gerulaitienė). 201
Fig. 3:   Images of a liquidator wearing a special protective green uniform and a gas-mask next to religious and spiritual symbols (Photograph by E. Gerulaitienė). 214

Magdalena Banaschkevicz
Fun in the Power Plant. Edutainment in the Chernobyl Exclusion Zone Tourism
Fig. 1:   Extract from the description of the trip from the "Zero Zone" website (screenshot), source: Zero Zone, https://strefazero.org/?id=104&ArticleId=145, date of access: 16.07.2019 235

Lina Kaminskienė
What We Find Outdoors: Discovering Nuclear Tourism Through Educational Pathways
Fig. 1:   Learning in a simulative pilot cabin at Aviation museum in Kaunas (picture from personal album of L. Kaminskienė). 246
Fig. 2:   Positive impacts in the five domains of child development (Kaminskienė, 2020, adapted from Malone, 2008). 255
Fig. 3:   Dimensions of place-based education and its links with educational tourism (Kaminskienė, 2020). 257

Judita Kasperiūnienė
Innovative Technological Solutions in Virtual Nuclear Education
Fig. 1:   The framework of serious gaming for non-edutainment purposes. 280
Fig. 2:   Experiential gaming tools, motivators and learning strategies for virtual nuclear education. 281
Fig. 3:   The number of articles from the five most popular scientific fields researching nuclear education and virtual tours in *Springer Link* online collection of scientific, technological and medical journals, books and reference works (N=634; the number of articles researching virtual tours in the rectangle; the number of articles researching nuclear education in the oval). 282

Fig. 4:	Most popular article topics, researching virtual reality solutions, virtual tours and formal and informal nuclear education and training in *Scopus* peer-reviewed journals (N=372).	283
Fig. 5:	Most popular article topics on a virtual tour and nuclear education-related topics in *Taylor & Francis* books and academic journals (N=165, number of articles researching virtual tours in the rectangle; the number of articles researching nuclear education in the oval).	284
Fig. 6:	Most popular article topics, researching virtual reality solutions, virtual tours and formal and informal nuclear education and training in *ACM* digital library (N=67).	285
Fig. 7:	An example of hands-on nuclear science workbook for secondary school children (7 to 10 years old), which could be used during a live class excursion to ANSTO (extracted from ANSTO files, 2019).	290
Fig. 8:	*TourBuilder* story gallery.	299
Fig. 9:	Teaching and learning materials for nuclear biology classes created with the *Tour Creator*.	299
Fig. 10:	A snippet of the *GeoGuessr* Tool website.	300
Fig. 11:	A snippet of *Space* and *AccessMars* tools.	302

List of Tables

Linara Dovydaitytė
The Pedagogy of Dissonant Heritage: Soviet Industry in Museums and Textbooks
Tab. 1:   History textbooks covering the period of Soviet Lithuania and approved by the Ministry of Education, Science and Sport of the Republic of Lithuania as valid for the academic year 2018–2019. Compiled by the author 93

Eglė Gerulaitienė and Natalija Mažeikienė
Energy Tourism at Nuclear Power Plants: Between Educational Mission and Retention of "Safety Myth"
Tab. 1:   The strategic directions of the Lithuanian energy sector (Lithuania National energy independence strategy, 2018) 160

Judita Kasperiūnienė
Innovative Technological Solutions in Virtual Nuclear Education
Tab. 1:   The challenges in virtual technology solutions and virtual tour technologies used within most popular scientific fields and educational settings 287
Tab. 2:   The challenges in nuclear education, nuclear information, and nuclear technologies used within most popular scientific fields and educational settings 289
Tab. 3:   The list of freely available nuclear educational apps and games ... 292
Tab. 4:   The list of popular freely available nuclear education tours and simulators 294
Tab. 5:   Concepts and most popular technological solutions for virtual nuclear education 295

Odeta Norkutė and Natalija Mažeikienė
Energy Literacy in Geography Curriculum: Redefining the Role of Nuclear Power
Tab. 1:   Areas of activity forming parts of the geography competency in different stages of education 325
Tab. 2:   Consistency in the gaining of skills and interaction of topics characterising the activity of "Cognition of Regions" in general programmes 329

Tab. 3:	Consistency in the gaining of skills and interaction of topics characterising the activity of "reading of geographical information" in general programmes 331
Tab. 4:	Consistency in the gaining of skills and interaction of topics characterising the activity of "orientation in the area and map" in general programmes 335
Tab. 5:	Consistency in the gaining of skills and interaction of topics characterising the activity of "cognition and investigation of environment" in general programmes 338

About the Authors

Mažeikienė, Natalija, prof., dr., Faculty of Social Sciences, Vytautas Magnus University. Leading researcher of the project "The Didactical Technology for the Development of Nuclear Educational Tourism in the Ignalina Nuclear Power Plant (INPP) Region (EDUATOM)". Research interests: critical theory, educational innovations, curriculum development, educational nuclear tourism. Contacts: Vytautas Magnus University, Jonavos g. 66–316, LT-44191 Kaunas, e-mail: natalija.mazeikiene@vdu.lt

Tandzegolskienė, Ilona, assoc. prof., dr., Vice-Chancellor of Education Academy at Vytautas Magnus University. Research interests: research on museum communication, educational tourism, critical discourse analysis and research methods in social sciences. She is one of the editors in the editorial board in the scientific journal "Vocational Training: Research and Realities". Contacts: Vytautas Magnus University, Education Academy, Jonavos str. 66–309, LT – 44191 Kaunas, e-mail: ilona.tandzegolskiene@vdu.lt

Dovydaitytė, Linara, assoc. prof., dr., Faculty of Arts, Head of the Department of Art History and Criticism, Vytautas Magnus University. Research in visual culture, museum studies, and memory politics. Contacts: Vytautas Magnus University, Muitinės str. 7, LT-44280 Kaunas, e-mail: linara.dovydaityte@vdu.lt

Dabašinskienė, Ineta, prof., dr., Department of Lithuanian Studies and Head of the Research Centre for Multilingualism, Vytautas Magnus University. Her research interests include, but are not limited to, psycho- and sociolinguistics, language acquisition, language policy and multilingualism, and spoken language and grammar. Contacts: Vytautas Magnus University, V. Putvinskio g. 23, 44248 Kaunas, e-mail: ineta.dabasinskiene@vdu.lt

Gerulaitienė, Eglė, assoc. prof., dr., Institute of Education, Head of the Project Management Unit in Department of the Research and Innovation, Vytautas Magnus University. Research interests: intercultural education, diversity and multiculturalism, implementation of innovations in the curriculum, application of innovative methods in the educational process, development of intercultural competence.

Contacts: Vytautas Magnus University, Education Academy, Institute of Educational Research, Jonavos str. 66, Kaunas, Lithuania. Email: egle.gerulaitiene@vdu.lt,

Banaszkiewicz, Magdalena, assoc. prof., dr., a cultural anthropologist, a professor in the Institute of Intercultural Studies at the Jagiellonian University in Krakow, Poland. Her research interests focus on the dissonances connected with tourism development and heritage sites. She co-edited the collective volume "Anthropology of Tourism in Central and Eastern Europe: Bridging Worlds" published in Lexington Books in 2018. Her newest monograph "Turystyka w miejscach kłopotliwego dziedzictwa" [Tourism in dissonant heritage sites] published in 2018 in the Jagiellonian University Press compares three cases (Lenin's Mausoleum, the Chernobyl Exclusion Zone and Nowa Huta district) through the lens of the tourist experience. Recently, she has been exploring tour guides' narratives in the Chernobyl Exclusion Zone (the research project 2016/23/D/HS3/01960 funded by the National Science Centre, Poland). She is the coordinator of the MA specialization "Intercultural relationships in tourism" at the Jagiellonian University and co-editor of the scientific journal "Cultural Tourism".

Kaminskienė, Lina, prof., dr., Chancellor of Education Academy and Head of the Institute of Educational Research, Vytautas Magnus University. Her research interests involve research on educational innovations, personalisation of learning, technology-enhanced learning, recognition of non-formal and informal learning, labour market and employability.

Contacts: Vytautas Magnus University, Education Academy, Institute of Educational Research, Jonavos str. 66–310, Kaunas, Lithuania. Email: lina.kaminskiene@vdu.lt,

Kasperiuniene, Judita, assoc. prof., dr., Faculty of Informatics, Vytautas Magnus University Research in: social media in education, serious games, technology enhanced learning, self-regulated learning. Contacts: Vytautas Magnus University, Vileikos str. 8, LT-44404, Kaunas, e-mail: judita.kasperiuniene@vdu.lt

Norkutė, Odeta, assoc. prof., dr., Education Academy, Institute of Teacher Training, Vytautas Magnus University. Research interests: curriculum design and development, subject integration (STEAM, PSHE), pedagogical competencies and training, nuclear tourism, educational innovations. Contacts: Vytautas Magnus University, Jonavos str. 66–314, LT-44191, Kaunas, e-mail: odeta.norkute@vdu.lt.

Baltische Studien zur Erziehungs- und Sozialwissenschaft
Baltic Studies in Educational and Social Sciences
Herausgegeben von Gerd-Bodo von Carlsburg, Natalija Mažeikienė und Airi Liimets
Edited by Gerd-Bodo von Carlsburg, Natalija Mažeikienė and Airi Liimets

Band	1	Gerd-Bodo Reinert / Irena Musteikienė (Hrsg.): Litauische Gespräche zur Pädagogik. Humanismus – Demokratie – Erziehung. 1999.
Band	2	Gerd-Bodo Reinert / Irena Musteikienė (Hrsg.): Litauische Gespräche zur Pädagogik II. Staat und Schule. 2000.
Band	3	Jaan Mikk: Textbook: Research and Writing. 2000.
Band	4	Gerd-Bodo Reinert / Irena Musteikienė (Hrsg.): Erziehungswandel und moderne pädagogische Verfahren. 2001.
Band	5	Airi Liimets (Hrsg.): Integration als Problem in der Erziehungswissenschaft. 2001.
Band	6	Gerd-Bodo Reinert / Irena Musteikienė (Hrsg.): Wissenschaft – Studium – Schule auf neuen Wegen. 2002.
Band	7	Palmira Jucevičienė / Gediminas Merkys / Gerd-Bodo Reinert: Towards the Learning Society. Educational Issues. 2002.
Band	8	Gerd-Bodo Reinert / Irena Musteikienė (Hrsg.): Bildung im Zeitalter der Informationsgesellschaft. 2003.
Band	9	Dalia Ohlinger: Argumentation in der Erst- und Fremdsprache. Pragmalinguistische und grammatikalische Aspekte anhand von Argumentationen deutscher und litauischer Studierender. 2003.
Band	10	Inga Šukevičiūtė: Interkulturelle Interferenzen im Bereich Wirtschaftskommunikation. Zur Analyse von Geschäftsbriefen im Vergleich Deutschland, Litauen und Russland. 2004.
Band	11	Gerd-Bodo Reinert von Carlsburg / Irena Musteikienė (Hrsg./eds.): Innovation durch Bildung. Innovation by Education. 2004.
Band	12	Gerd-Bodo von Carlsburg / Palmira Jucevičienė / Gediminas Merkys (eds.): Learning and Development for Innovation, Networking and Cohesion. 2005.
Band	13	Gerd-Bodo von Carlsburg / Irena Musteikienė (Hrsg./eds.): Bildungsreform als Lebensreform. Educational Systems Development as Development of Human Being. 2005.
Band	14	Gerd-Bodo von Carlsburg (Hrsg./ed.): Entwicklung erziehungswissenschaftlicher Paradigmen: Theorie und Praxis. Development of Educational Paradigms: Theory and Practice. 2007.
Band	15	Gerd-Bodo von Carlsburg (Hrsg./ed.): Bildungs- und Kulturmanagement. The Management of Education and Culture. 2008.
Band	16	Jaan Mikk / Marika Veisson / Piret Luik (eds.): Reforms and Innovations in Estonian Education. 2008.

Band 17 Gerd-Bodo von Carlsburg (Hrsg./ed.): Qualität von Bildung und Kultur. Theorie und Praxis. The Quality of Education and Culture. Theoretical and Practical Dimensions. 2009.

Band 18 Airi Liimets (Hrsg.): Denkkulturen. Selbstwerdung des Menschen. Erziehungskulturen. Festschrift für Professor Dr. Dr. h.c. Dr. h.c. Heino Liimets. 2010.

Band 19 Ellu Saar (ed.): Towards a Normal Stratification Order. Actual and Perceived Social Stratification in Post-Socialist Estonia. 2011.

Band 20 Marika Veisson / Eeva Hujala / Peter K. Smith/ Manjula Waniganayake / Eve Kikas (eds.): Global Perspectives in Early Childhood Education. Diversity, Challenges and Possibilities. 2011.

Band 21 Airi Liimets / Marit Mäesalu (eds.): Music Inside and Outside the School. 2011.

Band 22 Gerd-Bodo von Carlsburg (Hrsg./ed.): Enkulturation durch sozialen Kompetenzerwerb. Enculturation by Acquiring of Social Competences. 2011.

Band 23 Reet Liimets: Ich als raumzeitliches Konstrukt. Die Fiktionen vom Leben der estnischen und deutschen Jugendlichen. 2012.

Band 24 Raivo Vetik (ed.): Nation-Building in the Context of Post-Communist Transformation and Globalization. The Case of Estonia. 2012.

Band 25 Airi-Alina Allaste (ed.): ‚Back in the West'. Changing Lifestyles in Transforming Societies. 2013.

Band 26 Gerd-Bodo von Carlsburg (Hrsg./ed.): Bildungswissenschaft auf der Suche nach globaler Identität. Educational Sciences in Search of Global Identity. 2013.

Band 27 Airi Liimets / Marika Veisson (eds): Teachers and Youth in Educational Reality. 2014.

Band 28 Gerd-Bodo von Carlsburg / Thomas Vogel (Hrsg./eds.): Bildungswissenschaften und akademisches Selbstverständnis in einer globalisierten Welt. Education and Academic Self-Concept in the Globalized World. 2014.

Band 29 Marika Veisson / Airi Liimets / Pertti Kansanen / Edgar Krull (eds.): Tradition and Innovation in Education. 2015.

Band 30 Robertas Jucevičius / Jurgita Bruneckienė / Gerd-Bodo von Carlsburg (eds.): International Practices of Smart Development. 2015.

Band 31 Gerd-Bodo von Carlsburg (Hrsg./ed.): Strategien der Lehrerbildung. Zur Steigerung von Lehrkompetenzen und Unterrichtsqualität. Strategies for Teacher Training. Concepts for Improving Skills and Quality of Teaching. 2016.

Band 32 Gerd-Bodo von Carlsburg (Hrsg./ed.): Denk- und Lernkulturen im wissenschaftlichen Diskurs. Cultures of Thinking and Learning in the Scientific Discourse. 2017.

Band	33	Palmira Pečiuliauskienė / Aleksa Valdemaras: Motivation of New Generation Students for Learning Physics and Mathematics. 2018.
Band	34	Gerd-Bodo von Carlsburg (Hrsg./ed.): Transkulturelle Perspektiven in der Bildung. Transcultural Perspectives in Education. 2019.
Band	35	Giedre Kvieskienė / Vytautas Kvieska / Gerd-Bodo von Carlsburg: Social Clustering: Paradigm of Trust. 2021.
Band	36	Natalija Mažeikienė (ed.): Learning the Nuclear: Educational Tourism in (Post)Industrial Sites. 2021.

www.peterlang.com

www.ingramcontent.com/pod-product-compliance
Ingram Content Group UK Ltd.
Pitfield, Milton Keynes, MK11 3LW, UK
UKHW041924210426
5322IPUK00002B/49